OPTICAL RADIATION MEASUREMENTS

Volume 2

COLOR MEASUREMENT

OPTICAL RADIATION MEASUREMENTS

A Treatise

Edited by FRANC GRUM

OPTICAL RADIATION MEASUREMENTS

Volume 2

COLOR MEASUREMENT

Edited by

FRANC GRUM C. JAMES BARTLESON

Research Laboratories
Eastman Kodak Company
Rochester, New York

1980

ACADEMIC PRESS

A Subsidiary of Harcourt Brace Jovanovich, Publishers

New York London Toronto Sydney San Francisco

∨ 6337-8887

OPTOMETRY

ACADEMIC PRESS, INC.
111 Fifth Avenue, New York, New York 10003

United Kingdom Edition published by
ACADEMIC PRESS, INC. (LONDON) LTD.
24/28 Oval Road, London NW1 7DX

Library of Congress Cataloging in Publication Data

Main entry under title:

Optical radiation measurements.

    Vol. 2 edited by F. Grum and C. James Bartleson
    Includes bibliographies and index.
    CONTENTS: v. l. Grum, F., Becherer, R. Radi-
ometry.--v.2. Color measurement.
    1. Radiation--Measurement. 2. Optical measure-
ments. 3. Colorimetry. I. Grum, Franc
II. Becherer, Richard
QC475.Q67   539.2'028'7   78-31412
ISBN 0-12-304902-4 (v. 2)

PRINTED IN THE UNITED STATES OF AMERICA

80 81 82 83   9 8 7 6 5 4 3 2 1

Contents

v

4   Modern Illuminants

Frederick T. Simon

5   Color Order

Frederick T. Simon

6   Colorimetry of Fluorescent Materials

Franc Grum

7   Colorant Formulation and Shading

Eugene Allen

8   Modern Color-Measuring Instruments

M. Pearson

Index

List of Contributors

Numbers in parentheses indicate the pages on which the authors' contributions begin.

EUGENE ALLEN (289), Sinclair Laboratory, Lehigh University, Bethlehem, Pennsylvania 18015

C. J. BARTLESON (1, 33), Research Laboratories, Eastman Kodak Company, Rochester, New York 14650

FRANC GRUM (235), Research Laboratories, Eastman Kodak Company, Rochester, New York 14650

R. W. G. HUNT (11), Research Division, Kodak, Ltd., Harrow, United Kingdom HA1 4TY

M. PEARSON (337), Graphic Arts Research Center, Rochester Institute of Technology, Rochester, New York 14623

FREDERICK T. SIMON (149, 165), Clemson University, Clemson, South Carolina 29631

Preface

This volume of the treatise "Optical Radiation Measurements" deals with color measurement. It is an up-to-date, comprehensive treatment of the subject, combining historical, theoretical, and practical aspects in a single volume. The topics covered and the way they are addressed are intended to be both useful and generally informative. The book is directed to people who must deal with applications of color in solving practical problems, including color technologists, colorimetrists, students, and research workers; in short, anyone who faces the problem of measuring and specifying color. However, the volume is not simply a "cookbook" of application techniques. We believe that the intelligent worker will be able to apply techniques of color measurement more effectively if he has some understanding of the ways in which those techniques were developed over the years, the theoretical bases for them, and particularly, the strengths and limitations of various color measurement techniques. Accordingly, each chapter combines what we hope is a clear description of the most recently recommended methods of color measurement with a brief historical perspective and discussion of pertinent theoretical considerations that will help the reader to understand clearly the uses and constraints of each technique. We hope, in other words, that this volume of the treatise will help in explaining what color measurement is about and how it can be used effectively.

Acknowledgments

In preparing this book we had the pleasure of working with the authors who contributed individual chapters to the volume. We also have had the benefit of comments and suggestions from a number of our colleagues. We particularly wish to express our thanks to Professor F. W. Billmeyer, Jr. of Rensselaer Polytechnic Institute; Dr. D. L. MacAdam; Mr. Robert Peden of Eastman Kodak Company for reading and commenting on portions of the manuscript. Their comments have been very valuable to us in editing this manuscript.

The cooperation and encouragement of the Eastman Kodak Research Laboratories are gratefully acknowledged. Our thanks go to Mrs. Cathy Thayer for her skillful handling of many details in preparation of the volume.

1

Introduction

C. J. BARTLESON

Research Laboratories
Eastman Kodak Company
Rochester, New York

This book is about color measurement. It is intended to provide an introduction to and summary of the current theory and methodology of measuring color. Most readers will immediately understand what is meant by the phrase "color measurement." However, all readers will not necessarily understand the same thing by that phrase. Both "color" and "measurement" mean different things to different people. The purpose of this section is to make explicit the ways in which these terms are used in this book and to provide a guide to what is meant by the phrase color measurement as it is used here.

Color has meant different things when used throughout the literature. Some physicists have applied it to the variations in spectral power distributions of lights, either emitted directly by sources or indirectly reflected or transmitted by objects. It is not unusual to find chemists using the word color to refer to spectral differences arising from variations in molecular constitution or stereochemical arrangement of compounds. By color, colorimetrists and photometrists most often have meant that which is represented by a tristimulus measurement, that is, the relative or absolute amounts of three reference lights that, when additively mixed, elicit a color match to some sample when it is viewed in a particular way. When psychologists and sensory scientists speak of color, they usually mean an aspect of conscious response of a sensing human observer, a perception that takes place in the brain of an observer as a result of stimulation of his visual mechanism. In everyday language we usually associate color with objects; we expect the same object to have the same color whenever and wherever we encounter it.

We all tend to adopt different attitudes and to use the word color in different ways, depending on our interests of the moment. We might, for example, speak of a colored light (implying that color is a property of the light) and, at the same, of the color that we perceive when we look at the light (implying that color is a property of the visual response process) and of the light's color specification (implying that color is a property of some light mixture that meets a criterion of matching or equality of appearance under certain restricted conditions). Even under carefully controlled laboratory conditions it is not often possible to avoid the ambiguity with which we use the word color. An experimenter may mean one thing by the word when he instructs an observer to match colors; the observer may mean a different thing when he performs the match; and the same experimenter may adopt still another attitude when he reduces the experimental data by mathematical techniques.

There is, thus, no single, unique definition of what we mean by color. The Commission Internationale de l'Eclairage (CIE) has taken some cognizance of this fact by recommending more than one definition for the word color. Color appears in the CIE's "International Lighting Vocabulary" (CIE, 1970) in two different ways: perceived color and psychophysical color. Perceived color is said to be an "aspect of visual perception by which an observer may distinguish differences between two fields of view of the same size, shape, and structure, such as may be caused by differences in the spectral composition of the radiation concerned in the observation." Psychophysical color, on the other hand, is said to be a "characteristic of a visible radiation by which an observer may distinguish differences between two fields of view of the same size, shape, and structure, such as may be caused by differences in the spectral composition of the radiation concerned in the observation." According to these CIE recommendations, psychophysical color is something that may be specified by the tristimulus values of the radiation entering the observer's eyes, whereas perceived color is a more private matter, peculiar to the observer, for which no method of measurement or specification is proposed.

Color is used in both these senses throughout this book, either as an aspect of visual perception or as a characteristic of visible radiation. The choice of the sense in which the word is used is dictated by accepted conventions in dealing with the subject under discussion. For example, color-measuring instruments are almost universally designed to provide tristimulus specifications, and so it is logical to speak of color in the psychophysical sense when discussing such instruments. Similarly, measurement of fluorescent samples is generally a problem of attempting to derive unique specifications of color matches, and so the matter of prime interest in such discussions is psychophysical color. In formulating colorants an attempt is made to provide pigment or dye mixtures that will color-match some aim or reference; hence, psychophysical color is the sense in which the criterion can be said to be met or not. However, color order systems may be arranged in a number of ways, including structures that relate to differences among perceived colors. Accordingly, the reader will find that perceived color is what is meant by the word color throughout much of the chapter on color order. The chapter on colorimetry uses the word in both senses, depending on whether the colorimetric method under discussion is intended to measure psychophysical or perceived color. In all instances where the context does not make clear which meaning is used for color, the word is modified by appropriate adjectives so that the reader will be able to tell what is meant when the word is used. A chapter on color terms and their definitions is intended to help the reader understand the various ways in which the lexicon of color is structured to encompass the concepts that are part of both kinds of color. The implications of modifying adjectives are made explicit in that chapter.

We have chosen this approach, rather than to try to use consistently a single definition, for several reasons. As stated earlier, the primary reason relates to the way in which the word color is used in various departments of color measurement. However pleasing it might be to the philosopher or theorist to use a single definition, the fact is that the word is used in different ways for different kinds of color measurement tasks. To impose an artificial constraint of a singular definition would likely cause more confusion from treating various subjects in ways that are at variance with the ways in which they are usually discussed than the probably minor confusion that might arise from using two different definitions of color in ways that are consistent with those common to the subjects under discussion. Finally, each chapter has been written by a different author, each an expert in the subject of his chapter. It would be awkward and presumptuous to insist that each author use the word color in the same way, particularly when it might be a way in which his subject is never discussed. Our aim is to convey to the reader information about color measurement. We believe that can best be accomplished by allowing each expert author to set forth his subject in his own way.

A similar kind of ambivalence attends the term "measurement," although this creates a less obvious problem throughout the book. There are many different definitions proposed for what is meant by measurement. The literature of science and the philosophy of science abound with them. It has been said that "everyone would agree that, whatever else it consists of, measurement consists of the assignment of numerals or numbers to things" (Campbell, 1938). However, ubiquity of agreement seems to end here. There is no unanimous agreement about what things can be measured and what things cannot. Neither is there unanimous agreement about what constitutes a measurement, what operations are or are not properly operations of measurement. Generally, the various definitions of measurement that have been proposed fall into or closer to one of two broad categories; categories that have been called narrow and broad views of measurement. In the narrow view, measurement is the assignment of numbers to objects or events along some dimension by comparing them with units along some other dimension through the operation of adding these units together. This definition tends to be favored by physical scientists. The broader view defines measurement as the assignment of numbers to objects or events according to any consistent rule or set of rules. This definition tends to be favored by sensory scientists, those who study human and animal responses to stimulation. These statements may seem to differ only superficially, but their implications for the meaning of color measurements is fundamental. The narrow definition requires that scales of measurement consist of additive units. The broad definition requires only that measurement scales be isomorphic in some consistent way with the events under consideration. The ways in which we can properly interpret measurements are quite different in the two cases.

In the narrow view, additivity and its geometric consequence, distance, are fundamental to the question of what constitutes a measurement. The concept of metric distance requires that a valid measure is one that may be described by a mathematical function having all four of the following properties:

(1) the distance between any two points is never negative: $d(x, y) \geq 0$,
(2) the distance between two identical points is always zero: $d(x, x) = 0$,
(3) distance is symmetric: $d(x, y) = d(y, x)$,
(4) the sum of the distances between two points by way of a third point is always equal to or greater than the direct distance between these two points: $d(x, y) + d(y, z) \geq d(x, z)$.

Details of the implications of these requirements are treated in a number of textbooks (e.g., Coombs *et al.*, 1970; Savage, 1970). In general, they deal with the meaning of distance (differences, intervals). The most commonly used, but certainly not the only, class of distance functions that meet all of these requirements is that known as the "power metric." The general definition of distance between any two points $[d(x, y)]$ in a power metric is defined as

$$d(x, y) = \sum_{i}^{n} [|x_i - y_i|^r]^{1/r}, \tag{1}$$

where $r \geq 1$. The exponent r may be interpreted as a parameter of component weight. When $r = 1$, all components receive equal weight. As the value of the exponent increases, components become increasingly differentially weighted according to their sizes. All distances of the power metric are invariant over coordinate translation but only in the special case of $r = 2$ are they also invariant over rotation. That special case is called the Euclidean geometry, with which we are so familiar in physical science. Unfortunately, it does not always seem to apply to perceptual or psychological science. The implications here are that what the CIE has called psychophysical color may be specified according to physical quantities involving distances arrayed in Euclidean geometry, but perceptual color involves nonphysical quantities that may have distances arrayed in non-Euclidean geometry. From this fact alone it should be clear that we cannot hope to translate from psychophysical color specifications to perceived color with impunity.

There is also another way in which measurements of the two kinds of color may differ. This concerns the extent to which scales of measurements satisfy the requirements of the power metric of Eq. (1). The three basic properties of the power metric are (1) interdimensional additivity, (2) intradimensional subtractivity, and (3) power. In the first case, the distance between two points is a function of their component contributions. In the second case, the distance between two points is a function of the absolute values of their component-wise differences. When these two properties are satisfied, the model is one of additive differences. In the third case, that of power, all component-wise differences are transformed by the same convex power function. This involves a nonmetric or dimensional assumption that points may simply be represented in a dimensionally organized space. The narrow view of measurement alluded to earlier embraces all three properties but the broad view requires only the third. Thus, according to the broad view, any transform that preserves the invariance associated with any degree of isomorphism between two domains is properly a measurement.

The practical implications of these theoretical considerations are critically important to the understanding and interpretation of color measurements. Perceptual color measurements may not necessarily result in a scale that has *both* distance and ratio properties *at the same time*. In other words, a scale of perceived color differences (distances, intervals) will not necessarily yield information about ratios of the magnitudes of color appearances. Conversely, a scale of ratios of perceived color cannot be expected to provide information about perceived color differences, unlike our expectations for ratio scales in the physical sciences.

Perhaps this can be made somewhat more clear if we consider the kinds of scales that are involved in color measurement. Of the many different scale types that exist, there are four that need concern us here. They are usually called ratio, interval, ordinal, and nominal scales. Let us consider them in inverse order, the order in which their mathematical power increases:

(1) *Nominal scales* merely determine whether or not things are equal. They satisfy the invariance conditions $y = y$, $x = x$, $y \neq x$. The nominal scale is the least powerful scale that is even referred to as a scale of measurement.

(2) *Ordinal scales*, in addition to determining equality, determine the order in which things stand with respect to one another. They satisfy the invariance of monotonicity, that is, $y > x$, $x > z$, $y > z$.

(3) *Interval scales*, in addition, determine equality of intervals. They satisfy the invariance condition $y = ax + b$, which means that the ratios of intervals remain constant over any linear transform of the scale.

(4) *Ratio scales*, in addition, determine equality of magnitudes. They satisfy the invariance condition $y = ax$, which means that ratios of magnitudes remain invariant over any linear scale transform that passes through the origin.

Each of these scale types is found in physical science. The numbers placed on the jerseys of football players, to allow us to distinguish among the players, provide an example of a nominal scale. The Mohs scale of hardness of minerals is an example of an ordinal scale; a mineral with a higher Mohs value is harder than one with a lower value. Temperature scales of Celsius or Fahrenheit are examples of interval scales; we cannot say that a value of 40°C (or 104°F) is twice the temperature of 20°C (or 68°F), but we can say that the temperature interval from 20° to 40° (which is 20° on the Celsius scale or 36° on the Fahrenheit scale) is equal to the interval between 0°C (or 32°F) and 20° (which is, in fact, an interval of 20°C or 36°F). Length in meters and temperature in kelvins are examples of ratio scales; a length of 100 m is twice as long as one of 50 m, or a temperature of 4000 K is twice as high as one of 2000 K.

In color measurement we are generally concerned with all but the ordinal scale. The measurement criterion of equality is basic to color matches and conventional color specifications. The result is a nominal scale of perceived color; that is, a colorimetric tristimulus specification implies only nominal information about perceived color. By its definition, however, the tristimulus specification is set forth as a triad of values expressing amounts of physical stimulation on a ratio scale. This distinction between the *criterion* and the *specification* is critically important to our understanding of what is meant by a tristimulus color measurement. The physical stimuli leading to a criterion response (a color match) are described on the ratio scale of energy that applies to physics. The criterion itself, the equality of perceived color, provides only nominal information about perceived color.

We must be careful not to let the specification metric (a ratio scale of energy relations) mislead us into thinking that we know anything more than equality of color appearance (a nominal scale of perceived color) about the perceived color itself. Measurements of perceived color differences are also specified on scales that are derived from ratios of physical magnitudes, but here the criterion of measurement is the equality of intervals (differences). Thus, color difference measurement attempts to construct interval scales of perceived color. Finally, scales of color appearance attempt to provide measurements of perceived color magnitudes, ratios of perceived color.

These distinctions among scale types, and the information they convey, form one of the two basic distinctions that underly the structure of the chapter on color terms and concepts. Color measurement concepts are divided into the three groups there labeled "psychophysical color," "psychometric color," and "psychoquantitative color" according to whether they relate, respectively, to nominal, interval, or ratio scales of perceived color. The terms used to distinguish these categories may be provisional and open to criticism on the basis of philology, but the distinctions that they represent are incontrovertible and critical to our understanding of the meanings of various kinds of color measurements.

That chapter makes a second, or corollary, distinction among scales of color measurement as well. This is the distinction between "absolute" and "relative" attributes of perceived color. An absolute scale of perceived color is defined as one in which there is only a single perceptual anchor (or reference), which is assigned a perceptually meaningful modulus (or scale value); the anchor representing a complete lack of the attribute in question is assigned a modulus of zero. For example, perceived brightness (one of the dimensions of perceived color) is an absolute attribute in which complete lack of brightness is called "zero brightness." There are no other anchors and moduli on the scale of absolute perceived brightness. This means that brightness is a continuum forming a ratio scale. Lightness, on the other hand, forms a relative scale of perception. In addition to the same perceptual anchor as that of brightness (a modulus of zero for a complete lack of the attribute in question), the lightness scale has a second, upper scale, perceptually meaningful anchor; the brightness of a "white" (a hueless perception with no "gray content"), which is assigned a second modulus (e.g., a value of 100). In other words, lightness is anchored at both the low and high ends of the scale. The upper anchor may vary in brightness (the brightness of a white paper in moonlight is obviously lower than that of the same white paper in bright sunlight), but it nonetheless represents a perceptually meaningful scale value. In this sense, then, lightness is a form of *relative* brightness.

We have, then, a structure for distinguishing among forms of perceptual attributes and a structure for classifying the scales of perceptual color measurement. These distinctions are, we believe, used uniformly throughout the book. Each chapter is written in such a way that it can stand alone without suffering too much from the lack of context. However, we believe that there is greater utility in considering all the chapters as parts of a whole. The chapters are arranged in such a way that the reader acquires information and viewpoints as the book progresses that enhance understanding of the exposition of each succeeding chapter. The volume should be read from beginning to end, even though it is intended to be a kind of handbook of color measurement. Each chapter includes tabulations of fundamental data and is intended to be reasonably complete. Inevitably, however, a definitive treatment of a subject as broad and complex as color measurement would require a prohibitively large book; therefore, some compromises have been made between completeness and size. Wherever additional data are of interest but not included here, we have included references to primary or standard secondary sources. It is our hope that the treatment provided here will provide a useful reference to current theory and methodology of color measurement.

Following this introduction, the second chapter discusses color terms and symbols and their usage. This chapter, written by Hunt, presents the most current proposals for distinguishing among the various concepts and operational definitions of color attributes that are important in color measurement. Chapter 3, by Bartleson, sets forth a summary of theoretical considerations that are basic to colorimetry and addresses the most current CIE recommendations for the practice of colorimetry. The chapter on modern illuminants, by Simon, describes characteristics of various illuminants recommended for use in making color measurements and the nature of light sources that may be used to realize those illuminants in practice. Chapter 5, also by Simon, describes both historical and contemporary aspects of color order systems. The historical aspects help to place currently available collections and systems for color order in perspective, and the chapter is planned to provide the reader with an understanding of how such systems are organized as well as how they may be used for making color measurements. Chapter 6, by Grum, deals specifically with the troublesome case of making color measurements of materials that fluoresce. It describes the nature of the problems involved in such measurements and sets forth the most recent developments in the field, including recommended methods for making measurements that have practical utility. Another area of special problems in the general field of color measurement involves colorant formulation, predicting colorant mixtures that will produce a specified psychophysical color from color measurements of a standard or reference. This subject is treated in Chapter 7, by Allen. Recent advances in measurement and prediction theory are discussed and emphasis is placed on the use of computer assistance in making predictions from measurements. Finally, Pearson has provided a chapter on color-measuring instruments in which the various basic characteristics of instruments are described and actual devices classified according to that descriptive structure. The aim of this chapter is to provide the reader with information that is helpful in understanding what various instruments do and which of the several kinds of instruments are useful for different color measurement tasks. All these chapters combine to form a single volume that is reasonably complete and self-contained. Volume I in this treatise (*Optical Radiation Measurements*, Vol. 1, *Radiometry*) provides a reference source for radiometric and photometric measurement. The present volume extends the practice of light measurement to the domain of color measurement. In this manner the two volumes are companion pieces. We have not attempted to deal in detail with radiometric measurements in the present volume. We hope, however, that this volume will be a useful and reasonably complete reference for color measurement.

REFERENCES

Campbell, N. R. (1938). *Proc. Aristotelian Soc., Suppl.* **17,** 121–142.
CIE (1970). "International Lighting Vocabulary," 3rd ed., Publ. No. 17 (E-1.1). Bureau Central CIE, Paris.
Coombs, C. H., Dawes, R. M., and Tversky, A. (1970). "Mathematical Psychology." Prentice-Hall, Englewood Cliffs, New Jersey.
Savage, C. W. (1970). "The Measurement of Sensation." Univ. of California Press, Berkeley.

2

Color Terms, Symbols, and Their Usage*

R. W. G. HUNT

Research Division
Kodak Ltd.
Harrow, United Kingdom

I. INTRODUCTION

To some people, terminology may seem to be an unimportant subject: "Let's get on with the work and not fuss about words" may be their comment. But as in other subjects worthy of serious study, communication and discussion are vitally important in color, and these are seriously hampered if, like Humpty-Dumpty in "Alice through the Looking Glass," we each take the attitude that, "When I use a word it means just what I choose it to mean—neither more nor less."

It is therefore fortunate for the subject of color that its terminology has in the past received careful attention from some of its ablest workers and has been the subject of publications by prestigious bodies, including the International Commission on Illumination (CIE, 1970), the Colour Group of Great Britain (Physical Society, 1948), the Optical Society of America (Troland, 1922; Committee on Colorimetry, 1944), and the Inter-Society

* Adapted from the paper by R. W. G. Hunt entitled "Colour Terminology" in *Color Research and Application* (1978) by permission of John Wiley & Sons, Inc.

Color Council (Judd, 1939; Newhall and Brennan, 1949). The terms and definitions in this chapter are based mainly on the most recent work of the CIE Colorimetry Committee.

II. SUBJECTIVE TERMS

In color both the response of the observer (the subject) and the physical nature of the stimulus (the object) are important, and it is necessary to distinguish clearly between these *subjective* and *objective* aspects of color. It is convenient for the present discussion to start with the response of the observer (the subjective, or psychosensorial, or perceptual aspect), and where a term may be used both in this sense and in an objective sense, the adjective *perceived* is used to make it clear that it is the subjective aspect that is meant.

We therefore start with

(perceived) color: Attribute of visual perception that can be described by color names: white, gray, black, yellow, orange, brown, red, green, blue, purple, and so on, or by combinations of such names.

It will be noted that white, gray, and black are colors. The definition for *hue*, however, excludes white, gray, and black:

hue: Attribute of a visual sensation according to which an area appears to be similar to one, or to proportions of two, of the perceived colors red, yellow, orange, green, blue, and purple.

This then leads to the following two classes of perceived colors:

achromatic (perceived) color: Perceived color devoid of hue.
chromatic (perceived) color: Perceived color possessing a hue.

In a chromatic perceived color the hue may be exhibited weakly or strongly (as in pink and red, for instance) and hence we have

colorfulness: Attribute of a visual sensation according to which an area appears to exhibit more or less chromatic color.

This is a new term (Hunt, 1977); its relationship to the older terms, saturation and chroma, will be discussed later. Colorfulness and hue provide two of the three basic subjective color terms; the third term is *brightness*.

brightness: Attribute of a visual sensation according to which an area appears to exhibit more or less light.

These three terms—hue, colorfulness, and brightness—are all used in everyday speech, and this is a desirable feature for basic subjective terms.

Although all colors can be described in terms of hue, colorfulness, and brightness, the recognition of colored *objects* is often dependent not so much on colorfulness as on

saturation: The colorfulness of an area judged in proportion to its brightness.

Saturation is thus a *relative* colorfulness.

The following example illustrates the difference between colorfulness and saturation. Imagine a display of colored lights at night, seen both directly and reflected by a surface of still water or by a sheet of glass. The reflections of the lights will generally appear less bright and less colorful than when viewed directly, but the reflected lights will often be described as being of lower brightness but of the same so-called color (red and pink lights, for instance, still being seen as red and pink). In this case the colorfulness, although lower in the reflected lights, is being judged in proportion to the brightness, which is also lower, and the lights are seen as having the same saturation. Saturation is thus a very useful attribute in perceiving the nature of objects, enabling us, for instance, to recognize the difference between red and pink lights over a very wide range of brightnesses.

The terms hue, colorfulness, brightness, and saturation describe attributes that are always present in chromatic perceived colors, but there are some further attributes that are present only in some of these colors, and these will be considered next.

For related colors (seen in relation to other colors) the fraction of light perceived to be reflected (or transmitted) often provides a very important clue to the identity of an object, and this results in the following important perceptual attribute:

lightness (of a related color): The brightness of an area judged relative to the brightness of a similarly illuminated area that appears to be white or highly transmitting.

Lightness is thus a *relative* brightness, and the distinction between these two terms can be illustrated as follows. A diffusely reflecting gray object for which the fraction of reflected light is, say, 20%, when seen among other typical objects will have a certain lightness, often described as a medium gray. If now the level of illumination on the objects viewed is greatly increased, then, although the eye normally compensates for the change by reducing its sensitivity, such compensation is usually only partial, and the brightnesses of the objects increase. However, the object reflecting 20% of the light will still appear to be a medium gray: Its lightness will have remained constant, because although its brightness will have increased, so also will that of a white similarly illuminated. Similarly, the lightnesses of other reflecting objects will remain roughly the same. It is the tendency for the lightness of objects to be largely unaffected by changes in illumination level that makes it such an important attribute in the visual task of recognizing objects, and the use of the specific term *lightness*, in addition to *brightness*, is therefore useful. Lightness is applicable not only to diffusely reflecting objects, but also to transmitting objects.

Related colors require not only a special term for relative brightness, but also two special terms for relative colorfulness. One of these terms is *saturation* which has already been discussed and is also applicable to unrelated colors (colors seen in isolation from other colors); the other term is specific to related colors:

> **perceived chroma** (of a related color): The colorfulness of an area judged in proportion to the brightness of a similarly illuminated area that appears to be white or highly transmitting.

Perceived chroma is thus another form of *relative* colorfulness.

The distinction between saturation and perceived chroma can be illustrated as follows. Consider a scene that includes a person wearing a red coat. At a given level of illumination, the colorfulness of the coat will be perceived as at a certain level. But if we look carefully at different parts of the scene, we will usually note that some parts of it are receiving the light at a rather more glancing angle than others and that other parts are in shadows. In the case of the coat these more dimly lit areas will have lower brightnesses than the areas more highly illuminated, and they will also have lower colorfulness; but the coat will be perceived as being the same color all over. Hence, although the brightness and colorfulness decrease in the more dimly lit areas, there must be other attributes that remain constant: One of these constant attributes is the hue, and another is the *saturation*. The saturation is constant because the colorfulness, although lower in the more dimly lit areas, is then judged in proportion to the brightness, which is also lower.

Consider now that the person is also wearing a red scarf having a pattern containing areas of white and areas of red that are darker than the rest of the scarf; consider also that the sizes of the areas in the pattern are small compared to the nonuniformity of illumination on the scarf caused by its angles to the light and by shadows. Although the colorfulness of the dark red patterned areas could be judged in proportion to their brightnesses and could be seen to have the same saturation as the rest of the scarf, such a judgment would be less informative about the nature of the object than to judge the colorfulness in proportion to the brightnesses of the adjacent white areas. When this is done what is seen is *perceived chroma*. Judged in this way, the colorfulness is lower in the dark red areas of the scarf than in the rest of the scarf; hence, these areas have lower perceived chromas. However, the perceived chromas are the same in the dimly lit areas as in those more highly illuminated, because the colorfulnesses of the dark red areas and the brightnesses of the white areas adjacent to them will be changing together, both being low in the dimly lit areas and both being high in the brightly lit areas. Similarly, the lightnesses of the patterned areas are the same in the dimly and brightly lit areas, because their lightnesses will be judged as their brightnesses relative to those of the white patterned areas adjacent to them.

Because perceived chroma is colorfulness judged in proportion to the brightness of a similarly illuminated white, its assessment usually requires a color to have surroundings of similar illumination that contain a white or a near white. This is the case for each patterned area of the scarf, but the nonuniformity of illumination on the coat and on its surroundings may result in its perceived chroma being difficult or impossible to assess. But saturation, being assessed as colorfulness in proportion to the brightness of the color itself, does not suffer from this difficulty; hence, the coat saturation may be more in evidence than perceived chroma. Similar considerations apply to the perception of lightness. The nonuniform illumination on the coat and its surroundings may make its lightness difficult to assess, but the lightnesses in the scarf can be judged because of the presence of the adjacent white areas of the pattern.

Lightness and perceived chroma are most easily assessed when flat objects are viewed in uniform illumination in the presence of a white, as, for instance, when samples of fabrics or papers are seen in viewing booths. In these cases lightness and perceived chroma are usually more obvious than brightness and saturation. Conversely, lightness and perceived chroma are least easily assessed when intricate three-dimensional objects are viewed in nonuniform directional lighting, as, for instance, when flowers are seen in sunlight against a dark background. In these cases saturation is usually more obvious than perceived chroma, and lightness is assessed with more difficulty than brightness, but in these latter cases, lightness and perceived chroma can usually still be assessed in patterns on objects if the patterned areas are small compared to the nonuniformity of the illumination and contain areas of white.

Returning to the example of the coat and scarf again, if we now imagine that the *general* level of illumination is gradually reduced, the colorfulness of each area of the coat (and of other objects illuminated with it) will gradually decrease and will in fact reach zero when the illumination level is very low (below that corresponding to the level of moonlight). Conversely, if the general illumination level is increased, the colorfulness will increase, at least up to levels corresponding to that typical of clear sunlight. But over most of this range of illumination levels, although the colorfulness varies, the saturation of the coat will remain approximately constant. This is because its colorfulness will be judged in proportion to its brightness, which will be changing in a similar way, being low when the illumination level is dim and high when it is bright. Similarly, the perceived chromas of the patterned areas of the scarf will remain approximately constant, because their colorfulnesses will be judged in proportion to the brightnesses of the adjacent white areas, and both will be changing in a similar way, low at general illumination levels that are dim and high at those that are bright.

If we consider now the brightness and lightness perceived in the scarf we find that similar considerations apply. At general levels of illumination that are low, the brightnesses seen in the scarf will all be low; and at high levels, most areas of the scarf will be perceived as having high brightnesses. But in the patterned areas that contain the whites, lightness will be assessed, and it will be approximately the same for all levels of illumination.

It is clear from the preceding discussion that, although hue, brightness, and colorfulness are the three basic attributes of perceived color, when it comes to the recognition of objects, hue and saturation are important attributes of unrelated colors, and in the case of related colors, hue and either saturation or a combination of perceived chroma and lightness are important.

We shall conclude this section by considering a few other subjective terms of particular interest to our discussion in this chapter.

Another way in which perceived colors can be divided into two classes is

luminous (perceived) color: Color perceived to belong to an area that appears to be emitting light as a primary light source or that appears to be specularly reflecting such light.

nonluminous (perceived) color: Color perceived to belong to an area that appears to be transmitting or diffusely reflecting light as a secondary light source.

To clarify the nature of this distinction it is necessary to note that primary and secondary light sources are defined as follows:

primary light source: Surface or object emitting light produced by a transformation of energy.

secondary light source: Surface or object that is not self-emitting but receives light and redirects it, at least in part, by reflection or transmission.

It should be noted that the distinction between luminous and nonluminous perceived colors is made on the basis of whether the color *appears to be* emitting light or redirecting it; hence, the same object could appear to be luminous in one set of conditions and nonluminous in another. Thus, a gray cloud seen out-of-doors might appear to be nonluminous, but when seen through a small window from a rather dark room might appear to be luminous. Colors that are luminous but whose brightness is only modestly in excess of those that are nonluminous are commonly referred to as "fluorescent," but such colors may not be fluorescent in the physical sense of emitting at some wavelengths light that is in excess of that produced by thermal radiation.

Only related colors can have a lightness, and this is also true of gray content, which is defined as follows:

gray content: Attribute of a visual sensation that permits a judgment to be made of the apparent proportion of gray in a related color.

The distinction between luminous and nonluminous colors is usually quite a sharp one, and colors that are on the boundary between the two classes are defined as having

zero gray content: Attribute of a visual sensation according to which a perceived color is judged to have the highest possible brightness without being luminous.

A very widely used color name is *brown* and this is defined as follows:

brown: A color name used to describe nonluminous colors of yellowish, orange, or reddish hue that have an appreciable gray content.

Another class of color often encountered is

metallic (perceived) color: Color perceived to belong to an area that appears to be specularly reflecting from its surface a high proportion of light incident on it from another source, sometimes imparting to the light a characteristic hue.

Examples of metallic (perceived) colors are silver and gold, but these names are sometimes used rather loosely to describe colors whose specular reflection is quite low.

It is sometimes useful to consider the hue and saturation of a perceived color together; hence, we have

chromaticness: Attribute of a visual sensation consisting of the hue and the saturation.

III. OBJECTIVE TERMS

We must now turn our attention to some of the more important objective terms: in these terms it is the color stimulus, not the perceived color, that is considered.

It is convenient to divide some of the objective terms into *psychophysical* terms and *psychometric* terms. The distinction between these two sets of terms and subjective terms can be seen from the following definition:

Subjective

Perceptual terms: Terms denoting important attributes of sensations of light and color. Any measures of such attributes must indicate the subjective magnitudes of response in a visual process.

Objective

Psychophysical terms: Terms denoting objective measures of physical variables that are evaluated so as to relate to equality of magnitudes of important attributes of light and color. These measures identify stimuli that produce equal responses in a visual process in specified viewing conditions.

Psychometric terms: Terms denoting objective measures of physical variables that are evaluated so as to relate to differences between magnitudes of important attributes of light and color and such that equal scale intervals represent approximately equal perceived differences in the attribute considered. These measures identify pairs of stimuli that produce equally perceptible differences in response in a visual process in specified viewing conditions.

A. Psychophysical Terms

The psychophysical terms of major importance in color are luminance, luminance factor, dominant (or complementary) wavelength, purity, and chromaticity. They are defined as follows:

luminance (in a given direction, at a point on the surface of a source or a receptor, or at a point on the path of a beam): Quotient of the luminous flux leaving, arriving at, or passing through an element of surface at this point and propagated in directions defined by an elementary cone containing the given direction, by the product of the solid angle of the cone and the area of the orthogonal projection of the element of surface on a plane perpendicular to the given direction.

Symbol: L,

$$L = \frac{d^2\Phi}{d\Omega \, dA \cos \theta},\tag{1}$$

where Φ is the luminous flux, $d\Omega$ is the solid angle of the cone, dA the area of the element of the surface, and θ the angle between the direction of view and the normal to the surface.

Unit: candela per square meter cd/m^2

Luminance is the psychophysical measure that correlates approximately with *brightness*: two colors having equal luminance appear of approximately the same brightness when seen under the same conditions. [But if one of the colors is viewed in different conditions, for example against a much lighter or darker background, then the brightnesses would usually no longer be equal, although the luminances would not have changed.]

One reason that brightness correlates only approximately with luminance (even when the viewing conditions are the same) is that a given luminance produces slightly different brightnesses for various saturations and hues.

> **luminance factor:** Ratio of the luminance of a body to that of a perfect reflecting or transmitting diffuser identically illuminated.
> Symbol: β

Luminance factor correlates with *lightness* but does so only approximately for reasons similar to those just discussed in connection with luminance and brightness.

> **dominant wavelength** (of a color stimulus): Wavelength of the monochromatic light stimulus that, when additively mixed in suitable proportions with the specified achromatic light stimulus, matches the color stimulus considered.
> Symbol: λ_d
> **complementary wavelength** (of a color stimulus): Wavelength of the monochromatic light stimulus that, when additively mixed in suitable proportions with the color stimulus considered, matches the specified achromatic light stimulus.
> Symbol: λ_c

Complementary wavelength is used only for colors (purples) for which a dominant wavelength cannot be given. Dominant and complementary wavelength correlate approximately with *hue*, but the exact hue perceived depends on the viewing conditions and on the saturation.

> **purity:** A measure of the proportions of the amounts of the monochromatic light stimulus and of the specified achromatic light stimulus that, when additively mixed, match the color stimulus considered.
> Symbol: p

In the case of purples, a suitable mixture of light from the two ends of the spectrum is used instead of the monochromatic stimulus. Purity is related to *saturation*, but the saturation perceived for a given purity depends to a marked degree on the hue, as well as on the viewing conditions. If the proportions of the amounts of the stimuli are measured in luminances, then the measure is known as *colorimetric purity* p_c and if they are measured in terms of relative distances on an x, y (or x_{10}, y_{10}) chromaticity diagram, then the measure is known as *excitation purity* p_e.

chromaticity coordinates: Ratio of each of a set of three tristimulus values to their sum.

Tristimulus values are the amounts of three color-matching or reference color stimuli; they are always represented by capital letters, for example, R, G, B for a set of red, green, and blue color-matching stimuli or X, Y, Z for the CIE reference-color stimuli. Chromaticity coordinates are always represented by the corresponding small letters: r, g, b and x, y, z in these examples. Hence,

$$R/(R + G + B) = r,$$
$$G/(R + G + B) = g,$$
$$B/(R + G + B) = b,$$

and $r + g + b = 1$. Similarly,

$$X/(X + Y + Z) = x,$$
$$Y/(X + Y + Z) = y,$$
$$Z/(X + Y + Z) = z,$$

and $x + y + z = 1$.

chromaticity: Color quality of a color stimulus definable by its chromaticity coordinates or by its dominant (or complementary) wavelength and purity taken together.

Chromaticity correlates approximately with *chromaticness*, with the exact relationship depending very much on the viewing conditions, including the nature of the surround.

chromaticity diagram: A diagram in which distances along suitable axes represent chromaticity coordinates.

Two axes at right angles to one another are commonly used (the third chromaticity coordinate always being equal to the sum of the other two subtracted from unity). In the CIE 1931 Standard Colorimetric System y is normally plotted as ordinate and x as abcissa to obtain an x, y *chromaticity diagram* (or an x_{10}, y_{10} *chromaticity diagram* in the case of the CIE 1964 Supplementary Standard Colorimetric System).

B.   Psychometric Terms

The distinction between psychophysical and psychometric terms can be illustrated by the following example. Consider a series of flat gray samples viewed in uniform illumination. The most important subjective quality whereby the grays are distinguished is the *perceptual* attribute, *lightness*. The simplest objective measure of a physical variable that relates to the attribute of lightness is the *psychophysical* measure *luminance factor*. Two grays having equal luminance factors would have the same lightness when viewed in identical conditions. (But if one of the grays is viewed in different conditions—for example, against a much lighter or darker background—then the lightnesses would usually no longer be equal, although the luminance factors would not have changed.)

However, a scale of grays whose luminance factors are uniformly spaced, for example, 0.1, 0.3, 0.5, 0.7, and 0.9, appear to observers to be very unevenly spaced in lightness. A series of grays whose lightness spacing appears subjectively uniform might have luminance factors of about 0.10, 0.20, 0.36, 0.59, and 0.90. It is therefore convenient to have an objective measure that is *psychometric*, that is, one that relates to the attribute of lightness but whose scale intervals represent approximately equal perceived differences in lightness. Such a measure is called *psychometric lightness*:

psychometric lightness: Quantity defined by a suitable function of psychophysical measures (principally luminance factor) such that equal scale intervals correspond as closely as possible to equal differences in lightness for related colors.

One measure of psychometric lightness (in which the luminance factor is written as Y/Y_n) is

CIE 1976 psychometric lightness: Quantity L^* defined by the relation

$$L^* = 116(Y/Y_n)^{1/3} - 16, \qquad Y/Y_n > 0.008856. \qquad (2)$$

Y is the Y tristimulus value of the color stimulus considered, and Y_n that of a specified white object-color stimulus.

Values of L^* can be calculated when Y/Y_n is less than or equal to 0.008856 by using the formula

$$L^* = 903.3(Y/Y_n), \qquad Y/Y_n \lessgtr 0.008856.$$

The choice of 0.008856 as the value at which the functions change was made so that the first differential is smooth when transferring from one function to another.

Pairs of grays having equal differences in CIE 1976 psychometric lightness have approximately equal differences in lightness when viewed in identical conditions. (But if one of the pairs of grays is viewed in different conditions —for example, against a much lighter or darker background—then the perceived difference in lightness might no longer equal that of the other pairs of grays, although the differences in psychometric lightness would not have changed.)

In Table I are shown psychometric terms relating not only to lightness but also to brightness, hue, colorfulness, saturation, perceived chroma, and chromaticness. Table I also includes the psychophysical terms luminance, luminance factor, dominant wavelength, purity, and chromaticity, already described.

The psychometric terms in Table I are all defined in ways similar to that already given for psychometric lightness. Thus, psychometric brightness is defined as

psychometric brightness: Quantity defined by a suitable function of psychophysical measures (principally luminance and also, in the case of related colors, luminance factor) such that equal scale intervals correspond as closely as possible to equal differences in brightness.

In the definition of each of these psychometric terms mention is made of the principal psychophysical measure or measures involved and of the perceptual concept concerned, as shown in Table II.

Of a character slightly different from the psychometric terms in Table II is *psychometric chromaticness*, which is defined with reference to a *uniform-chromaticity-scale (UCS) diagram*:

TABLE I.   Perceptual, Psychometric, and Psychophysical Terms

Subjective Perceptual term	Objective Psychometric term	Psychophysical term
Brightness	Psychometric brightness	Luminance
Lightness[a]	Psychometric lightness[a]	Luminance factor[a]
Hue	Psychometric hue	Dominant wavelength
Colorfulness	Psychometric colorfulness	—
Saturation	Psychometric saturation	—
—		Purity
Perceived chroma[a]	Psychometric chroma[a]	—
Chromaticness	Psychometric chromaticness	Chromaticity

[a] These terms are only applicable to related colors.

TABLE II. Psychometric Terms, Psychophysical Measures, and Perceptual Concept

Psychometric term	Psychophysical measures	Perceptual concept
Psychometric brightness	Luminance, luminance factor	Brightness
Psychometric lightness	Luminance factor	Lightness
Psychometric hue	Dominant wavelength	Hue
Psychometric colorfulness	Chromaticity coordinates, luminance, luminance factor	Colorfulness
Psychometric saturation	Chromaticity coordinates	Saturation
Psychometric chroma	Chromaticity coordinates, luminance	Perceived chroma

uniform-chromaticity-scale diagram; UCS diagram: Chromaticity diagram in which the coordinate scales are chosen with the intention of making equal intervals represent as nearly as possible equal steps of discrimination for colors of the same luminance at all parts of the diagram.

psychometric chromaticness: Color quality of a color stimulus definable by its coordinates in a uniform-chromaticity-scale (UCS) diagram.

There are at present no agreed formulas for calculating measures of psychometric brightness and psychometric colorfulness, although formulas have been proposed for discussion (Hunt, 1977). But formulas for calculating measures of all the other psychometric terms given in Table I are available (CIE, 1977). These measures are denoted by terms with the prefix *CIE 1976* to distinguish them from other similar measures that may be introduced at a future date.

Thus, *CIE 1976 psychometric chromaticness* consists of the coordinates in the *CIE 1976 uniform-chromaticity-scale (UCS) diagram*; in this diagram v' is plotted against u' where

$$u' = 4X/(X + 15Y + 3Z), \tag{3}$$

$$v' = 9Y/(X + 15Y + 3Z), \tag{4}$$

and where X, Y, Z are the tristimulus values of the color considered. This diagram is shown in Fig. 1. *CIE 1976 psychometric saturation* s_{uv} is calculated as 13 times the distance on this diagram between the color stimulus considered (at u', v') and the specified achromatic light stimulus (at u'_n, v'_n), as also shown in Fig. 1. Thus,

$$s_{uv} = 13[(u' - u'_n)^2 + (v' - v'_n)^2]^{1/2}. \tag{5}$$

The reason for the constant 13 will be explained later. The diagram also

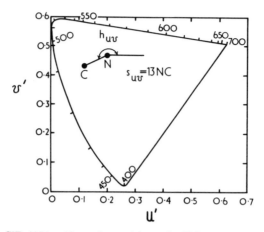

Fig. 1. The CIE 1976 uniform-chromaticity-scale (UCS) diagram showing CIE 1976 psychometric saturation s_{uv} and CIE 1976 u, v hue angle, h_{uv}. H represents the specified achromatic light stimulus and C the color considered.

provides a psychometric measure of hue, *CIE 1976 u, v hue angle, h_{uv}*, which, as shown in Fig. 1, is given by

$$h_{uv} = \tan^{-1}[(v' - v_n)/(u' - u'_n)]$$
$$= \arctan[(v' - v'_n)/(u' - u'_n)].$$

C. Color Solids

It is often convenient to regard triads of psychometric measures as being plotted in such a way as to form a three-dimensional *color solid* or *color space*. In such color spaces it is usually the intention that the perceptual difference between any two colors should be proportional to the distance between them. The Munsell system, which consists of small painted color chips, can be regarded as generating a space of this type. Munsell colors are denoted by their Munsell hue, which correlates with hue; by their Munsell value, which correlates with lightness; and by their Munsell chroma, which correlates with perceived chroma. For each of these Munsell variables scales are used that are intended to be perceptually uniform. For each Munsell sample a CIE color specification is available in terms of the usual X, Y, Z tristimulus values. However, the relationship between these specifications and Munsell hue, value, and chroma is very complicated. The CIE has therefore established two other color spaces that are similar to that generated by the Munsell system but that are more simply related to the CIE color specifications.

Both the CIE color spaces use *CIE 1976 psychometric lightness* L^* (the formula for which has already been given) as the variable that correlates with lightness. The L^* axis is usually regarded as being vertical in the spaces. The other two axes are

$$u^* = 13L^*(u' - u'_n),\tag{6}$$

$$v^* = 13L^*(v' - v'_n),\tag{7}$$

for the CIE 1976 ($L^*u^*v^*$) color space, and

$$a^* = 500[(X/X_n)^{1/3} - (Y/Y_n)^{1/3}],\tag{8}$$

$$b^* = 200[(Y/Y_n)^{1/3} - (Z/Z_n)^{1/3}],\tag{9}$$

where

$$X/X_n > 0.008856,$$

$$Y/Y_n > 0.008856,$$

$$Z/Z_n > 0.008856,$$

for the CIE 1976 ($L^*a^*b^*$) color space. (The suffix n in all cases denotes that the quantity is for the specified white object color stimulus.) These variables (u^*, v^* or a^*, b^*) are usually regarded as being plotted on a horizontal plane. The multiplication of the chromaticity differences ($u' - u'_n$) and ($v' - v'_n$) by L^* is to allow for the fact that a given difference in chromaticity becomes more difficult to see as the lightness is reduced. In the case of a^* and b^* this effect is allowed for by using functions of tristimulus values instead of chromaticity coordinates. The constant 13 in the formulae for u^* and v^* provides the appropriate scaling relative to L^*; the constants 500 and 200 in the formulae for a^* and b^* provide the appropriate scaling relative to each other and to L^*. Values of a^* and b^* can be calculated when X/X_n, Y/Y_n, or Z/Z_n are not greater than 0.008856 by using the more elaborate formulae

$$a^* = 500[f(X/X_n) - f(Y/Y_n)],$$

$$b^* = 200[f(Y/Y_n) - f(Z/Z_n)],$$

where

$$f(X/X_n) = (X/X_n)^{1/3}, \qquad X/X_n > 0.008856,$$

$$f(X/X_n) = 7.787(X/X_n) + 16/116, \qquad X/X_n \leq 0.008856,$$

and similarly for $f(Y/Y_n)$ amd $f(Z/Z_n)$.

Correlates of perceived chroma in these spaces are given by *CIE 1976 u, v chroma* C_{uv}^* and *CIE 1976 a, b chroma* C_{ab}^*. Correlates of hue are given by *CIE 1976 u, v hue angle* h_{uv}, and *CIE 1976 a, b hue angle* h_{ab}. The formulae for these variables are as follows (the formula for h_{uv} now being given again, but in terms of u^* and v^*):

$$C_{uv}^* = (u^{*2} + v^{*2})^{1/2}, \tag{10}$$

$$C_{ab}^* = (a^{*2} + b^{*2})^{1/2}, \tag{11}$$

$$h_{uv} = \tan^{-1}(v^*/u^*) = \arctan(v^*/u^*), \tag{12}$$

$$h_{ab} = \tan^{-1}(b^*/a^*) = \arctan(b^*/a^*). \tag{13}$$

The constant 13 is used in the formula for s_{uv} so that

$$s_{uv} = C_{uv}^*/L^* \tag{14}$$

The formulae given provide psychometric measures that are intended to approximate uniform scales of the concepts concerned. In the case of C^* and h it is interesting to plot the chroma contours and hue lines of the Munsell system in u^*, v^* and a^*, b^* diagrams. This is done (for samples of Munsell value 5) in Figs. 2 and 3. If the Munsell system is regarded as a perceptually uniform color space, then the degree to which C^* and h approximate psychometric measures of perceived chroma and hue can be

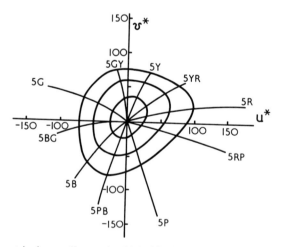

Fig. 2. Psychometric chroma diagram in which v^* is plotted against u^*, where v^* and u^* are two of the variables used in the CIE 1976 ($L^*u^*v^*$) color space. The approximately straight radial lines are loci of Munsell hues, 5R, 5YR, 5Y, 5GY, 5G, 5BG, 5B, 5PB, 5P, and 5RP at Munsell value 5. The approximately circular concentric contours are loci of Munsell chroma 4, 8, and 12 at Munsell value 5.

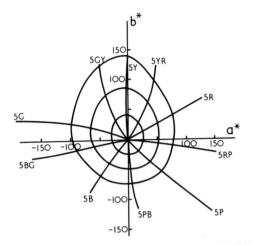

Fig. 3.   Same as Fig. 2 but for b^* and a^* of the CIE 1976 ($L^*a^*b^*$) color space.

gauged by noting departures of the Munsell hue loci in Figs. 2 and 3 from straight radial lines having equal angular spacing and departures of the Munsell chroma loci from equally spaced concentric circles.

Figures in which pairs of variables such as u^*, v^* and a^*, b^* are plotted against one another do not generate chromaticity diagrams, because u^*, v^* and a^*, b^* are not ratios of tristimulus values; they are related to tristimulus values by nonlinear functions, as shown in Eq. (6), (7), (8), and (9). It is therefore helpful to define

> **psychometric chroma coordinates:** Measures of psychometric chroma along two directions at right angles to one another and in planes normal to the achromatic axis in a uniform color space.
> **psychometric chroma diagram:** A diagram in which distances along two axes at right angles to one another represent psychometric chroma coordinates.

Thus, u^*, v^* and a^*, b^* plots are *psychometric chroma diagrams*.

The CIE 1976 ($L^*u^*v^*$) color space can be used to show the relationships between psychometric lightness, hue angle, chroma, and saturation, and this is done in Fig. 4. If the L^* axis is regarded as vertical and the u^* and v^* axes are regarded as lying in a horizontal plane, then vertical planes containing the L^* axis as one edge are planes of constant CIE 1976 u, v hue angle h_{uv}, cylinders having L^* as their axes are surfaces of constant CIE 1976 u, v chroma C^*_{uv}, and cones having L^* as their axes and their apices at the origin are surfaces of constant CIE 1976 u, v saturation s_{uv}. Horizontal planes through the solid are planes of constant CIE 1976 psycho-

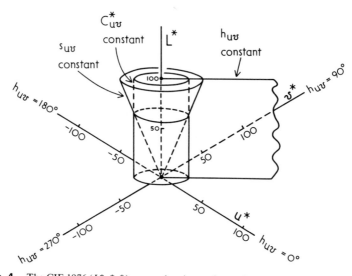

Fig. 4. The CIE 1976 ($L^*u^*v^*$) space, showing surfaces of constant CIE 1976 u,v hue angle h_{uv}, CIE 1976 u,v chroma C^*_{uv}, and CIE 1976 psychometric saturation s_{uv}. Similar surfaces occur in the CIE 1976 ($L^*a^*b^*$) space except in the case of psychometric saturation.

metric lightness L^*. They contain CIE 1976 uniform-chromaticity-scale diagrams in the sense that, on any given plane, if points in the solid are projected onto it by means of straight lines passing through the origin, then the projected points generate a CIE 1976 UCS diagram, multiplied by an overall scale factor. [Similar properties belong to the CIE 1976

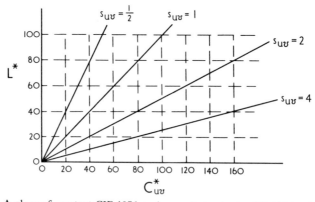

Fig. 5. A plane of constant CIE 1976 u,v hue angle h_{uv}, in the CIE 1976 ($L^*u^*v^*$) color space, showing horizontal lines of constant CIE 1976 psychometric lightness L^*, vertical lines of constant CIE 1976 u,v chroma C^*_{uv}, and radial lines of constant CIE 1976 psychometric saturation s_{uv}. Similar lines occur in the CIE 1976 ($L^*a^*b^*$) space except in the case of psychometric saturation.

($L^*a^*b^*$) color space except that no simple psychometric correlate of saturation exists, and there is no simple relationship to any chromaticity diagram.]

Figure 5 shows a plane containing a single CIE 1976 u, v hue angle; on such a plane horizontal lines are loci of constant CIE 1976 psychometric lightness L^*, vertical lines are loci of constant CIE 1976 u, v chroma C_{uv}^*, and lines through the origin are lines of constant CIE 1976 psychometric saturation s_{uv}.

It is not possible to represent measures of psychometric brightness and psychometric colorfulness in CIE 1976 ($L^*u^*v^*$) and ($L^*a^*b^*$) color spaces, because only correlates of *relative* brightness (lightness) and *relative* color-fulness (saturation and perceived chroma) are considered. But the definition of saturation as colorfulness judged in proportion to brightness is represented psychometrically by the relationship

$$s_{uv} = C_{uv}^*/L^*. \tag{15}$$

This arises because perceived chroma and lightness are, respectively, color-fulness and brightness both judged in proportion to a white or highly transmitting object color similarly illuminated.

The CIE 1976 ($L^*u^*v^*$) and ($L^*a^*b^*$) color spaces provide the basis for two different color difference formulae (Robertson, 1977):

$$\Delta E_{uv}^* = [(\Delta L^*)^2 + (\Delta u^*)^2 + (\Delta v^*)^2]^{1/2}, \tag{16}$$

$$\Delta E_{ab}^* = [(\Delta L^*)^2 + (\Delta a^*)^2 + (\Delta b^*)^2]^{1/2}, \tag{17}$$

where ΔL^*, Δu^*, Δv^*, Δa^*, and Δb^* indicate the difference in L^*, u^*, v^*, a^*, and b^*, respectively, between the two colors concerned. The term ΔE_{uv}^* is thus the distance in the ($L^*u^*v^*$) space, and ΔE_{ab}^* is that in the ($L^*a^*b^*$) space between points representing the two colors.

When it is desired to identify the components of color differences in terms of approximate correlates of lightness, perceived chroma, and hue, the two color differences may be rewritten as

$$\Delta E_{uv}^* = [(\Delta L^*)^2 + (\Delta C_{uv}^*)^2 + (\Delta H_{uv}^*)^2]^{1/2}, \tag{18}$$

$$\Delta E_{ab}^* = [(\Delta L^*)^2 + (\Delta C_{ab}^*)^2 + (\Delta H_{ab}^*)^2]^{1/2}, \tag{19}$$

where ΔH_{uv}^* is the *CIE 1976 u, v hue difference* and ΔH_{ab}^* is the *CIE 1976 a, b hue difference*:

$$\Delta H_{uv}^* = [(\Delta E_{uv}^*)^2 - (\Delta L^*)^2 - (\Delta C_{uv}^*)^2]^{1/2}, \tag{20}$$

$$\Delta H_{ab}^* = [(\Delta E_{ab}^*)^2 - (\Delta L^*)^2 - (\Delta C_{ab}^*)^2]^{1/2}. \tag{21}$$

It is necessary to use ΔH_{uv}^* and ΔH_{ab}^* in these formulas rather than Δh_{uv} and Δh_{ab}, because differences in these latter measures, being angular in nature, correspond to color differences that are proportional to C^*. For small color differences away from the L^* axis, if h is expressed in degrees,

$$\Delta H^* = C^* \Delta h(\pi/180). \tag{22}$$

CIE 1976 psychometric saturation s_{uv} is also an angular measure; in this case differences in s_{uv} correspond to color differences that are proportional to L^*:

$$\Delta C_{uv}^* = L^* \Delta s_{uv}. \tag{23}$$

D. Correlated Color Temperature

The replacement of the CIE 1960 UCS diagram (in which v is plotted against u) by the CIE 1976 UCS diagram (in which $v' = 1.5v$ is plotted against $u' = u$) (CIE, 1977) raised the question as to whether the method of determining *correlated color temperature*, which depended on the u, v diagram (CIE, 1974), should be replaced by a similar method depending on the u', v' diagram. It has been decided not to change the method, for two reasons. First, experimental studies (Clarke, 1977; Grum, 1977; Terstiege, 1977) showed that correlated color temperature could only be judged very imprecisely by a single observer from one occasion to another, and among several observers, and neither diagram appeared to predict the results significantly better than the other. Second, a change in the method would occasion great inconvenience in connection with the calculation of color-rendering indices of lamps. Accordingly, the definition of correlated color temperature has been reworded in such a way as to retain the existing method of calculation:

correlated color temperature: The temperature of the Planckian radiator whose perceived color most closely resembles that of a given illuminant at the same brightness and under specified viewing conditions.
Symbol: T_{cp}
Unit: kelvin (K)
Note: Correlated color temperatures determined experimentally exhibit a considerable spread among observers. The agreed method of calculating the correlated color temperature of an illuminant is to determine on a chromaticity diagram the temperature on the Planckian locus that is intersected by the agreed isotemperature line containing the point representing the illuminant. The agreed isotemperature lines to be used are those that are normal to the Planckian locus on a chromaticity diagram in which $\frac{2}{3}v'$ is plotted against u', where v' and u' are the coordinates of the CIE 1976 uniform-chromaticity-scale diagram.

E. Abbreviations

When the context is such that no confusion is likely, the abbreviation *metric* for *psychometric* is suggested.

Abbreviations suggested for CIE 1976 ($L^*u^*v^*$) and CIE 1976 ($L^*a^*b^*$) are CIELUV and CIELAB, respectively. Thus, CIE 1976 u, v chroma can be referred to as CIELUV chroma, and CIE 1976 a, b hue angle as CIELAB hue angle.

REFERENCES

CIE (1970). "International Lighting Vocabulary," 3rd Ed., Publ. No. 17 (E-1.1). Bureau Central CIE, Paris.

CIE (1974). "Method of Measuring and Specifying Colour Rendering Properties of Light Sources," 2nd Ed., Publ. No. 13.2. Bureau Central CIE, Paris.

CIE (1977). "Official Recommendations on Uniform Color Spaces, Color Difference Equations, and Metric Color Terms," Suppl. No. 2 to "Colorimetry," Publ. No. 15. Bureau Central CIE, Paris.

Clarke, F. J. J. (1977). Personal communication.

Committee on Colorimetry (1944). *J. Opt. Soc. Am.* **34**, 245–266.

Grum, F. (1977). Personal communication.

Hunt, R. W. G., (1977). *Color Res. Appl.* **2**, 55–68, 109–120.

Judd, D. B., ed. (1939). "Comparative List of Color Terms," 1st Ed. Inter-Soc. Color Counc., Washington, D.C.

Newhall, S. M., and Brennan, J. G., eds. (1949). "Comparative List of Color Terms," 2nd Ed. Inter-Soc. Color Counc., Washington, D.C.

Physical Society (1948). "Report on Colour Terminology." Phys. Soc., London.

Robertson, A. R. (1977). *Color Res. Appl.* **2**, 7–11.

Terstiege, H. (1977). Personal communication.

Troland, L. T. (1922). *J. Opt. Soc. Am.* **6**, 527–596.

3

Colorimetry

C. J. BARTLESON

Research Laboratories
Eastman Kodak Comapny
Rochester, New York

I. INTRODUCTION

"Colorimetry" is a word that stands for methods of measuring or evaluating the colors of objects. We commonly use the term "color" to refer to an aspect of visual experience. When we speak of seeing the colors of objects, we generally mean that we perceive the object to be dark or light; to be white, gray, or black; or to have some hue such as red, green, yellow, blue, or some combination of those attributes; and to be more or less colorful.

33

These perceptions are private phenomena; they take place in our own minds. We alone know what we see. Such private or personal events are often called "psychological phenomena" because only the individual experiencing the event knows what it is. If the individual wants to communicate to others something about what he sees, then he must try to describe the psychological phenomenon in more or less formal ways. Certainly the oldest and probably least formal way of communicating psychological phenomena such as perceived color is by the use of descriptive words. People who share a common language also share a generally common set of associations for the words of that language. When I say that I see a red object, all readers of English will immediately understand something of the color that I experience; certainly they will not conjure up a vision of something blue, purple, or green. If I further qualify my statement by saying that I see a colorful red object, I will have conveyed a somewhat more precise description of what I see. By saying that I see a moderately light, colorful, red tomato, the reader will probably form a mental image that is reasonably congruent with my own perception; the precision of my communication will have been enhanced by the addition of more descriptive words. Such descriptions of color served mankind for many thousands of years. The descriptive process is a crude and imprecise method of measurement, or as some would say, a substitute for measurement. The process embodies the necessary elements of successful communication. It attempts to convey information about an event according to a convention of rules. The rules that apply are those of the language convention used. I would not use the word "blue" when trying to describe an ordinary tomato, for example. To communicate successfully, I am constrained to use only a limited number of words to describe the color of any one object that I see.

In the most general sense, measurement is simply a process for description of events according to a convention of rules. Usually measurement is considered to involve the assignment of numerals or numbers to things according to some set of rules. In the narrow sense, measurement may be said to be the assignment of numerals to events that may vary along some dimension by a process of comparing them with other events that vary along a different dimension according to the operation of addition, for example, a process that involves explicit or implicit additive combining of objects. For example, if one large tomato weighs twice as much as one small tomato, then we assume that two small tomatoes will weigh the same as the one large one. This is a view of measurement that is based on the primitive operations of counting things, the same operations that gave rise to arithmetic and algebra. It is the concept of combination involving additivity that is the basic element of such a definition of measurement and that is responsible for its being called a narrow view. By contrast, a broader view of measurement is one that involves the assignment of numerals to objects or events within a dimension of variation according to any consistent, non-random rule or set of rules. That is, a process of measurement is one that involves assignment of numbers along a scale that is in any way isomorphic with some dimension of variation such as to involve some criterion of invariance. This broader view of measurement differs from the narrower view in a number of ways, but it differs most importantly in not requiring additivity. This distinction is an important one to which we shall return shortly.

Regardless of which view of measurement is adopted, it follows that "color measurement" must involve the assignment of numbers to represent attributes of the psychological phenomena that we call color. Most often, color measurement does not attempt directly to describe color perceptions. Instead, color measurement generally tries to relate psychological phenomena (color) to the physical phenomena (intensity of light, its wavelength, etc.) that elicit the perception. Color measurement usually consists of the process for determining what physical conditions give rise to a particular psychological (perceptual) condition; it is a process of relating psychological phenomena to physical phenomena according to some criterion. The name for such a process is "psychophysics," and measurements thus made are called "psychophysical measurements." All color measurements are psychophysical measurements of one kind or another.

The task of psychophysics is to elucidate numerical relationships between psychological dimensions and physical dimensions. We may say that psychophysics determines what function $\mathbf{f}$ of physical stimulation Φ relates to perception Ψ according to what rules, and we may represent the process in mathematical notation as

$$\Psi = \mathbf{f}(\Phi). \tag{1}$$

There are a number of criteria that may be used to establish the function $\mathbf{f}$ of Eq. (1). Color measurement usually involves one of three criteria, each implying an invariance of Ψ, and together defining three different kinds of color measurement. The first, and most common, is the criterion of equality or identity: $\Psi \equiv \Psi$; this is the basis for what is called "color matching." The second invariance criterion is one of ratios of difference: $\Delta\Psi = a\Delta\Psi$; this is the basis for "color differences." The third, and least common, criterion involves the invariance of ratios of magnitudes: $\Psi = a\Psi$, which is the basis for measurement of "color appearances."

The oldest form of colorimetry, and the one still most heavily practiced today, is that of determining equality of colors, the process of color matching. It requires the fewest assumptions about measurement of perception. Probably for that reason it is acceptable to the largest number of people. Chemists, physicists, psychologists, technologists, and artists all use this form of colorimetry without considering the question of what constitutes measurement and whether or not a private phenomenon such as the perception of color is amenable of measurement in the same sense that we think of measuring length or weight. When two colors are matched, they appear exactly the same. There is no difference. In short, they are equal (in color appearance). We may not know how to express their singular color appearance, but we can quite simply express the physical or stimulus conditions that give rise to the equality of appearance. In this way we can communicate something about color in a way that is useful. By duplicating the stimulus conditions, another person can recreate the color match; he can see for himself what the color looks like. Often the mere information that two stimulus conditions elicit the same color appearances is all that is required; we need not bother to recreate the color appearances if we are only interested in specifying the physical conditions for a color match. Suppose that a manufacturer of plastic tableware decides to produce large quantities of a particular yellow-appearing plate. He must convey to his plant manager by some means the color of the plate that is to be produced. In other words, an aim must be specified. That aim may take the form of a color specification, derived by measurement, for a condition of color match. Under a particular set of viewing conditions, a given concentration of yellow colorant sufficient to yield the required colorimetric specification should mean that all plates will have the same color appearance (to the extent that they each do have the same colorimetric specification and also to the extent that all people see color in the same way). It will be immediately evident that such a method of color specification can be broadly useful. For that reason there is wide international agreement about the methodology by which such specfications should be made. These conventions relate to attempts to standardize the conditions and assumptions that underly the colorimetry of color match specfication. The principal international body to make recommendations for such colorimetry is the Commission Internationale de l'Eclairage, usually abbreviated as CIE.

The second form of color measurement referred to previously is that concerned with differences in color. The invariance underlying such a scheme of measurement is one involving ratios of difference in color sensation. Here we must face squarely the philosophical and operational problems of measuring sensations, a process that some scientists consider impossible. It is argued that we cannot isolate a unit of sensation from which to build up an additively quantitative scale of sensation magnitude as required by the narrower view of measurement. There is, in other words, no *unit* of sensation. Accordingly, proponents of the narrow view of measurement dispute the possibility of measuring sensation in the manner that, say, length can be measured in physical science.

This problem is one that has preoccupied many scientists for a long time. It continues to do so today. Noteworthy among those who have been concerned with the problem of finding a unit for sensation is Gustav Theodor Fechner, whose enquiries led to the establishment of an indirect unit. This in turn provided a means for "measuring" sensations according to a schema not unlike that used in physical science (Fechner, 1860). Fechner considered the implications and possible extensions of the evidence collected by Ernst Heinrich Weber (1834) describing a physical invariance associated with sensory thresholds. Weber studied the relationship between stimulus quantities and the perception of just noticeably different weights and also lengths. Weber experimented on only the two sense modalities, over rather limited ranges of stimulation, and by methods that would be considered careless by today's standards. Nonetheless, he proposed a general law of sensory thresholds that says the difference between one stimulus and another that is just noticeably different is a constant fraction of the first. That fraction, now called the "Weber fraction," is constant within any one sense modality for a given set of observing conditions. If we symbolize the amount by which a stimulus Φ must be increased or decreased to produce another stimulus just noticeably different from the first as $\delta\Phi$, then Weber's law may be written as

$$\delta\Phi/\Phi = c. \tag{2}$$

Fechner reasoned that the criterion on which Eq. (2) is based, the just noticeable difference (usually abbreviated JND), must represent a change in sensation (which we may represent here as $\delta\Theta$).* Thus, he extended Eq. (2) to read

$$\delta\Theta = c[\delta\Phi/\Phi]. \tag{3}$$

Fechner then went on to make an *assumption* that is now both famous and controversial. That assumption is that since $\delta\Theta$ represents a least perceptible difference in sensation, no other sensation can be smaller than $\delta\Theta$, and therefore all $\delta\Theta$'s must be equal. Accordingly, $\delta\Theta$ may serve as a unit of sensation. Here, then, was the elusive unit of sensation needed to satisfy the requirement for additivity. He then simply treated Eq. (3) as a differential equation that may be integrated to yield what was called the "measurement formula":

$$\Theta = c \ln \Phi + C. \tag{4}$$

The constant of integration C may be eliminated by further assuming that

---

* The symbol Θ is used rather than ψ to indicate that the perceptual threshold is not necessarily the same kind of quantity as the perceptual magnitude.

the magnitude of sensation must be zero at the absolute threshold where $\Phi = t$. Thus, $C = -c \ln(t)$ at the absolute threshold and if t is arbitrarily stipulated as unity, Eq. (4) becomes what is now known as "Fechner's law":

$$\Theta = k \log \Phi. \tag{5}$$

The implication of Eq. (5) is that a perceptual magnitude may be determined by summing JNDs. Fechner thus addressed the philosophical problem of measuring sensations. He also treated the operational problem by setting forth (not necessarily for the first time) three methods by which JND's can be determined experimentally. These are now known as the methods of (1) limits, (2) constant stimuli, and (3) average error. The method of limits is an experimental procedure by which the stimulus is changed in successive serial increments until a point is reached where the observer's response changes. Usually the boundary between "no change" and "change" is approached from opposite directions and the data are averaged. The method of constant stimuli involves presentation of stimuli to which the observer responds with either of two categories (for absolute thresholds) or three categories of judgment (for differential thresholds). By treating each stimulus as a constant and recording the frequency with which it is assigned to categories, one obtains a "psychometric curve" (Urban, 1909) on which the 50% point is generally taken as indicating threshold. Finally, the method of average error provides a standard stimulus that the observer tries to match with an adjustable stimulus. The standard or average error of matching is then assumed to represent the threshold, although there are logical bases for considering such measures to be subthreshold (Herbart, 1824). In one way or another each of these three methods leads to a measure of the stimulus that corresponds to the JND. Any scale of perception must be built up by cumulative addition of JNDs according to Fechner's law.

In colorimetry, methods for specifying color differences are based on just such summations of JNDs. The experimental methods that form the base of data from which such expressions have been derived typically have been ones in which JNDs were determined by the method of average error. That is, most equations for calculating color differences are based on some normalization of the standard errors of color matching, although there are some important exceptions.

Let us return to the example of our plastic tableware manufacturer. He has specified the colorimetric aim for his yellow-appearing plate and now starts to produce samples. One of the first things he finds is that his factory is unable to produce plastic plates all with exactly the same aim colorimetric specifications. He is now in a businessman's quandary, for if he accepts only those few plates that are exactly on aim, his cost of manufacturing will be very high; after all, he must recover the cost of producing all those plates that do not meet the aim, so the price of those that he will sell must be high enough to keep his business solvent. He soon decides that he must have a *tolerance* about the aim. That is, there must be a gamut of color differences representing colorimetric specifications that differ from the aim by some amount but for which the product is accepted and marketed. The range of color difference found acceptable involves business decisions, not science or technology. If the tolerance is too great, his product will not find acceptance in the marketplace and his business will be ruined. On the other hand, if the tolerance is too small, either the plates will have prohibitively high prices or, if they are priced competitively, the manufacturer will not realize a satisfactory profit; in either case his business will be ruined. However, the *specification* of the tolerance is a matter for technology and it is accomplished through some easily communicated specification of color difference from aim. Since the very livelihood of the manufacturer may depend on the precision with which such color differences can be measured and communicated, it is easy to see that colorimetric specification of color differences is as important as the specification of matching colors.

In situations where complex products are involved, such as television or photographic images, and where design of products is practiced, there is a need to be able to measure not only color matches and differences, but also color appearances. As we have seen, color matches and color differences are measured and specified by methods that can be traced back to the matching process. The data provided are stimulus data. They tell us something about the stimulus conditions for a match or for a specified difference in color appearance, but they do not specify the color appearance itself. To measure color appearance, we must become even more deeply embroiled in the philosophical and methodological questions of measuring sensations than was the case for extending matching data to color difference derivations.

To measure color appearances we must first ask whether or not it is possible to measure sensations. Colorimetry of matches and differences may be made without assuming that sensations can be measured. Specifications of both color matches and color differences are based on data from color matching experiments. The color match is simply the average stimulus adjustment for a color match. The color difference is a sum of the average errors (JNDs) of a color match. Both kinds of specification may be made even though it is assumed that sensations cannot be measured directly.

Not all workers have made the assumption that sensations cannot be measured, however. Franz Brentano (1874) implied his belief that sensations could be measured by suggesting that Weber's law applied equally to sensations and stimuli. That is, for any two sensations produced by just noticeably different stimuli, the difference between one sensation and the other is a constant fraction of the first: $\Pi = k\Pi$, where Π stands for the magnitude of sensation. Accordingly, Brentano would derive a different "fundamental formula" than that of Fechner shown in Eq. (3). Brentano's equation is instead

$$\delta\Pi/\Pi = \delta\Phi/\Phi. \tag{6}$$

If Eq. (6) is treated as a differential equation and integrated, the result is a power function, rather than the Fechnernerian logarithmic law of Eq. (5):

$$\Pi = k\Phi^p. \tag{7}$$

It follows from Eq. (7) that ratios of sensation are proportional to ratios of stimulation, and this in turn implies that sensations can be measured.

Merkel (1888) proposed a method for experimentally determining sensation ratios by factors of two. He called it *die Methode der doppelten Reize* (the method of doubling stimuli), but it is clear from the context that he meant doubling *sensations* rather than stimuli. Four years later Fullerton and Cattell (1892) suggested an extension of the method in which observers were instructed to produce stimuli that elicited both multiples and fractions of the sensation for a standard stimulus. Beginning in the 1930s, Stevens (1946, 1975) extended and developed these methods to the extent that they are now common experimental practices by sensory scientists.

Unfortunately for those who insist that measurement involve additivity, these methods of direct magnitude scaling of sensations do not generally yield additive scales. If we have three color stimuli (A, B, C) where the color differences between A and B and between B and C are each 20 JNDs, then by the principle of additivity we assume that the difference between A and C is twice as large (40 JNDs) as that between either A and B or B and C. However, when these same stimuli are scaled by direct methods, the ratio of magnitudes is not twice; more often it is around the square root of 2. Color differences and color ratios seem to be perceived in different ways. Ratio scales do not seem to have symmetry of distances. The sum of the distances between any two points and a third is not additive. In short, perceptual ratio scales do not obey the requirements for measurement according to the narrower view of measurement. They do, however, obey the requirements for the broader view of measurement, for example, that measurement is the assignment of number to objects or events according to any consistent, nonrandom rule. This is not surprising since the liberal view was proposed by Stevens (1946) purposely to address the problem of nonadditivity of ratio scaling data. He points out [as have others, even opponents of the idea of measuring sensations directly (Guild, 1932)] that much can be learned and specified about vision by direct estimation, even if the scales are not additive. In the final analysis we have no way of knowing which scale is correct. We do not know how the visual mechanism operates as yet. Fechner's assumption of equality of JNDs may be as much in error as the assumption of Brentano that ratios of sensation are obvious and unitary to observers.

Accordingly, although there is broad international agreement about measurement of color equality (color matching), there is only provisional agreement regarding measurement of color differences and, as yet, no international agreement on ways to measure color appearances. The bulk of the conventions for colorimetry are, then, agreed methods for specifying conditions of color matching. A much smaller and more recent body of agreement exists for methods by which color differences may be measured and specified. These agreements stem from the basic data of color matching that have been reviewed elsewhere in this book. Those basic principles will not be repeated here. In the following sections of this chapter we will summarize the methods for colorimetry that have been recommended by the CIE and will sketch briefly the nature of some continuing work toward standardization that is presently being carried out by that organization. In all of what follows, it is important to recognize what kind of color measurement is implied by any standard colorimetric method and to be aware of the limitations and constraints that must be observed with each kind of color measurement if it is to be maximally useful. Most of the problems encountered in the application of colorimetry can be traced to attempts to use colorimetric specifications for predicting or expressing sensory color relationships for which the colorimetric specification is not appropriate. When used correctly, colorimetry provides us with powerful tools for evaluating color.

II. DERIVATION OF THE CIE 1931 (X, Y, Z) METHOD OF TRISTIMULUS SPECIFICATION

The CIE method of colorimetric specification is based on the rules of color matching by additive color mixture. Observations made over many years of additive color mixture experimentation were elucidated by Hermann Günter Grassman (1853, 1854) and have come to be known as Grassman's laws of color mixture. By additive mixture we mean the combination of two stimuli acting in such a manner that they enter the eye simultaneously or in rapid succession and are incident on the same area of the retina or are incident in the form of a mosaic too fine to be resolved as such. Grassman's laws for such color mixture are three in number:

(1) Three independent variables are necessary and sufficient for specifying a color mixture.

(2) Stimuli evoking the same color appearance produce identical results in additive color mixtures, regardless of their spectral compositions.

(3) If one component of a color mixture changes, the color of the mixture changes in a corresponding manner.

These three principles of additive color mixture imply certain analogies with mathematical relations that are very powerful. The first establishes what is called "trichromacy," for example, the principle that all *hues* of color experience can be matched by a suitable mixture of three different stimuli.* Mathematically, this principle may be stated as the following equation:

$$\mathfrak{C} = X\mathfrak{X} + Y\mathfrak{Y} + Z\mathfrak{Z}, \tag{8}$$

where X, Y, Z are the tristimulus values of color stimulus $\mathfrak{C}$ and $\mathfrak{X}$, $\mathfrak{Y}$, $\mathfrak{Z}$ are unit vectors of the reference stimuli (usually called "primaries") used in the matching mixture.

The second principle means that stimuli with different physical characteristics (for example, different spectral radiance distributions) may provide the same color match. Such physically dissimilar stimuli that elicit the same color matches are called "metamers" and the phenomenon is said to be "metamerism" because an identical color match may consist of different mixture components. Metamerism may be noted mathematically by extending Eq. (8) as follows: If

$$\mathfrak{C}_1 = \mathfrak{C}_2,$$

$$\mathfrak{C}_1 = X_1\mathfrak{X}_1 + Y_1\mathfrak{Y}_1 + Z_1\mathfrak{Z}_1,$$

and

$$\mathfrak{C}_2 = X_2\mathfrak{X}_2 + Y_2\mathfrak{Y}_2 + Z_2\mathfrak{Z}_2$$

then

$$X_1\mathfrak{X}_1 + Y_1\mathfrak{Y}_1 + Z_1\mathfrak{Z}_1 = X_2\mathfrak{X}_2 + Y_2\mathfrak{Y}_2 + Z_2\mathfrak{Z}_2. \tag{9}$$

The third principle establishes additivity of the stimulus metric for color specification. That is, Grassman's third law is analogous to the algebraic axioms: (1) if equals are added to or subtracted from equals, the results are equal, and (2) if equals are multiplied or divided by equals, the results are equal. In other words, if Eq. (8) is true, then we may also take as true

---

* The only restriction attendent to the choice of primary stimuli for use in color matching is that none of them may be matched in color by any mixture of the others.

the expression

$$k\mathfrak{C} = kX\mathfrak{X} + kY\mathfrak{Y} + kZ\mathfrak{Z}, \tag{10}$$

where k is a constant.

The implications of these principles have been tested many times over the years and found to be generally (if not strictly) true. In particular, during the 1920s and 1930s, many measurements were made to establish quantitative expressions for Eqs. (8)–(10). For reasons of experimental convenience, the color-matching primaries were generally chosen as monochromatic lights. The most important experimental choices were monochromatic lights with spectral centroids of 700.0, 546.1, and 435.8 nm. Symbolizing the unit vectors of this primary system as $\mathfrak{R}, \mathfrak{G}, \mathfrak{B}$, we rewrite Eq. (8) as

$$\mathfrak{C} = R\mathfrak{R} + G\mathfrak{G} + B\mathfrak{B}. \tag{11}$$

The tristimulus values R, G, B may be determined as the radiances of the primaries necessary to match each, in turn, of a large series of other monochromatic lights. These quantities are called "spectral tristimulus values," and if the power of each spectral stimulus is the same, the unit normalized spectral tristimulus values are symbolized as $\bar{r}$, $\bar{g}$, $\bar{b}$ at each wavelength. They are plotted as a function of wavelength in Fig. 1 for

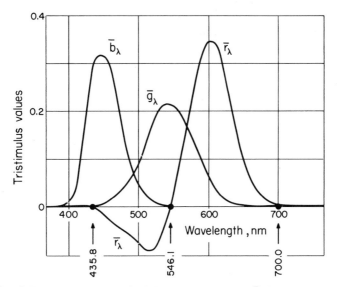

Fig. 1. CIE 2° standard spectral tristimulus values $\bar{r}_\lambda, \bar{g}_\lambda, \bar{b}_\lambda$ for color matching with monochromatic lights of 435.8, 546.1, and 700.0 nm.

monochromatic primaries of 700.0, 546.1, and 435.8 nm having radiances that stand in the ratios of 72.1 : 1.4 : 1.0 and representing matching by an average observer of normal color vision.

From Grassman's laws we see that the tristimulus values R, G, B of any stimulus function of wavelength $\varphi(\lambda)$ may be obtained by integrating the stimulus function with the spectral tristimulus functions over wavelength. That is,

$$R = k \int_{\lambda} \varphi(\lambda)\bar{r}(\lambda)\, d\lambda,$$

$$G = k \int_{\lambda} \varphi(\lambda)\bar{g}(\lambda)\, d\lambda, \qquad (12)$$

$$B = k \int_{\lambda} \varphi(\lambda)\bar{b}(\lambda)\, d\lambda,$$

where for object colors the color stimulus function may be the product of the spectral reflectance $\rho(\lambda)$, or the spectral radiance factor $\beta(\lambda)$, or the spectral transmittance $\tau(\lambda)$, of the object with the spectral power distribution of the illuminant irradiating the object $S(\lambda)$:

$$\varphi(\lambda) = \rho(\lambda)S(\lambda), \qquad \varphi(\lambda) = \beta(\lambda)S(\lambda), \qquad \text{or} \qquad \varphi(\lambda) = \tau(\lambda)S(\lambda).$$

The tristimulus values may be expressed as dimensionless ratios called "chromaticity coordinates." They are related to the tristimulus values simply as

$$r = R/(R + G + B), \qquad g = G/(R + G + B), \qquad b = B/(R + G + B). \quad (13)$$

Chromaticity coordinates provide the advantage that a knowledge of any two determines the value of the third as well; for example, $r + g + b = 1.0$.

Grassman's laws together with the spectral tristimulus functions illustrated in Fig. 1 form the basis for the conventions adopted by the CIE for specifications of the conditions of color matching. These fundamental standards were first adopted by the CIE in 1931 and have been refined over a period of more than four decades. They are summarized in CIE Publication No. 15 and its supplements (CIE, 1971a,b, 1978a).

In addition to adopting the tristimulus specification system based on the real, monochromatic primaries of spectral centroids 700.0, 546.1, and 435.8 nm, the CIE also adopted a tristimulus specification based on imaginary primaries that are a linear transformation of the former. This was done as a matter of computational and interpretive convenience. The tristimulus values for these imaginary primaries are symbolized X, Y, Z, and they have become the fundamental specification metric of CIE colorimetry.

Grassman's second law states that one set of primaries may be substituted for another. In particular, different triads of primaries yield different tristimulus values that are linearly related. Thus, if the tristimulus values for a given color sample are R, G, B for the first triad of primaries and the tristimulus values X, Y, Z are those corresponding to the same color sample with respect to the second set of primaries, then the relationship between the two triads of tristimulus values may be expressed as

$$X = a_{11}R + a_{12}G + a_{13}B,$$
$$Y = a_{21}R + a_{22}G + a_{23}B, \tag{14}$$
$$Z = a_{31}R + a_{32}G + a_{33}B,$$

where the coefficients a_{1j} are the amounts of the second set of primaries required to match the color of the first set mixture having tristimulus values $R = 1$, $G = 0$, $B = 0$; the coefficients a_{2j} are the amounts required to match $R = 0$, $G = 1$, $B = 0$; and the coefficients a_{3j} are those required to match $R = 0$, $G = 0$, $B = 1$. Together the coefficients a_{ij} form the transformation matrix from the first primary set to the second. This unique description of the relationship stems from Grassman's third law, which says that we may take the sum of any stimuli and consider it to be a separate, new, stimulus. In matrix notation the relationship is

$$\begin{bmatrix} X \\ Y \\ Z \end{bmatrix} = \mathbf{A} \begin{bmatrix} R \\ G \\ B \end{bmatrix}, \tag{15}$$

where the matrix $\mathbf{A}$ is the transpose of a matrix $\mathbf{A}^*$ that describes the transformation between the two underlying linearly independent sets of primaries. The inverse of matrix $\mathbf{A}$, written $\mathbf{A}^{-1}$, provides the reverse transformation from the second set of tristimulus values to the first:

$$\begin{bmatrix} R \\ G \\ B \end{bmatrix} = \mathbf{A}^{-1} \begin{bmatrix} X \\ Y \\ Z \end{bmatrix}. \tag{16}$$

The linear equations represented by matrix Eq. (16) are as follows:

$$R = \frac{a_{22}a_{33} - a_{23}a_{32}}{\Delta} X + \frac{a_{13}a_{32} - a_{12}a_{33}}{\Delta} Y + \frac{a_{12}a_{23} - a_{13}a_{22}}{\Delta} Z,$$

$$G = \frac{a_{23}a_{31} - a_{21}a_{33}}{\Delta} X + \frac{a_{11}a_{33} - a_{13}a_{31}}{\Delta} Y + \frac{a_{13}a_{21} - a_{11}a_{33}}{\Delta} Z, \tag{17}$$

$$B = \frac{a_{21}a_{32} - a_{22}a_{31}}{\Delta} X + \frac{a_{12}a_{31} - a_{11}a_{32}}{\Delta} Y + \frac{a_{11}a_{22} - a_{12}a_{21}}{\Delta} Z,$$

where Δ is the determinant of matrix $\mathbf{A}$; that is,

$$\Delta = a_{11}(a_{22}a_{33} - a_{23}a_{32}) + a_{12}(a_{23}a_{31} - a_{21}a_{33})$$
$$+ a_{13}(a_{21}a_{32} - a_{22}a_{31}). \tag{18}$$

Equations (14)–(18) may be used to state the transformation for chromaticity coordinates as well, since these are simply expressions of relative proportions of the tristimulus values. Equation (13) defines the r, g, b chromaticity coordinates of the R, G, B tristimulus values. An equivalent expression may be written for chromaticity coordinates x, y, z of the tristimulus values X, Y, Z. Since we need know only two chromaticity coordinates to determine all three uniquely, the transformation of chromaticity coordinates may be represented as

$$x = \frac{c_{11}r + c_{12}g + c_{13}}{c_{31}r + c_{32}g + c_{33}},$$

$$y = \frac{c_{21}r + c_{22}g + c_{23}}{c_{31}r + c_{32}g + c_{33}}, \tag{19}$$

where, referring back to Eq. (14), the values of c_{ij} are

$$c_{11} = a_{11} - a_{13}, \qquad c_{12} = a_{12} - a_{13}, \qquad c_{13} = a_{13};$$

$$c_{21} = a_{21} - a_{23}, \qquad c_{22} = a_{22} - a_{23}, \qquad c_{23} = a_{23};$$

$$c_{31} = (a_{11} - a_{13}) + (a_{21} - a_{23}) + (a_{31} - a_{33}),$$

$$c_{32} = (a_{12} - a_{13}) + (a_{22} - a_{23}) + (a_{32} - a_{33}),$$

$$c_{33} = a_{13} + a_{23} + a_{33}.$$

In determining the coefficients of Eqs. (14)–(19), the CIE set down five requirements. These were (1) to avoid negative tristimulus values, such as those obtained with the R, G, B system as illustrated in Fig. 1; (2) to make the new X, Y, Z tristimulus values equal for a stimulus with equal power at all wavelengths; (3) to develop a chromaticity diagram in which the locus of chromaticities corresponding to spectral stimuli are as nearly as possible equally distant from the chromaticity point for an equienergy stimulus; (4) to make the dominant wavelength (which will be defined in the following paragraphs) of the Z reference stimulus for normal viewing conditions as nearly that of the perceptually unitary blue wavelength as possible (for example, the wavelength of the spectral stimulus that, on average, would appear to an observer of normal color vision to contain no reddish or greenish tint), and (5) to set the luminances [as defined by the CIE in 1924 (CIE, 1970)] of X, Y, Z at 0, 1, 0 as unit quantities. This last requirement results in the tristimulus value Y being equal to luminance. In addition, the spectral tristimulus function $\bar{y}(\lambda)$ becomes identical to the standard relative luminous efficacy function $V(\lambda)$.

The foregoing sets forth the bases for derivation of the CIE recommendations for tristimulus specification. Such a specification represents the amounts of primary stimuli required to provide a color match for the standard observer [whose color matching characteristics are defined by the spectral tristimulus functions $\bar{x}(\lambda)$, $\bar{y}(\lambda)$, $\bar{z}(\lambda)$] under a standard set of viewing conditions. It remained to specify the conditions of viewing for which such specifications would be considered appropriate. In addition, in order to enhance communication, a set of recommended methods for expressing the tristimulus specifications were gradually evolved over the years, as were a set of illuminants to be used or assumed for the tristimulus matches. The following paragraphs describe these additional CIE recommendations.

III. CIE STANDARD OBSERVERS

The CIE has recommended two standard observers for different purposes. The CIE 1931 Standard Colorimetric Observer is intended to represent an average observer of normal color vision when attending a stimulus that subtends 2° diameter in visual subtense. The CIE 1964 Supplementary Standard Colorimetric Observer is intended to represent an average observer of normal color vision when attending a stimulus that subtends 10° diameter in visual subtense. In practice these two CIE observers are referred to as the "2° observer" and the "10° observer," but the former is satisfactorily used for stimuli whose diameters subtend visual angles from about 1° to 4° and the latter is used for stimuli with diameters larger than 4°.

The CIE 1931 Standard Colorimetric Observer is defined by the spectral tristimulus values listed in Table I. Those values are determined as a linear transform of the $\bar{r}$, $\bar{g}$, $\bar{b}$ spectral tristimulus values adopted by the CIE in 1931. The chromaticity coordinates of spectral stimuli for the 2° observer are related to the original r, g, b chromaticity coordinates (with respect to spectral primaries of 700.0, 546.1, and 435.8 nm) according to the following projective transformation:

$$x(\lambda) = \frac{0.49000r(\lambda) + 0.31000g(\lambda) + 0.20000b(\lambda)}{0.66697r(\lambda) + 1.13240g(\lambda) + 1.20063b(\lambda)},$$

$$y(\lambda) = \frac{0.17697r(\lambda) + 0.81240g(\lambda) + 0.01063b(\lambda)}{0.66697r(\lambda) + 1.13240g(\lambda) + 1.20063b(\lambda)}, \qquad (20)$$

$$z(\lambda) = \frac{0.00000r(\lambda) + 0.01000g(\lambda) + 0.99000b(\lambda)}{0.66697r(\lambda) + 1.13240g(\lambda) + 1.20063b(\lambda)},$$

C. J. Bartleson

TABLE I.   CIE 1931 Spectral Tristimulus Values $\bar{x}(\lambda)$, $\bar{y}(\lambda)$, $\bar{z}(\lambda)$[a]

λ (nm)	Spectral Tristimulus Values			Chromaticity Coordinates		
	$\bar{x}(\lambda)$	$\bar{y}(\lambda)$	$\bar{z}(\lambda)$	$x(\lambda)$	$y(\lambda)$	$z(\lambda)$
360	0.000 129 900 0	0.000 003 917 000	0.000 606 100 0	0.175 56	0.005 29	0.819 15
61	0.000 145 847 0	0.000 004 393 581	0.000 680 879 2	0.175 48	0.005 29	0.819 23
62	0.000 163 802 1	0.000 004 929 604	0.000 765 145 6	0.175 40	0.005 28	0.819 32
63	0.000 184 003 7	0.000 005 532 136	0.000 860 012 4	0.175 32	0.005 27	0.819 41
64	0.000 206 690 2	0.000 006 208 245	0.000 966 592 8	0.175 24	0.005 26	0.819 50
365	0.000 232 100 0	0.000 006 965 000	0.001 086 000	0.175 16	0.005 26	0.819 58
66	0.000 260 728 0	0.000 007 813 219	0.001 220 586	0.175 09	0.005 25	0.819 66
67	0.000 293 075 0	0.000 008 767 336	0.001 372 729	0.175 01	0.005 24	0.819 75
68	0.000 329 388 0	0.000 009 839 844	0.001 543 579	0.174 94	0.005 23	0.819 83
69	0.000 369 914 0	0.000 011 043 23	0.001 734 286	0.174 88	0.005 22	0.819 90
370	0.000 414 900 0	0.000 012 390 00	0.001 946 000	0.174 82	0.005 22	0.819 96
71	0.000 464 158 7	0.000 013 886 41	0.002 177 777	0.174 77	0.005 23	0.820 00
72	0.000 518 986 0	0.000 015 557 28	0.002 435 809	0.174 72	0.005 24	0.820 04
73	0.000 581 854 0	0.000 017 442 96	0.002 731 953	0.174 66	0.005 24	0.820 10
74	0.000 655 234 7	0.000 019 583 75	0.003 078 064	0.174 59	0.005 22	0.820 19
375	0.000 741 600 0	0.000 022 020 00	0.003 486 000	0.174 51	0.005 18	0.820 31
76	0.000 845 029 6	0.000 024 839 65	0.003 975 227	0.174 41	0.005 13	0.820 46
77	0.000 964 526 8	0.000 028 041 26	0.004 540 880	0.174 31	0.005 07	0.820 62
78	0.001 094 949	0.000 031 531 04	0.005 158 320	0.174 22	0.005 02	0.820 76
79	0.001 231 154	0.000 035 215 21	0.005 802 907	0.174 16	0.004 98	0.820 86
380	0.001 368 000	0.000 039 000 00	0.006 450 001	0.174 11	0.004 96	0.820 93
81	0.001 502 050	0.000 042 826 40	0.007 083 216	0.174 09	0.004 96	0.820 95
82	0.001 642 328	0.000 046 914 60	0.007 745 488	0.174 07	0.004 97	0.820 96
83	0.001 802 382	0.000 051 589 60	0.008 501 152	0.174 06	0.004 98	0.820 96
84	0.001 995 757	0.000 057 176 40	0.009 414 544	0.174 04	0.004 98	0.820 98
385	0.002 236 000	0.000 064 000 00	0.010 549 99	0.174 01	0.004 98	0.821 01
86	0.002 535 385	0.000 072 344 21	0.011 965 80	0.173 97	0.004 97	0.821 06
87	0.002 892 603	0.000 082 212 24	0.013 655 87	0.173 93	0.004 94	0.821 13
88	0.003 300 829	0.000 093 508 16	0.015 588 05	0.173 89	0.004 93	0.821 18
89	0.003 753 236	0.000 106 136 1	0.017 730 15	0.173 84	0.004 92	0.821 24
390	0.004 243 000	0.000 120 000 0	0.020 050 01	0.173 80	0.004 92	0.821 28
91	0.004 762 389	0.000 134 984 0	0.022 511 36	0.173 76	0.004 92	0.821 32
92	0.005 330 048	0.000 151 492 0	0.025 202 88	0.173 70	0.004 94	0.821 36
93	0.005 978 712	0.000 170 208 0	0.028 279 72	0.173 66	0.004 94	0.821 40
94	0.006 741 117	0.000 191 816 0	0.031 897 04	0.173 61	0.004 94	0.821 45
395	0.007 650 000	0.000 217 000 0	0.036 210 00	0.173 56	0.004 92	0.821 52
96	0.008 751 373	0.000 246 906 7	0.041 437 71	0.173 51	0.004 90	0.821 59
97	0.010 028 88	0.000 281 240 0	0.047 503 72	0.173 47	0.004 86	0.821 67
98	0.011 421 70	0.000 318 520 0	0.054 119 88	0.173 42	0.004 84	0.821 74
99	0.012 869 01	0.000 357 266 7	0.060 998 03	0.173 38	0.004 81	0.821 81
400	0.014 310 00	0.000 396 000 0	0.067 850 01	0.173 34	0.004 80	0.821 86
01	0.015 704 43	0.000 433 714 7	0.074 486 32	0.173 29	0.004 79	0.821 92
02	0.017 147 44	0.000 473 024 0	0.081 361 56	0.173 24	0.004 78	0.821 98
03	0.018 781 22	0.000 517 876 0	0.089 153 64	0.173 17	0.004 78	0.822 05
04	0.020 748 01	0.000 572 218 7	0.098 540 48	0.173 10	0.004 77	0.822 13
405	0.023 190 00	0.000 640 000 0	0.110 200 0	0.173 02	0.004 78	0.822 20
06	0.026 207 36	0.000 724 560 0	0.124 613 3	0.172 93	0.004 78	0.822 29
07	0.029 782 48	0.000 825 500 0	0.141 701 7	0.172 84	0.004 79	0.822 37
08	0.033 880 92	0.000 941 160 0	0.161 303 5	0.172 75	0.004 80	0.822 45
09	0.038 468 24	0.001 069 880	0.183 256 8	0.172 66	0.004 80	0.822 54

λ (nm)	Spectral Tristimulus Values			Chromaticity Coordinates		
	$\bar{x}(\lambda)$	$\bar{y}(\lambda)$	$\bar{z}(\lambda)$	$x(\lambda)$	$y(\lambda)$	$z(\lambda)$
410	0.043 510 00	0.001 210 000	0.207 400 0	0.172 58	0.004 80	0.822 62
11	0.048 995 60	0.001 362 091	0.233 692 1	0.172 49	0.004 80	0.822 71
12	0.055 022 60	0.001 530 752	0.262 611 4	0.172 39	0.004 80	0.822 81
13	0.061 718 80	0.001 720 368	0.294 774 6	0.172 30	0.004 80	0.822 90
14	0.069 212 00	0.001 935 323	0.330 798 5	0.172 19	0.004 82	0.822 99
415	0.077 630 00	0.002 180 000	0.371 300 0	0.172 09	0.004 83	0.823 08
16	0.086 958 11	0.002 454 800	0.416 209 1	0.171 98	0.004 86	0.823 16
17	0.097 176 72	0.002 764 000	0.465 464 2	0.171 87	0.004 89	0.823 24
18	0.108 406 3	0.003 117 800	0.519 694 8	0.171 74	0.004 94	0.823 32
19	0.120 767 2	0.003 526 400	0.579 530 3	0.171 59	0.005 01	0.823 40
420	0.134 380 0	0.004 000 000	0.645 600 0	0.171 41	0.005 10	0.823 49
21	0.149 358 2	0.004 546 240	0.718 483 8	0.171 21	0.005 21	0.823 58
22	0.165 395 7	0.005 159 320	0.796 713 3	0.170 99	0.005 33	0.823 68
23	0.181 983 1	0.005 829 280	0.877 845 9	0.170 77	0.005 47	0.823 76
24	0.198 611 0	0.006 546 160	0.959 439 0	0.170 54	0.005 62	0.823 84
425	0.214 770 0	0.007 300 000	1.039 050 1	0.170 30	0.005 79	0.823 91
26	0.230 186 8	0.008 086 507	1.115 367 3	0.170 05	0.005 97	0.823 98
27	0.244 879 7	0.008 908 720	1.188 497 1	0.169 78	0.006 18	0.824 04
28	0.258 777 3	0.009 767 680	1.258 123 3	0.169 50	0.006 40	0.824 10
29	0.271 807 9	0.010 664 43	1.323 929 6	0.169 20	0.006 64	0.824 16
430	0.283 900 0	0.011 600 00	1.385 600 0	0.168 88	0.006 90	0.824 22
31	0.294 943 8	0.012 573 17	1.442 635 2	0.168 53	0.007 18	0.824 29
32	0.304 896 5	0.013 582 72	1.494 803 5	0.168 15	0.007 49	0.824 36
33	0.313 787 3	0.014 629 68	1.542 190 3	0.167 75	0.007 82	0.824 43
34	0.321 645 4	0.015 715 09	1.584 880 7	0.167 33	0.008 17	0.824 50
435	0.328 500 0	0.016 840 00	1.622 960 0	0.166 90	0.008 55	0.824 55
36	0.334 351 1	0.018 007 36	1.656 404 8	0.166 45	0.008 96	0.824 59
37	0.339 210 1	0.019 214 48	1.685 295 9	0.165 98	0.009 40	0.824 62
38	0.343 121 3	0.020 453 92	1.709 874 5	0.165 48	0.009 87	0.824 65
39	0.346 129 6	0.021 718 24	1.730 382 1	0.164 96	0.010 35	0.824 69
440	0.348 280 0	0.023 000 00	1.747 060 0	0.164 41	0.010 86	0.824 73
41	0.349 599 9	0.024 294 61	1.760 044 6	0.163 83	0.011 38	0.824 79
42	0.350 147 4	0.025 610 24	1.769 623 3	0.163 21	0.011 94	0.824 85
43	0.350 013 0	0.026 958 57	1.776 263 7	0.162 55	0.012 52	0.824 93
44	0.349 287 0	0.028 351 25	1.780 433 4	0.161 85	0.013 14	0.825 01
445	0.348 060 0	0.029 800 00	1.782 600 0	0.161 11	0.013 79	0.825 10
46	0.346 373 3	0.031 310 83	1.782 968 2	0.160 31	0.014 49	0.825 20
47	0.344 262 4	0.032 883 68	1.781 699 8	0.159 47	0.015 23	0.825 30
48	0.341 808 8	0.034 521 12	1.779 198 2	0.158 57	0.016 02	0.825 41
49	0.339 094 1	0.036 225 71	1.775 867 1	0.157 63	0.016 84	0.825 53
450	0.336 200 0	0.038 000 00	1.772 110 0	0.156 64	0.017 71	0.825 65
51	0.333 197 7	0.039 846 67	1.768 258 9	0.155 60	0.018 61	0.825 79
52	0.330 041 1	0.041 768 00	1.764 039 0	0.154 52	0.019 56	0.825 92
53	0.326 635 7	0.043 766 00	1.758 943 8	0.153 40	0.020 55	0.826 05
54	0.322 886 8	0.045 842 67	1.752 466 3	0.152 22	0.021 61	0.826 17
455	0.318 700 0	0.048 000 00	1.744 100 0	0.150 99	0.022 74	0.826 27
56	0.314 025 1	0.050 243 68	1.733 559 5	0.149 69	0.023 95	0.826 36
57	0.308 884 0	0.052 573 04	1.720 858 1	0.148 34	0.025 25	0.826 41
58	0.303 290 4	0.054 980 56	1.705 936 9	0.146 93	0.026 63	0.826 44
59	0.297 257 9	0.057 458 72	1.688 737 2	0.145 47	0.028 12	0.826 41

TABLE I. (*Continued*)

λ (nm)	Spectral Tristimulus Values			Chromaticity Coordinates		
	$\bar{x}(\lambda)$	$\bar{y}(\lambda)$	$\bar{z}(\lambda)$	$x(\lambda)$	$y(\lambda)$	$z(\lambda)$
460	0.290 800 0	0.060 000 00	1.669 200 0	0.143 96	0.029 70	0.826 34
61	0.283 970 1	0.062 601 97	1.647 528 7	0.142 41	0.031 39	0.826 20
62	0.276 721 4	0.065 277 52	1.623 412 7	0.140 80	0.033 21	0.825 99
63	0.268 917 8	0.068 042 08	1.596 022 3	0.139 12	0.035 20	0.825 68
64	0.260 422 7	0.070 911 09	1.564 528 0	0.137 37	0.037 40	0.825 23
465	0.251 100 0	0.073 900 00	1.528 100 0	0.135 50	0.039 88	0.824 62
66	0.240 847 5	0.077 016 00	1.486 111 4	0.133 51	0.042 69	0.823 80
67	0.229 851 2	0.080 266 40	1.439 521 5	0.131 37	0.045 88	0.822 75
68	0.218 407 2	0.083 666 80	1.389 879 9	0.129 09	0.049 45	0.821 46
69	0.206 811 5	0.087 232 80	1.338 736 2	0.126 66	0.053 43	0.819 91
470	0.195 360 0	0.090 980 00	1.287 640 0	0.124 12	0.057 80	0.818 08
71	0.184 213 6	0.094 917 55	1.237 422 3	0.121 47	0.062 59	0.815 94
72	0.173 327 3	0.099 045 84	1.187 824 3	0.118 70	0.067 83	0.813 47
73	0.162 688 1	0.103 367 4	1.138 761 1	0.115 81	0.073 58	0.810 61
74	0.152 283 3	0.107 884 6	1.090 148 0	0.112 78	0.079 89	0.807 33
475	0.142 100 0	0.112 600 0	1.041 900 0	0.109 60	0.086 84	0.803 56
76	0.132 178 6	0.117 532 0	0.994 197 6	0.106 26	0.094 49	0.799 25
77	0.122 569 6	0.122 674 4	0.947 347 3	0.102 78	0.102 86	0.794 36
78	0.113 275 2	0.127 992 8	0.901 453 1	0.099 13	0.112 01	0.788 86
79	0.104 297 9	0.133 452 8	0.856 619 3	0.095 31	0.121 94	0.782 75
480	0.095 640 00	0.139 020 0	0.812 950 1	0.091 29	0.132 70	0.776 01
81	0.087 299 55	0.144 676 4	0.770 517 3	0.087 08	0.144 32	0.768 60
82	0.079 308 04	0.150 469 3	0.729 444 8	0.082 68	0.156 87	0.760 45
83	0.071 717 76	0.156 461 9	0.689 913 6	0.078 12	0.170 42	0.751 46
84	0.064 580 99	0.162 717 7	0.652 104 9	0.073 44	0.185 03	0.741 53
485	0.057 950 01	0.169 300 0	0.616 200 0	0.068 71	0.200 72	0.730 57
86	0.051 862 11	0.176 243 1	0.582 328 6	0.063 99	0.217 47	0.718 54
87	0.046 281 52	0.183 558 1	0.550 416 2	0.059 32	0.235 25	0.705 43
88	0.041 150 88	0.191 273 5	0.520 337 6	0.054 67	0.254 09	0.691 24
89	0.036 412 83	0.199 418 0	0.491 967 3	0.650 03	0.274 00	0.675 97
490	0.032 010 00	0.208 020 0	0.465 180 0	0.045 39	0.294 98	0.659 63
91	0.027 917 20	0.217 119 9	0.439 924 6	0.040 76	0.316 98	0.642 26
92	0.024 144 40	0.226 734 5	0.416 183 6	0.036 20	0.339 90	0.623 90
93	0.020 687 00	0.236 857 1	0.393 882 2	0.031 76	0.363 60	0.604 64
94	0.017 540 40	0.247 481 2	0.372 945 9	0.027 49	0.387 92	0.584 59
495	0.014 700 00	0.258 600 0	0.353 300 0	0.023 46	0.412 70	0.563 84
96	0.012 161 79	0.270 184 9	0.334 857 8	0.019 70	0.437 76	0.542 54
97	0.009 919 960	0.282 293 9	0.317 552 1	0.016 27	0.462 95	0.520 78
98	0.007 967 240	0.295 050 5	0.301 337 5	0.013 18	0.488 21	0.498 61
99	0.006 296 346	0.308 578 0	0.286 168 6	0.010 48	0.513 40	0.476 12
500	0.004 900 000	0.323 000 0	0.272 000 0	0.008 17	0.538 42	0.453 41
01	0.003 777 173	0.338 402 1	0.258 817 1	0.006 28	0.563 07	0.430 65
02	0.002 945 320	0.354 685 8	0.246 483 8	0.004 87	0.587 12	0.408 01
03	0.002 424 880	0.371 698 6	0.234 771 8	0.003 98	0.610 45	0.385 57
04	0.002 236 293	0.389 287 5	0.223 453 3	0.003 64	0.633 01	0.363 35
505	0.002 400 000	0.407 300 0	0.212 300 0	0.003 86	0.654 82	0.341 32
06	0.002 925 520	0.425 629 9	0.201 169 2	0.004 64	0.675 90	0.319 46
07	0.003 836 560	0.444 309 6	0.190 119 6	0.006 01	0.696 12	0.297 87
08	0.005 174 840	0.463 394 4	0.179 225 4	0.007 99	0.715 34	0.276 67
09	0.006 982 080	0.482 939 5	0.168 560 8	0.010 60	0.733 41	0.255 99

λ (nm)	Spectral Tristimulus Values			Chromaticity Coordinates		
	$\bar{x}(\lambda)$	$\bar{y}(\lambda)$	$\bar{z}(\lambda)$	$x(\lambda)$	$y(\lambda)$	$z(\lambda)$
510	0.009 300 000	0.503 000 0	0.158 200 0	0.013 87	0.750 19	0.235 94
11	0.012 149 49	0.523 569 3	0.148 138 3	0.017 77	0.765 61	0.216 62
12	0.015 535 88	0.544 512 0	0.138 375 8	0.022 24	0.779 63	0.198 13
13	0.019 477 52	0.565 690 0	0.128 994 2	0.027 27	0.792 11	0.180 62
14	0.023 992 77	0.586 965 3	0.120 075 1	0.032 82	0.802 93	0.164 25
515	0.029 100 00	0.608 200 0	0.111 700 0	0.038 85	0.812 02	0.149 13
16	0.034 814 85	0.629 345 6	0.103 904 8	0.045 33	0.819 39	0.135 28
17	0.041 120 16	0.650 306 8	0.096 667 48	0.052 18	0.825 16	0.122 66
18	0.047 985 04	0.670 875 2	0.089 982 72	0.059 32	0.829 43	0.111 25
19	0.055 378 61	0.690 842 4	0.083 845 31	0.066 72	0.832 27	0.101 01
520	0.063 270 00	0.710 000 0	0.078 249 99	0.074 30	0.833 80	0.091 90
21	0.071 635 01	0.728 185 2	0.073 208 99	0.082 05	0.834 09	0.083 86
22	0.080 462 24	0.745 463 6	0.068 678 16	0.089 94	0.833 29	0.076 77
23	0.089 739 96	0.761 969 4	0.064 567 84	0.097 94	0.831 59	0.070 47
24	0.099 456 45	0.777 836 8	0.060 788 35	0.106 02	0.829 18	0.064 80
525	0.109 600 0	0.793 200 0	0.057 250 01	0.114 16	0.826 21	0.059 63
26	0.120 167 4	0.808 110 4	0.053 904 35	0.122 35	0.822 77	0.054 88
27	0.131 114 5	0.822 496 2	0.050 746 64	0.130 55	0.818 93	0.050 52
28	0.142 367 9	0.836 306 8	0.047 752 76	0.138 70	0.814 78	0.046 52
29	0.153 854 2	0.849 491 6	0.044 898 59	0.146 77	0.810 40	0.042 83
530	0.165 500 0	0.862 000 0	0.042 160 00	0.154 72	0.805 86	0.039 42
31	0.177 257 1	0.873 810 8	0.039 507 28	0.162 53	0.801 24	0.036 23
32	0.189 140 0	0.884 962 4	0.036 935 64	0.170 24	0.796 52	0.033 24
33	0.201 169 4	0.895 493 6	0.034 458 36	0.177 85	0.791 69	0.030 46
34	0.213 365 8	0.905 443 2	0.032 088 72	0.185 39	0.786 73	0.027 88
535	0.225 749 9	0.914 850 1	0.029 840 00	0.192 88	0.781 63	0.025 49
36	0.238 320 9	0.923 734 8	0.027 711 81	0.200 31	0.776 40	0.023 29
37	0.251 066 8	0.932 092 4	0.025 694 44	0.207 69	0.771 05	0.021 26
38	0.263 992 2	0.939 922 6	0.023 787 16	0.215 03	0.765 59	0.019 38
39	0.277 101 7	0.947 225 2	0.021 989 25	0.222 34	0.760 02	0.017 64
540	0.290 400 0	0.954 000 0	0.020 300 00	0.229 62	0.754 33	0.016 05
41	0.303 891 2	0.960 256 1	0.018 718 05	0.236 89	0.748 52	0.014 59
42	0.317 572 6	0.966 007 4	0.017 240 36	0.244 13	0.742 62	0.013 25
43	0.331 438 4	0.971 260 6	0.015 863 64	0.251 36	0.736 61	0.012 03
44	0.345 482 8	0.976 022 5	0.014 584 61	0.258 58	0.730 51	0.010 91
545	0.359 700 0	0.980 300 0	0.013 400 00	0.265 78	0.724 32	0.009 90
46	0.374 083 9	0.984 092 4	0.012 307 23	0.272 96	0.718 06	0.008 98
47	0.388 639 6	0.987 418 2	0.011 301 88	0.280 13	0.711 72	0.008 15
48	0.403 378 4	0.990 312 8	0.010 377 92	0.287 29	0.705 32	0.007 39
49	0.418 311 5	0.992 811 6	0.009 529 306	0.294 45	0.698 84	0.006 71
550	0.433 449 9	0.994 950 1	0.008 749 999	0.301 60	0.692 31	0.006 09
51	0.448 795 3	0.996 710 8	0.008 035 200	0.308 76	0.685 71	0.005 53
52	0.464 336 0	0.998 098 3	0.007 381 600	0.315 92	0.679 06	0.005 02
53	0.480 064 0	0.999 112 0	0.006 785 400	0.323 06	0.672 37	0.004 57
54	0.495 971 3	0.999 748 2	0.006 242 800	0.330 21	0.665 63	0.004 16
555	0.512 050 1	1.000 000 0	0.005 749 999	0.337 36	0.658 85	0.003 79
56	0.528 295 9	0.999 856 7	0.005 303 600	0.344 51	0.652 03	0.003 46
57	0.544 691 6	0.999 304 6	0.004 899 800	0.351 67	0.645 17	0.003 16
58	0.561 209 4	0.998 325 5	0.004 534 200	0.358 81	0.638 29	0.002 90
59	0.577 821 5	0.996 898 7	0.004 202 400	0.365 96	0.631 38	0.002 66

TABLE I. (*Continued*)

λ (nm)	Spectral Tristimulus Values			Chromaticity Coordinates		
	$\bar{x}(\lambda)$	$\bar{y}(\lambda)$	$\bar{z}(\lambda)$	$x(\lambda)$	$y(\lambda)$	$z(\lambda)$
560	0.594 500 0	0.995 000 0	0.003 900 000	0.373 10	0.624 45	0.002 45
61	0.611 220 9	0.992 600 5	0.003 623 200	0.380 24	0.617 50	0.002 26
62	0.627 975 8	0.989 742 6	0.003 370 600	0.387 38	0.610 54	0.002 08
63	0.644 760 2	0.986 444 4	0.003 141 400	0.394 51	0.603 57	0.001 92
64	0.661 569 7	0.982 724 1	0.002 934 800	0.401 63	0.596 59	0.001 78
565	0.678 400 0	0.978 600 0	0.002 749 999	0.408 73	0.589 61	0.001 66
66	0.695 239 2	0.974 083 7	0.002 585 200	0.415 83	0.582 62	0.001 55
67	0.712 058 6	0.969 171 2	0.002 438 600	0.422 92	0.575 63	0.001 45
68	0.728 828 4	0.963 856 8	0.002 309 400	0.429 99	0.568 65	0.001 36
69	0.745 518 8	0.958 134 9	0.002 196 800	0.437 04	0.561 67	0.001 29
570	0.762 100 0	0.952 000 0	0.002 100 000	0.444 06	0.554 72	0.001 22
71	0.778 543 2	0.945 450 4	0.002 017 733	0.451 06	0.547 77	0.001 17
72	0.794 825 6	0.938 499 2	0.001 948 200	0.458 04	0.540 84	0.001 12
73	0.810 926 4	0.931 162 8	0.001 889 800	0.464 99	0.533 93	0.001 08
74	0.826 824 8	0.923 457 6	0.001 840 933	0.471 90	0.527 05	0.001 05
575	0.842 500 0	0.915 400 0	0.001 800 000	0.478 78	0.520 20	0.001 02
76	0.857 932 5	0.907 006 4	0.001 766 267	0.485 61	0.513 39	0.001 00
77	0.873 081 6	0.898 277 2	0.001 737 800	0.492 41	0.506 61	0.000 98
78	0.887 894 4	0.889 204 8	0.001 711 200	0.499 15	0.499 89	0.000 96
79	0.902 318 1	0.879 781 6	0.001 683 067	0.505 85	0.493 21	0.000 94
580	0.916 300 0	0.870 000 0	0.001 650 001	0.512 49	0.486 59	0.000 92
81	0.929 799 5	0.859 861 3	0.001 610 133	0.519 07	0.480 03	0.000 90
82	0.942 798 4	0.849 392 0	0.001 564 400	0.525 60	0.473 53	0.000 87
83	0.955 277 6	0.838 622 0	0.001 513 600	0.532 07	0.467 09	0.000 84
84	0.967 217 9	0.827 581 3	0.001 458 533	0.538 46	0.460 73	0.000 81
585	0.978 600 0	0.816 300 0	0.001 400 000	0.544 79	0.454 43	0.000 78
86	0.989 385 6	0.804 794 7	0.001 336 667	0.551 03	0.448 23	0.000 74
87	0.999 548 8	0.793 082 0	0.001 270 000	0.557 19	0.442 10	0.000 71
88	1.009 089 2	0.781 192 0	0.001 205 000	0.563 27	0.436 06	0.000 67
89	1.018 006 4	0.769 154 7	0.001 146 667	0.569 26	0.430 10	0.000 64
590	1.026 300 0	0.757 000 0	0.001 100 000	0.575 15	0.424 23	0.000 62
91	1.033 982 7	0.744 754 1	0.001 068 800	0.580 94	0.418 46	0.000 60
92	1.040 986 0	0.732 422 4	0.001 049 400	0.586 65	0.412 76	0.000 59
93	1.047 188 0	0.720 003 6	0.001 035 600	0.592 22	0.407 19	0.000 59
94	1.052 466 7	0.707 496 5	0.001 021 200	0.597 66	0.401 76	0.000 58
595	1.056 700 0	0.694 900 0	0.001 000 000	0.602 93	0.396 50	0.000 57
96	1.059 794 4	0.682 219 2	0.000 968 640 0	0.608 03	0.391 41	0.000 56
97	1.061 799 2	0.669 471 6	0.000 929 920 0	0.612 98	0.386 48	0.000 54
98	1.062 806 8	0.656 674 4	0.000 886 880 0	0.617 78	0.381 71	0.000 51
99	1.062 909 6	0.643 844 8	0.000 842 560 0	0.622 46	0.377 05	0.000 49
600	1.062 200 0	0.631 000 0	0.000 800 000 0	0.627 04	0.372 49	0.000 47
01	1.060 735 2	0.618 155 5	0.000 760 960 0	0.631 52	0.368 03	0.000 45
02	1.058 443 6	0.605 314 4	0.000 723 680 0	0.635 90	0.363 67	0.000 43
03	1.055 224 4	0.592 475 6	0.000 685 920 0	0.640 16	0.359 43	0.000 41
04	1.050 976 8	0.579 637 9	0.000 645 440 0	0.644 27	0.355 33	0.000 40
605	1.045 600 0	0.566 800 0	0.000 600 000 0	0.648 23	0.351 40	0.000 37
06	1.039 036 9	0.553 961 1	0.000 547 866 7	0.652 03	0.347 63	0.000 34
07	1.031 360 8	0.541 137 2	0.000 491 600 0	0.655 67	0.344 02	0.000 31
08	1.022 666 2	0.528 352 8	0.000 435 400 0	0.659 17	0.340 55	0.000 28
09	1.013 047 7	0.515 632 3	0.000 383 466 7	0.662 53	0.337 22	0.000 25

λ (nm)	Spectral Tristimulus Values			Chromaticity Coordinates		
	$\bar{x}(\lambda)$	$\bar{y}(\lambda)$	$\bar{z}(\lambda)$	$x(\lambda)$	$y(\lambda)$	$z(\lambda)$
610	1.002 600 0	0.503 000 0	0.000 340 000 0	0.665 76	0.334 01	0.000 23
11	0.991 367 5	0.490 468 8	0.000 307 253 3	0.668 87	0.330 92	0.000 21
12	0.979 331 4	0.478 030 4	0.000 283 160 0	0.671 86	0.327 95	0.000 19
13	0.966 491 6	0.465 677 6	0.000 265 440 0	0.674 72	0.325 09	0.000 19
14	0.952 847 9	0.453 403 2	0.000 251 813 3	0.677 46	0.322 36	0.000 18
615	0.938 400 0	0.441 200 0	0.000 240 000 0	0.680 08	0.319 75	0.000 17
16	0.923 194 0	0.429 080 0	0.000 229 546 7	0.682 58	0.317 25	0.000 17
17	0.907 244 0	0.417 036 0	0.000 220 640 0	0.684 97	0.314 86	0.000 17
18	0.890 502 0	0.405 032 0	0.000 211 960 0	0.687 25	0.312 59	0.000 16
19	0.872 920 0	0.393 032 0	0.000 202 186 7	0.689 43	0.310 41	0.000 16
620	0.854 449 9	0.381 000 0	0.000 190 000 0	0.691 51	0.308 34	0.000 15
21	0.835 084 0	0.368 918 4	0.000 174 213 3	0.693 49	0.306 37	0.000 14
22	0.814 946 0	0.356 827 2	0.000 155 640 0	0.695 39	0.304 48	0.000 13
23	0.794 186 0	0.344 776 8	0.000 135 960 0	0.697 21	0.302 67	0.000 12
24	0.772 954 0	0.332 817 6	0.000 116 853 3	0.698 94	0.300 95	0.000 11
625	0.751 400 0	0.321 000 0	0.000 100 000 0	0.700 61	0.299 30	0.000 09
26	0.729 583 6	0.309 338 1	0.000 086 133 33	0.702 19	0.297 73	0.000 08
27	0.707 588 8	0.297 850 4	0.000 074 600 00	0.703 71	0.296 22	0.000 07
28	0.685 602 2	0.286 593 6	0.000 065 000 00	0.705 16	0.294 77	0.000 07
29	0.663 810 4	0.275 624 5	0.000 056 933 33	0.706 56	0.293 38	0.000 06
630	0.642 400 0	0.265 000 0	0.000 049 999 99	0.707 92	0.292 03	0.000 05
31	0.621 514 9	0.254 763 2	0.000 044 160 00	0.709 23	0.290 72	0.000 05
32	0.601 113 8	0.244 889 6	0.000 039 480 00	0.710 50	0.289 45	0.000 05
33	0.581 105 2	0.235 334 4	0.000 035 720 00	0.711 73	0.288 23	0.000 04
34	0.561 397 7	0.226 052 8	0.000 032 640 00	0.712 90	0.287 06	0.000 04
635	0.541 900 0	0.217 000 0	0.000 030 000 00	0.714 03	0.285 93	0.000 04
36	0.522 599 5	0.208 161 6	0.000 027 653 33	0.715 12	0.284 84	0.000 04
37	0.503 546 4	0.199 548 8	0.000 025 560 00	0.716 16	0.283 80	0.000 04
38	0.484 743 6	0.191 155 2	0.000 023 640 00	0.717 16	0.282 81	0.000 03
39	0.466 193 9	0.182 974 4	0.000 021 813 33	0.718 12	0.281 85	0.000 03
640	0.447 900 0	0.175 000 0	0.000 020 000 00	0.719 03	0.280 94	0.000 03
41	0.429 861 3	0.167 223 5	0.000 018 133 33	0.719 91	0.280 06	0.000 03
42	0.412 098 0	0.159 646 4	0.000 016 200 00	0.720 75	0.279 22	0.000 03
43	0.394 644 0	0.152 277 6	0.000 014 200 00	0.721 55	0.278 42	0.000 03
44	0.377 533 3	0.145 125 9	0.000 012 133 33	0.722 32	0.277 66	0.000 02
645	0.360 800 0	0.138 200 0	0.000 010 000 00	0.723 03	0.276 95	0.000 02
46	0.344 456 3	0.131 500 3	0.000 007 733 333	0.723 70	0.276 28	0.000 02
47	0.328 516 8	0.125 024 8	0.000 005 400 000	0.724 33	0.275 66	0.000 01
48	0.313 019 2	0.118 779 2	0.000 003 200 000	0.724 91	0.275 08	0.000 01
49	0.298 001 1	0.112 769 1	0.000 001 333 333	0.725 47	0.274 53	0.000 00
650	0.283 500 0	0.107 000 0	0.000 000 000 000	0.725 99	0.274 01	0.000 00
51	0.269 544 8	0.101 476 2		0.726 49	0.273 51	
52	0.256 118 4	0.096 188 64		0.726 98	0.273 02	
53	0.243 189 6	0.091 122 96		0.727 43	0.272 57	
54	0.230 727 2	0.086 264 85		0.727 86	0.272 14	

54

C. J. Bartleson

TABLE I. (*Continued*)

λ (nm)	x̄(λ)	ȳ(λ)	z̄(λ)	x(λ)	y(λ)	z(λ)
655	0.218 700 0	0.081 600 00	0.000 000 0	0.728 27	0.271 73	0.000 00
56	0.207 097 1	0.077 120 64		0.728 66	0.271 34	
57	0.195 923 2	0.072 825 52		0.729 02	0.270 98	
58	0.185 170 8	0.068 710 08		0.729 36	0.270 64	
59	0.174 832 3	0.064 769 76		0.729 68	0.270 32	
660	0.164 900 0	0.061 000 00		0.729 97	0.270 03	
61	0.155 366 7	0.057 396 21		0.730 23	0.269 77	
62	0.146 230 0	0.053 955 04		0.730 47	0.269 53	
63	0.137 490 0	0.050 673 76		0.730 69	0.269 31	
64	0.129 146 7	0.047 549 65		0.730 90	0.269 10	
665	0.121 200 0	0.044 580 00		0.731 09	0.268 91	
66	0.113 639 7	0.041 758 72		0.731 28	0.268 72	
67	0.106 465 0	0.039 084 96		0.731 47	0.268 53	
68	0.099 690 44	0.036 563 84		0.731 65	0.268 35	
69	0.093 330 61	0.034 200 48		0.731 83	0.268 17	
670	0.087 400 00	0.032 000 00		0.731 99	0.268 01	
71	0.081 900 96	0.029 962 61		0.732 15	0.267 85	
72	0.076 804 28	0.028 076 64		0.732 30	0.267 70	
73	0.072 077 12	0.026 329 36		0.732 44	0.267 56	
74	0.067 686 64	0.024 708 05		0.732 58	0.267 42	
675	0.063 600 00	0.023 200 00		0.732 72	0.267 28	
76	0.059 806 85	0.021 800 77		0.732 86	0.267 14	
77	0.056 282 16	0.020 501 12		0.733 00	0.267 00	
78	0.052 971 04	0.019 281 08		0.733 14	0.266 86	
79	0.049 818 61	0.018 120 69		0.733 28	0.266 72	
680	0.046 770 00	0.017 000 00		0.733 42	0.266 58	
81	0.043 784 05	0.015 903 79		0.733 55	0.266 45	
82	0.040 875 36	0.014 837 18		0.733 68	0.266 32	
83	0.038 072 64	0.013 810 68		0.733 81	0.266 19	
84	0.035 404 61	0.012 834 78		0.733 94	0.266 06	
685	0.032 900 00	0.011 920 00		0.734 05	0.265 95	
86	0.030 564 19	0.011 068 31		0.734 14	0.265 86	
87	0.028 380 56	0.010 273 39		0.734 22	0.265 78	
88	0.026 344 84	0.009 533 311		0.734 29	0.265 71	
89	0.024 452 75	0.008 846 157		0.734 34	0.265 66	
690	0.022 700 00	0.008 210 000		0.734 39	0.265 61	
91	0.021 084 29	0.007 623 781		0.734 44	0.265 56	
92	0.019 599 88	0.007 085 424		0.734 48	0.265 52	
93	0.018 237 32	0.006 591 476		0.734 52	0.265 48	
94	0.016 987 17	0.006 138 485		0.734 56	0.265 44	
695	0.015 840 00	0.005 723 000		0.734 59	0.265 41	
96	0.014 790 64	0.005 343 059		0.734 62	0.265 38	
97	0.013 831 32	0.004 995 796		0.734 65	0.265 35	
98	0.012 948 68	0.004 676 404		0.734 67	0.265 33	
99	0.012 129 20	0.004 380 075		0.734 69	0.265 31	
700	0.011 359 16	0.004 102 000		0.734 69	0.265 31	
01	0.010 629 35	0.003 838 453		0.734 69	0.265 31	
02	0.009 938 846	0.003 589 099		0.734 69	0.265 31	
03	0.009 288 422	0.003 354 219		0.734 69	0.265 31	
04	0.008 678 854	0.003 134 093		0.734 69	0.265 31	

λ (nm)	$\bar{x}(\lambda)$	$\bar{y}(\lambda)$	$\bar{z}(\lambda)$	$x(\lambda)$	$y(\lambda)$	$z(\lambda)$
		Spectral Tristimulus Values			Chromaticity Coordinates	
705	0.008 110 916	0.002 929 000	0.000 000 0	0.734 69	0.265 31	0.000 00
06	0.007 582 388	0.002 738 139		0.734 69	0.265 31	
07	0.007 088 746	0.002 559 876		0.734 69	0.265 31	
08	0.006 627 313	0.002 393 244		0.734 69	0.265 31	
09	0.006 195 408	0.002 237 275		0.734 69	0.265 31	
710	0.005 790 346	0.002 091 000		0.734 69	0.265 31	
11	0.005 409 826	0.001 953 587		0.734 69	0.265 31	
12	0.005 052 583	0.001 824 580		0.734 69	0.265 31	
13	0.004 717 512	0.001 703 580		0.734 69	0.265 31	
14	0.004 403 507	0.001 590 187		0.734 69	0.265 31	
715	0.004 109 457	0.001 484 000		0.734 69	0.265 31	
16	0.003 833 913	0.001 384 496		0.734 69	0.265 31	
17	0.003 575 748	0.001 291 268		0.734 69	0.265 31	
18	0.003 334 342	0.001 204 092		0.734 69	0.265 31	
19	0.003 109 075	0.001 122 744		0.734 69	0.265 31	
720	0.002 899 327	0.001 047 000		0.734 69	0.265 31	
21	0.002 704 348	0.000 976 589 6		0.734 69	0.265 31	
22	0.002 523 020	0.000 911 108 8		0.734 69	0.265 31	
23	0.002 354 168	0.000 850 133 2		0.734 69	0.265 31	
24	0.002 196 616	0.000 793 238 4		0.734 69	0.265 31	
725	0.002 049 190	0.000 740 000 0		0.734 69	0.265 31	
26	0.001 910 960	0.000 690 082 7		0.734 69	0.265 31	
27	0.001 781 438	0.000 643 310 0		0.734 69	0.265 31	
28	0.001 660 110	0.000 599 496 0		0.734 69	0.265 31	
29	0.001 546 459	0.000 558 454 7		0.734 69	0.265 31	
730	0.001 439 971	0.000 520 000 0		0.734 69	0.265 31	
31	0.001 340 042	0.000 483 913 6		0.734 69	0.265 31	
32	0.001 246 275	0.000 450 052 8		0.734 69	0.265 31	
33	0.001 158 471	0.000 418 345 2		0.734 69	0.265 31	
34	0.001 076 430	0.000 388 718 4		0.734 69	0.265 31	
735	0.000 999 949 3	0.000 361 100 0		0.734 69	0.265 31	
36	0.000 928 735 8	0.000 335 383 5		0.734 69	0.265 31	
37	0.000 862 433 2	0.000 311 440 4		0.734 69	0.265 31	
38	0.000 800 750 3	0.000 289 165 6		0.734 69	0.265 31	
39	0.000 743 396 0	0.000 268 453 9		0.734 69	0.265 31	
740	0.000 690 078 6	0.000 249 200 0		0.734 69	0.265 31	
41	0.000 640 515 6	0.000 231 301 9		0.734 69	0.265 31	
42	0.000 594 502 1	0.000 214 685 6		0.734 69	0.265 31	
43	0.000 551 864 6	0.000 199 288 4		0.734 69	0.265 31	
44	0.000 512 429 0	0.000 185 047 5		0.734 69	0.265 31	
745	0.000 476 021 3	0.000 171 900 0		0.734 69	0.265 31	
46	0.000 442 453 6	0.000 159 778 1		0.734 69	0.265 31	
47	0.000 411 511 7	0.000 148 604 4		0.734 69	0.265 31	
48	0.000 382 981 4	0.000 138 301 6		0.734 69	0.265 31	
49	0.000 356 649 1	0.000 128 792 5		0.734 69	0.265 31	
750	0.000 332 301 1	0.000 120 000 0		0.734 69	0.265 31	
51	0.000 309 758 6	0.000 111 859 5		0.734 69	0.265 31	
52	0.000 288 887 1	0.000 104 322 4		0.734 69	0.265 31	
53	0.000 269 539 4	0.000 097 335 60		0.734 69	0.265 31	
54	0.000 251 568 2	0.000 090 845 87		0.734 69	0.265 31	

TABLE I. (*Continued*)

λ (nm)	Spectral Tristimulus Values			Chromaticity Coordinates		
	$\bar{x}(\lambda)$	$\bar{y}(\lambda)$	$\bar{z}(\lambda)$	$x(\lambda)$	$y(\lambda)$	$z(\lambda)$
755	0.000 234 826 1	0.000 084 800 00	0.000 000 0	0.734 69	0.265 31	0.000 00
56	0.000 219 171 0	0.000 079 146 67		0.734 69	0.265 31	
57	0.000 204 525 8	0.000 073 858 00		0.734 69	0.265 31	
58	0.000 190 840 5	0.000 068 916 00		0.734 69	0.265 31	
59	0.000 178 065 4	0.000 064 302 67		0.734 69	0.265 31	
760	0.000 166 150 5	0.000 060 000 00		0.734 69	0.265 31	
61	0.000 155 023 6	0.000 055 981 87		0.734 69	0.265 31	
62	0.000 144 621 9	0.000 052 225 60		0.734 69	0.265 31	
63	0.000 134 909 8	0.000 048 718 40		0.734 69	0.265 31	
64	0.000 125 852 0	0.000 045 447 47		0.734 69	0.265 31	
765	0.000 117 413 0	0.000 042 400 00		0.734 69	0.265 31	
66	0.000 109 551 5	0.000 039 561 04		0.734 69	0.265 31	
67	0.000 102 224 5	0.000 036 915 12		0.734 69	0.265 31	
68	0.000 095 394 45	0.000 034 448 68		0.734 69	0.265 31	
69	0.000 089 023 90	0.000 032 148 16		0.734 69	0.265 31	
770	0.000 083 075 27	0.000 030 000 00		0.734 69	0.265 31	
71	0.000 077 512 69	0.000 027 991 25		0.734 69	0.265 31	
72	0.000 072 313 04	0.000 026 113 56		0.734 69	0.265 31	
73	0.000 067 457 78	0.000 024 360 24		0.734 69	0.265 31	
74	0.000 062 928 44	0.000 022 724 61		0.734 69	0.265 31	
775	0.000 058 706 52	0.000 021 200 00		0.734 69	0.265 31	
76	0.000 054 770 28	0.000 019 778 55		0.734 69	0.265 31	
77	0.000 051 099 18	0.000 018 452 85		0.734 69	0.265 31	
78	0.000 047 676 54	0.000 017 216 87		0.734 69	0.265 31	
79	0.000 044 485 67	0.000 016 064 59		0.734 69	0.265 31	
780	0.000 041 509 94	0.000 014 990 00		0.734 69	0.265 31	
81	0.000 038 733 24	0.000 013 987 28		0.734 69	0.265 31	
82	0.000 036 142 03	0.000 013 051 55		0.734 69	0.265 31	
83	0.000 033 723 52	0.000 012 178 18		0.734 69	0.265 31	
84	0.000 031 464 87	0.000 011 362 54		0.734 69	0.265 31	
785	0.000 029 353 26	0.000 010 600 00		0.734 69	0.265 31	
86	0.000 027 375 73	0.000 009 885 877		0.734 69	0.265 31	
87	0.000 025 524 33	0.000 009 217 304		0.734 69	0.265 31	
88	0.000 023 793 76	0.000 008 592 362		0.734 69	0.265 31	
89	0.000 022 178 70	0.000 008 009 133		0.734 69	0.265 31	
790	0.000 020 673 83	0.000 007 465 700		0.734 69	0.265 31	
91	0.000 019 272 26	0.000 006 959 567		0.734 69	0.265 31	
92	0.000 017 966 40	0.000 006 487 995		0.734 69	0.265 31	
93	0.000 016 749 91	0.000 006 048 699		0.734 69	0.265 31	
94	0.000 015 616 48	0.000 005 639 396		0.734 69	0.265 31	
795	0.000 014 559 77	0.000 005 257 800		0.734 69	0.265 31	
96	0.000 013 573 87	0.000 004 901 771		0.734 69	0.265 31	
97	0.000 012 654 36	0.000 004 569 720		0.734 69	0.265 31	
98	0.000 011 797 23	0.000 004 260 194		0.734 69	0.265 31	
99	0.000 010 998 44	0.000 003 971 739		0.734 69	0.265 31	
800	0.000 010 253 98	0.000 003 702 900		0.734 69	0.265 31	
01	0.000 009 559 646	0.000 003 452 163		0.734 69	0.265 31	
02	0.000 008 912 044	0.000 003 218 302		0.734 69	0.265 31	
03	0.000 008 308 358	0.000 003 000 300		0.734 69	0.265 31	
04	0.000 007 745 769	0.000 002 797 139		0.734 69	0.265 31	

λ (nm)	Spectral Tristimulus Values			Chromaticity Coordinates		
	$\bar{x}(\lambda)$	$\bar{y}(\lambda)$	$\bar{z}(\lambda)$	$x(\lambda)$	$y(\lambda)$	$z(\lambda)$
805	0.000 007 221 456	0.000 002 607 800	0.000 000 0	0.734 69	0.265 31	0.000 00
06	0.000 006 732 475	0.000 002 431 220		0.734 69	0.265 31	
07	0.000 006 276 423	0.000 002 266 531		0.734 69	0.265 31	
08	0.000 005 851 304	0.000 002 113 013		0.734 69	0.265 31	
09	0.000 005 455 118	0.000 001 969 943		0.734 69	0.265 31	
810	0.000 005 085 868	0.000 001 836 600		0.734 69	0.265 31	
11	0.000 004 741 466	0.000 001 712 230		0.734 69	0.265 31	
12	0.000 004 420 236	0.000 001 596 228		0.734 69	0.265 31	
13	0.000 004 120 783	0.000 001 488 090		0.734 69	0.265 31	
14	0.000 003 841 716	0.000 001 387 314		0.734 69	0.265 31	
815	0.000 003 581 652	0.000 001 293 400		0.734 69	0.265 31	
16	0.000 003 339 127	0.000 001 205 820		0.734 69	0.265 31	
17	0.000 003 112 949	0.000 001 124 143		0.734 69	0.265 31	
18	0.000 002 902 121	0.000 001 048 009		0.734 69	0.265 31	
19	0.000 002 705 645	0.000 000 977 057 8		0.734 69	0.265 31	
820	0.000 002 522 525	0.000 000 910 930 0		0.734 69	0.265 31	
21	0.000 002 351 726	0.000 000 849 251 3		0.734 69	0.265 31	
22	0.000 002 192 415	0.000 000 791 721 2		0.734 69	0.265 31	
23	0.000 002 043 902	0.000 000 738 090 4		0.734 69	0.265 31	
24	0.000 001 905 497	0.000 000 688 109 8		0.734 69	0.265 31	
825	0.000 001 776 509	0.000 000 641 530 0		0.734 69	0.265 31	
26	0.000 001 656 215	0.000 000 598 089 5		0.734 69	0.265 31	
27	0.000 001 544 022	0.000 000 557 574 6		0.734 69	0.265 31	
28	0.000 001 439 440	0.000 000 519 808 0		0.734 69	0.265 31	
29	0.000 001 341 977	0.000 000 484 612 3		0.734 69	0.265 31	
830	0.000 001 251 141	0.000 000 451 810 0		0.734 69	0.265 31	

$^a \sum \bar{x}(\lambda) = 106.865\ 469\ 489\ 595, \sum \bar{y}(\lambda) = 106.856\ 917\ 101\ 172, \sum \bar{z}(\lambda) = 106.892\ 251\ 278\ 636.$

These chromaticity coordinates are converted to spectral tristimulus values $\bar{x}(\lambda)$, $\bar{y}(\lambda)$, $\bar{z}(\lambda)$ as follows:

$$\bar{x}(\lambda) = [x(\lambda)/y(\lambda)]V(\lambda), \qquad \bar{y}(\lambda) = V(\lambda), \qquad \bar{z}(\lambda) = [z(\lambda)/y(\lambda)]V(\lambda), \quad (21)$$

where $V(\lambda)$ is the photopic luminous efficacy function.

The basic data on which the CIE 1931 Standard Colorimetric Observer's spectral tristimulus values rest derive from experiments with 2° stimulus fields by Guild (1931) for 7 observers together with similar data from Wright (1928–1929) for 10 observers and for 35 observers (Wright 1928–1929) whose data served to define more precisely the absorptances of ocular pigmentation. These experiments actually determined only matching *chromaticities* for spectral stimuli. Spectral tristimulus values were determined according to Eq. (21). Wright and Guild used the $V(\lambda)$ function adopted by the CIE in 1924 and did not actually measure values of $\bar{y}(\lambda)$. That practice has two consequences. The first has a salutary effect on the consistency of CIE standards; it means that the 1931 Standard Colorimetric Observer agrees exactly with the CIE 1924 Standard Photometric Observer in specifying luminance. However, to the extent that the 1924 Photometric Observer's $V(\lambda)$ data are inappropriate for typical observers of normal color vision, the luminance specifications of the colorimetric observer are also inappropriate.

Over the years, since 1931, some important instances of the inappropriateness of the $V(\lambda)$ function have come to light. Although the function provides acceptable data in the vast majority of cases where it has been used, a number of workers have found that the values of $V(\lambda)$ and, from Eq. (21), values of $\bar{r}(\lambda)$, $\bar{g}(\lambda)$, $\bar{b}(\lambda)$ are somewhat too low throughout the wavelength region of 380–460 nm. Judd (1951) has used newer determinations of luminous efficacy for this region of wavelengths and has computed a modified set of spectral tristimulus values that are sometimes used where the small errors of the original functions may be important. However, Judd's revised data are not presently recommended by the CIE.

Instead, the CIE has addressed two problems at once in recommending the CIE 1964 Supplementary Standard Colorimetric Observer. The 2° observer data relate to stimuli whose retinal images cover an area that is essentially free of rod receptors; for example, the data represent color mixture characteristics based almost exclusively on cone receptor characteristics. However, there are many instances where it is of interest to include some effects of rod receptors. This is done by determining color mixture characteristics for a retinal area larger than 2°. Extensive experiments of this kind were carried out by Stiles and Burch (1959) and by Speranskaya (1959). They measured directly the $\bar{r}(\lambda)$, $\bar{g}(\lambda)$, $\bar{b}(\lambda)$ functions, with no appeal to the $V(\lambda)$ function, using primaries with mean wavenumbers (v) of 15,500, 19,000, and 22,500 cm^{-1} (approximately wavelengths of 645.2, 526.3, and 444.4 nm). These measurements were made for stimuli subtending a visual angle of 10°. The experimental data were reduced to derive a coordinate system similar to that of the CIE 1931 Standard Colorimetric Observer's. After a period of field trials, the CIE adopted the new data and coordinate system as the CIE 1964 Supplementary Standard Colorimetric Observer. The following transformation from the original set of primaries to an imaginary set defines the supplementary 10° observer:

$$\bar{x}_{10}(\bar{v}) = 0.341080\bar{r}_{10}(\bar{v}) + 0.189145\bar{g}_{10}(\bar{v}) + 0.387529\bar{b}_{10}(\bar{v}),$$

$$\bar{y}_{10}(\bar{v}) = 0.139058\bar{r}_{10}(\bar{v}) + 0.837460\bar{g}_{10}(\bar{v}) + 0.073316\bar{b}_{10}(\bar{v}), \quad (22)$$

$$\bar{z}_{10}(\bar{v}) = 0.000000\bar{r}_{10}(\bar{v}) + 0.039553\bar{g}_{10}(\bar{v}) + 2.026200\bar{b}_{10}(\bar{v}).$$

Values of $\bar{x}_{10}(\lambda)$, $\bar{y}_{10}(\lambda)$, $\bar{z}_{10}(\lambda)$ are listed in Table II. The properties of these spectral tristimulus values are similar to those of the 1931 observer's

TABLE II. CIE 1964 Supplementary Spectral Tristimulus Values $\bar{x}_{10}(\lambda)$, $\bar{y}_{10}(\lambda)$, $\bar{z}_{10}(\lambda)^a$

λ (nm)	Spectral Tristimulus Values			Chromaticity Coordinates		
	$\bar{x}_{10}(\lambda)$	$\bar{y}_{10}(\lambda)$	$\bar{z}_{10}(\lambda)$	$x_{10}(\lambda)$	$y_{10}(\lambda)$	$z_{10}(\lambda)$
360	0.000 000 122 200	0.000 000 013 398	0.000 000 535 027	0.182 22	0.019 98	0.797 80
361	0.000 000 185 138	0.000 000 020 294	0.000 000 810 720	0.182 20	0.019 97	0.797 83
362	0.000 000 278 83	0.000 000 030 56	0.000 001 221 20	0.182 '7	0.019 97	0.797 86
363	0.000 000 417 47	0.000 000 045 74	0.000 001 828 70	0.182 15	0.019 96	0.797 89
364	0.000 000 621 33	0.000 000 068 05	0.000 002 722 20	0.182 12	0.019 95	0.797 93
365	0.000 000 919 27	0.000 000 100 65	0.000 004 028 30	0.182 10	0.019 94	0.797 96
366	0.000 001 351 98	0.000 000 147 98	0.000 005 925 70	0.182 07	0.019 93	0.798 00
367	0.000 001 976 54	0.000 000 216 27	0.000 008 665 10	0.182 04	0.019 92	0.798 04
368	0.000 002 872 5	0.000 000 314 2	0.000 012 596 0	0.182 00	0.019 91	0.798 09
369	0.000 004 149 5	0.000 000 453 7	0.000 018 201 0	0.181 96	0.019 90	0.798 14
370	0.000 005 958 6	0.000 000 651 1	0.000 026 143 7	0.181 92	0.019 88	0.798 20
371	0.000 008 505 6	0.000 000 928 8	0.000 037 330 0	0.181 88	0.019 86	0.798 26
372	0.000 012 068 6	0.000 001 317 5	0.000 052 987 0	0.181 83	0.019 85	0.798 32
373	0.000 017 022 6	0.000 001 857 2	0.000 074 764 0	0.181 78	0.019 83	0.798 39
374	0.000 023 868	0.000 002 602	0.000 104 870	0.181 73	0.019 81	0.798 46
375	0.000 033 266	0.000 003 625	0.000 146 220	0.181 67	0.019 80	0.798 53
376	0.000 046 087	0.000 005 019	0.000 202 660	0.181 61	0.019 78	0.798 61
377	0.000 063 472	0.000 006 907	0.000 279 230	0.181 55	0.019 76	0.798 69
378	0.000 086 892	0.000 009 449	0.000 382 450	0.181 48	0.019 74	0.798 78
379	0.000 118 246	0.000 012 848	0.000 520 720	0.181 41	0.019 71	0.798 88
380	0.000 159 952	0.000 017 364	0.000 704 776	0.181 33	0.019 69	0.798 98
381	0.000 215 080	0.000 023 327	0.000 948 230	0.181 25	0.019 66	0.799 09
382	0.000 287 49	0.000 031 15	0.001 268 20	0.181 17	0.019 63	0.799 20
383	0.000 381 99	0.000 041 35	0.001 686 10	0.181 09	0.019 60	0.799 31
384	0.000 504 55	0.000 054 56	0.002 228 50	0.181 00	0.019 57	0.799 43
385	0.000 662 44	0.000 071 56	0.002 927 80	0.180 91	0.019 54	0.799 55
386	0.000 864 50	0.000 093 30	0.003 823 70	0.180 80	0.019 51	0.799 69
387	0.001 121 50	0.000 120 87	0.004 964 20	0.180 70	0.019 47	0.799 83
388	0.001 446 16	0.000 155 64	0.006 406 70	0.180 58	0.019 43	0.799 99
389	0.001 853 59	0.000 199 20	0.008 219 30	0.180 45	0.019 39	0.800 16
390	0.002 361 6	0.000 253 4	0.010 482 2	0.180 31	0.019 35	0.800 34
391	0.002 990 6	0.000 320 2	0.013 289 0	0.180 16	0.019 29	0.800 55
392	0.003 764 5	0.000 402 4	0.016 747 0	0.180 00	0.019 24	0.800 76
393	0.004 710 2	0.000 502 3	0.020 980 0	0.179 83	0.019 18	0.800 99
394	0.005 858 1	0.000 623 2	0.026 127 0	0.179 65	0.019 11	0.801 24
395	0.007 242 3	0.000 768 5	0.032 344 0	0.179 47	0.019 04	0.801 49
396	0.008 899 6	0.000 941 7	0.039 802 0	0.179 27	0.018 97	0.801 76
397	0.010 870 9	0.001 147 8	0.048 691 0	0.179 06	0.018 91	0.802 03
398	0.013 198 9	0.001 390 3	0.059 210 0	0.178 85	0.018 84	0.802 31
399	0.015 929 2	0.001 674 0	0.071 576 0	0.178 62	0.018 77	0.802 61
400	0.019 109 7	0.002 004 4	0.086 010 9	0.178 39	0.018 71	0.802 90
401	0.022 788	0.002 386	0.102 740	0.178 15	0.018 65	0.803 20
402	0.027 011	0.002 822	0.122 000	0.177 90	0.018 59	0.803 51
403	0.031 829	0.003 319	0.144 020	0.177 65	0.018 52	0.803 83
404	0.037 278	0.003 880	0.168 990	0.177 39	0.018 46	0.804 15
405	0.043 400	0.004 509	0.197 120	0.177 12	0.018 40	0.804 48
406	0.050 223	0.005 209	0.228 570	0.176 84	0.018 34	0.804 82
407	0.057 764	0.005 985	0.263 470	0.176 53	0.018 29	0.805 18
408	0.066 038	0.006 833	0.301 900	0.176 21	0.018 23	0.805 56
409	0.075 033	0.007 757	0.343 870	0.175 86	0.018 18	0.805 96
410	0.084 736	0.008 756	0.389 366	0.175 49	0.018 13	0.806 38

TABLE II. (*Continued*)

λ (nm)	Spectral Tristimulus Values			Chromaticity Coordinates		
	$\bar{x}_{10}(\lambda)$	$\bar{y}_{10}(\lambda)$	$\bar{z}_{10}(\lambda)$	$x_{10}(\lambda)$	$y_{10}(\lambda)$	$z_{10}(\lambda)$
411	0.095 041	0.009 816	0.437 970	0.175 09	0.018 08	0.806 83
412	0.105 836	0.010 918	0.489 220	0.174 65	0.018 02	0.807 33
413	0.117 066	0.012 058	0.542 900	0.174 20	0.017 94	0.807 86
414	0.128 682	0 013 237	0.598 810	0.173 72	0.017 87	0.808 41
415	0.140 638	0.014 456	0.656 760	0.173 23	0.017 81	0.808 96
416	0.152 893	0.015 717	0.716 580	0.172 72	0.017 76	0.809 52
417	0.165 416	0.017 025	0.778 120	0.172 21	0.017 72	0.810 07
418	0.178 191	0.018 399	0.841 310	0.171 68	0.017 73	0.810 59
419	0.191 214	0.019 848	0.906 110	0.171 16	0.017 77	0.811 07
420	0.204 492	0.021 391	0.972 542	0.170 63	0.017 85	0.811 52
421	0.217 650	0.022 992	1.038 90	0.170 10	0.017 97	0.811 93
422	0.230 267	0.024 598	1.103 10	0.169 57	0.018 11	0.812 32
423	0.242 311	0.026 213	1.165 10	0.169 02	0.018 28	0.812 70
424	0.253 793	0.027 841	1.224 90	0.168 46	0.018 48	0.813 06
425	0.264 737	0.029 497	1.282 50	0.167 90	0.018 71	0.813 39
426	0.275 195	0.031 195	1.338 20	0.167 33	0.018 97	0.813 70
427	0.285 301	0.032 927	1.392 60	0.166 76	0.019 25	0.813 99
428	0.295 143	0.034 738	1.446 10	0.166 19	0.019 56	0.814 25
429	0.304 869	0.036 654	1.499 40	0.165 61	0.019 91	0.814 48
430	0.314 679	0.038 676	1.553 48	0.165 03	0.020 28	0.814 69
431	0.324 355	0.040 792	1.607 20	0.164 45	0.020 68	0.814 87
432	0.333 570	0.042 946	1.658 90	0.163 88	0.021 10	0.815 02
433	0.342 243	0.045 114	1.708 20	0.163 32	0.021 53	0.815 15
434	0.350 312	0.047 333	1.754 80	0.162 75	0.021 99	0.815 26
435	0.357 719	0.049 602	1.798 50	0.162 17	0.022 49	0.815 34
436	0.364 482	0.051 934	1.839 20	0.161 59	0.023 02	0.815 39
437	0.370 493	0.054 337	1.876 60	0.160 98	0.023 61	0.815 41
438	0.375 727	0.056 822	1.910 50	0.160 36	0.024 25	0.815 39
439	0.380 158	0.059 399	1.940 80	0.159 71	0.024 95	0.815 34
440	0.383 734	0.062 077	1.967 28	0.159 02	0.025 73	0.815 25
441	0.386 327	0.064 737	1.989 10	0.158 32	0.026 53	0.815 15
442	0.387 858	0.067 285	2.005 70	0.157 61	0.027 34	0.815 05
443	0.388 396	0.069 764	2.017 40	0.156 89	0.028 18	0.814 93
444	0.387 978	0.072 218	2.024 40	0.156 15	0.029 07	0.814 78
445	0.386 726	0.074 704	2.027 30	0.155 39	0.030 02	0.814 59
446	0.384 696	0.077 272	2.026 40	0.154 60	0.031 05	0.814 35
447	0.382 006	0.079 979	2.022 30	0.153 77	0.032 19	0.814 04
448	0.378 709	0.082 874	2.015 30	0.152 90	0.033 46	0.813 64
449	0.374 915	0.086 000	2.006 00	0.151 98	0.034 86	0.813 16
450	0.370 702	0.089 456	1.994 80	0.151 00	0.036 44	0.812 56
451	0.366 089	0.092 947	1.981 40	0.150 01	0.038 09	0.811 90
452	0.361 045	0.096 275	1.965 30	0.149 03	0.039 74	0.811 23
453	0.355 518	0.099 535	1.946 40	0.148 04	0.041 45	0.810 51
454	0.349 486	0.102 829	1.924 80	0.147 02	0.043 26	0.809 72
455	0.342 957	0.106 256	1.900 70	0.145 94	0.045 22	0.808 84
456	0.335 893	0.109 901	1.874 10	0.144 79	0.047 37	0.807 84
457	0.328 284	0.113 835	1.845 10	0.143 53	0.049 77	0.806 70
458	0.320 150	0.118 167	1.813 90	0.142 15	0.052 47	0.805 38
459	0.311 475	0.122 932	1.780 60	0.140 62	0.055 50	0.803 88
460	0.302 273	0.128 201	1.745 37	0.138 92	0.058 92	0.802 16

λ (nm)		Spectral Tristimulus Values			Chromaticity Coordinates	
	$\bar{x}_{10}(\lambda)$	$\bar{y}_{10}(\lambda)$	$\bar{z}_{10}(\lambda)$	$x_{10}(\lambda)$	$y_{10}(\lambda)$	$z_{10}(\lambda)$
461	0.292 858	0.133 457	1.709 10	0.137 14	0.062 50	0.800 36
462	0.283 502	0.138 323	1.672 30	0.135 38	0.066 05	0.798 57
463	0.274 044	0.143 042	1.634 70	0.133 56	0.069 72	0.796 72
464	0.264 263	0.147 787	1.595 60	0.131 63	0.073 61	0.794 76
465	0.254 085	0.152 761	1.554 90	0.129 52	0.077 87	0.792 61
466	0.243 392	0.158 102	1.512 20	0.127 18	0.082 62	0.790 20
467	0.232 187	0.163 941	1.467 30	0.124 60	0.087 98	0.787 42
468	0.220 488	0.170 362	1.419 90	0.121 77	0.094 08	0.784 15
469	0.208 198	0.177 425	1.370 00	0.118 59	0.101 06	0.780 35
470	0.195 618	0.185 190	1.317 56	0.115 18	0.109 04	0.775 78
471	0.183 034	0.193 025	1.262 40	0.111 71	0.117 81	0.770 48
472	0.170 222	0.200 313	1.205 00	0.108 04	0.127 14	0.764 82
473	0.157 348	0.207 156	1.146 60	0.104 13	0.137 09	0.758 78
474	0.144 650	0.213 644	1.088 00	0.100 01	0.147 72	0.752 27
475	0.132 349	0.219 940	1.030 20	0.095 73	0.159 09	0.745 18
476	0.120 584	0.226 170	0.973 830	0.091 31	0.171 27	0.737 42
477	0.109 456	0.232 467	0.919 430	0.086 78	0.184 30	0.728 92
478	0.099 042	0.239 025	0.867 460	0.082 16	0.198 27	0.719 57
479	0.089 388	0.245 997	0.818 280	0.077 48	0.213 23	0.709 29
480	0.080 507	0.253 589	0.772 125	0.072 78	0.229 24	0.697 98
481	0.072 034	0.261 876	0.728 290	0.067 82	0.246 54	0.685 64
482	0.063 710	0.270 643	0.686 040	0.062 44	0.265 23	0.672 33
483	0.055 694	0.279 645	0.645 530	0.056 78	0.285 10	0.658 12
484	0.048 117	0.288 694	0.606 850	0.050 99	0.305 93	0.643 08
485	0.041 072	0.297 665	0.570 060	0.045 19	0.327 54	0.627 27
486	0.034 642	0.306 469	0.535 220	0.039 53	0.349 72	0.610 75
487	0.028 896	0.315 035	0.502 340	0.034 15	0.372 26	0.593 59
488	0.023 876	0.323 335	0.471 400	0.029 17	0.394 98	0.575 85
489	0.019 628	0 331 366	0.442 390	0.024 74	0.417 66	0.557 60
490	0.016 172	0.339 133	0.415 254	0.020 99	0.440 11	0.538 90
491	0.013 300	0.347 860	0.390 024	0.017 71	0.463 08	0.519 21
492	0.010 759	0.358 326	0.366 399	0.014 63	0.487 20	0.498 17
493	0.008 542	0.370 001	0.344 015	0.011 82	0.512 07	0.476 11
494	0.006 661	0.382 464	0.322 689	0.009 36	0.537 31	0.453 33
495	0.005 132	0.395 379	0.302 356	0.007 30	0.562 52	0.430 18
496	0.003 982	0.408 482	0.283 036	0.005 73	0.587 32	0.406 95
497	0.003 239	0.421 588	0.264 816	0.004 70	0.611 31	0.383 99
498	0.002 934	0.434 619	0.247 848	0.004 28	0.634 11	0.361 61
499	0.003 114	0.447 601	0.232 318	0.004 56	0.655 31	0.340 13
500	0.003 816	0.460 777	0.218 502	0.005 59	0.674 54	0.319 87
501	0.005 095	0.474 340	0.205 851	0.007 43	0.692 18	0.300 39
502	0.006 936	0.488 200	0.193 596	0.010 07	0.708 84	0.281 09
503	0.009 299	0.502 340	0.181 736	0.013 41	0.724 49	0.262 10
504	0.012 147	0.516 740	0.170 281	0.017 37	0.739 08	0.243 55
505	0.015 444	0.531 360	0.159 249	0.021 87	0.752 58	0.225 55
506	0.019 156	0.546 190	0.148 673	0.026 83	0.764 95	0.208 22
507	0.023 250	0.561 180	0.138 609	0.032 16	0.776 14	0.191 70
508	0.027 690	0.576 290	0.129 096	0.037 77	0.786 13	0.176 10
509	0.032 444	0.591 500	0.120 215	0.043 60	0.794 86	0.161 54
510	0.037 465	0.606 741	0.112 044	0.049 54	0.802 30	0.148 16

TABLE II. (*Continued*)

λ (nm)		Spectral Tristimulus Values			Chromaticity Coordinates	
	$\bar{x}_{10}(\lambda)$	$\bar{y}_{10}(\lambda)$	$\bar{z}_{10}(\lambda)$	$x_{10}(\lambda)$	$y_{10}(\lambda)$	$z_{10}(\lambda)$
511	0.042 956	0.622 150	0.104 710	0.055 80	0.808 18	0.136 02
512	0.049 114	0.637 830	0.098 196	0.062 55	0.812 38	0.125 07
513	0.055 920	0.653 710	0.092 361	0.069 73	0.815 11	0.115 16
514	0.063 349	0.669 680	0.087 088	0.077 24	0.816 57	0.106 19
515	0.071 358	0.685 660	0.082 248	0.085 02	0.816 98	0.098 00
516	0.079 901	0.701 550	0.077 744	0.093 00	0.816 52	0.090 48
517	0.088 909	0.717 230	0.073 456	0.101 08	0.815 41	0.083 51
518	0.098 293	0.732 570	0.069 268	0.109 20	0.813 85	0.076 95
519	0.107 949	0.747 460	0.065 060	0.117 28	0.812 04	0.070 68
520	0.117 749	0.761 757	0.060 709	0.125 24	0.810 19	0.064 57
521	0.127 839	0.775 340	0.056 457	0.133 22	0.807 95	0.058 83
522	0.138 450	0.788 220	0.052 609	0.141 38	0.804 90	0.053 72
523	0.149 516	0.800 460	0.049 122	0.149 65	0.801 18	0.049 17
524	0.161 041	0.812 140	0.045 954	0.158 02	0.796 89	0.045 09
525	0.172 953	0.823 330	0.043 050	0.166 41	0.792 17	0.041 42
526	0.185 209	0.834 120	0.040 368	0.174 78	0.787 13	0.038 09
527	0.197 755	0.844 600	0.037 839	0.183 07	0.781 90	0.035 03
528	0.210 538	0.854 870	0.035 384	0.191 26	0.776 60	0.032 14
529	0.223 460	0.865 040	0.032 949	0.199 26	0.771 36	0.029 38
530	0.236 491	0.875 211	0.030 451	0.207 06	0.766 28	0.026 66
531	0.249 633	0.885 370	0.028 029	0.214 64	0.761 26	0.024 10
532	0.262 972	0.895 370	0.025 862	0.222 07	0.756 09	0.021 84
533	0.276 515	0.905 150	0.023 920	0.229 36	0.750 80	0.019 84
534	0.290 269	0.914 650	0.022 174	0.236 55	0.745 38	0.018 07
535	0.304 213	0.923 810	0.020 584	0.243 64	0.739 87	0.016 49
536	0.318 361	0.932 550	0.019 127	0.250 67	0.734 27	0.015 06
537	0.332 705	0.940 810	0.017 740	0.257 66	0.728 60	0.013 74
538	0.347 232	0.948 520	0.016 403	0.264 63	0.722 87	0.012 50
539	0.361 926	0.955 600	0.015 064	0.271 60	0.717 10	0.011 30
540	0.376 772	0.961 988	0.013 676	0.278 59	0.711 30	0.010 11
541	0.391 683	0.967 540	0.012 308	0.285 58	0.705 45	0.008 97
542	0.406 594	0.972 230	0.011 056	0.292 54	0.699 51	0.007 95
543	0.421 539	0.976 170	0.009 915	0.299 47	0.693 49	0.007 04
544	0.436 517	0.979 460	0.008 872	0.306 36	0.687 41	0.006 23
545	0.451 584	0.982 200	0.007 918	0.313 23	0.681 28	0.005 49
546	0.466 782	0.984 520	0.007 030	0.320 08	0.675 10	0.004 82
547	0.482 147	0.986 520	0.006 223	0.326 90	0.668 88	0.004 22
548	0.497 738	0.988 320	0.005 453	0.333 71	0.662 63	0.003 66
549	0.513 606	0.990 020	0.004 714	0.340 51	0.656 36	0.003 13
550	0.529 826	0.991 761	0.003 988	0.347 30	0.650 09	0.002 61
551	0.546 440	0.993 530	0.003 289	0.354 08	0.643 79	0.002 13
552	0.563 426	0.995 230	0.002 646	0.360 87	0.637 44	0.001 69
553	0.580 726	0.996 770	0.002 063	0.367 65	0.631 04	0.001 31
554	0.598 290	0.998 090	0.001 533	0.374 42	0.624 62	0.000 96
555	0.616 053	0.999 110	0.001 091	0.381 16	0.618 16	0.000 68
556	0.633 948	0.999 770	0.000 711	0.387 87	0.611 69	0.000 44
557	0.651 901	1.000 000	0.000 407	0.394 54	0.605 21	0.000 25
558	0.669 824	0.999 710	0.000 184	0.401 16	0.598 73	0.000 11
559	0.687 632	0.998 850	0.000 047	0.407 72	0.592 25	0.000 03
560	0.705 224	0.997 340	0.000 000	0.414 21	0.585 79	0.000 00

λ (nm)	Spectral Tristimulus Values			Chromaticity Coordinates		
	$\bar{x}_{10}(\lambda)$	$\bar{y}_{10}(\lambda)$	$\bar{z}_{10}(\lambda)$	$x_{10}(\lambda)$	$y_{10}(\lambda)$	$z_{10}(\lambda)$
561	0.722 773	0.995 260	0.000 000	0.420 70	0.579 30	0.000 00
562	0.740 483	0.992 740		0.427 23	0.572 77	
563	0.758 273	0.989 750		0.433 79	0.566 21	
564	0.776 083	0.986 300		0.440 36	0.559 64	
565	0.793 832	0.982 380		0.446 92	0.553 08	
566	0.811 436	0.977 980		0.453 46	0.546 54	
567	0.828 822	0.973 110		0.459 96	0.540 04	
568	0.845 879	0.967 740		0.466 40	0.533 60	
569	0.862 525	0.961 890		0.472 77	0.527 23	
570	0.878 655	0.955 552		0.479 04	0.520 96	
571	0.894 208	0.948 601		0.485 24	0.514 76	
572	0.909 206	0.940 981		0.491 41	0.508 59	
573	0.923 672	0.932 798		0.497 54	0.502 46	
574	0.937 638	0.924 158		0.503 62	0.496 38	
575	0.951 162	0.915 175		0.509 64	0.490 36	
576	0.964 283	0.905 954		0.515 59	0.484 41	
577	0.977 068	0.896 608		0.521 47	0.478 53	
578	0.989 590	0.887 249		0.527 26	0.472 74	
579	1.001 91	0.877 986		0.532 96	0.467 04	
580	1.014 16	0.868 934		0.538 56	0.461 44	
581	1.026 50	0.860 164		0.544 08	0.455 92	
582	1.038 80	0.851 519		0.549 54	0.450 46	
583	1.051 00	0.842 963		0.554 92	0.445 08	
584	1.062 90	0.834 393		0.560 22	0.439 78	
585	1.074 30	0.825 623		0.565 44	0.434 56	
586	1.085 20	0.816 764		0.570 57	0 429 43	
587	1.095 20	0.807 544		0.575 59	0.424 41	
588	1.104 20	0.797 947		0.580 50	0.419 50	
589	1.112 00	0.787 893		0.585 30	0.414 70	
590	1.118 52	0.777 405		0.589 96	0.410 04	
591	1.123 80	0.766 490		0.594 51	0.405 49	
592	1.128 00	0.755 309		0.598 95	0.401 05	
593	1.131 10	0.743 845		0.603 27	0.396 73	
594	1.133 20	0.732 190		0.607 49	0.392 51	
595	1.134 30	0.720 353		0.611 60	0.388 40	
596	1.134 30	0.708 281		0.615 60	0.384 40	
597	1.133 30	0.696 055		0.619 51	0.380 49	
598	1.131 20	0.683 621		0.623 31	0.376 69	
599	1.128 10	0.671 048		0.627 02	0.372 98	
600	1.123 99	0.658 341		0.630 63	0.369 37	
601	1.118 90	0.645 545		0.634 14	0.365 86	
602	1.112 90	0.632 718		0.637 54	0.362 46	
603	1.105 90	0.619 815		0.640 84	0.359 16	
604	1.098 00	0.606 887		0.644 03	0.355 97	
605	1.089 10	0.593 878		0.647 13	0.352 87	
606	1.079 20	0.580 781		0.650 13	0.349 87	
607	1.068 40	0.567 653		0.653 04	0.346 96	
608	1.056 70	0.554 490		0.655 85	0.344 15	
609	1.044 00	0.541 228		0.658 58	0.341 42	
610	1.030 48	0.527 963		0.661 22	0.338 78	

TABLE II. *(Continued)*

λ (nm)	Spectral Tristimulus Values			Chromaticity Coordinates		
	$\bar{x}_{10}(\lambda)$	$\bar{y}_{10}(\lambda)$	$\bar{z}_{10}(\lambda)$	$x_{10}(\lambda)$	$y_{10}(\lambda)$	$z_{10}(\lambda)$
611	1.016 00	0.514 634	0.000 000	0.663 78	0.336 22	0.000 00
612	1.000 80	0.501 363		0.666 24	0.333 76	
613	0.984 790	0.488 124		0.668 60	0.331 40	
614	0.968 080	0.474 935		0.670 87	0.329 13	
615	0.950 740	0.461 834		0.673 06	0.326 94	
616	0.932 800	0.448 823		0.675 15	0.324 85	
617	0.914 340	0.435 917		0.677 16	0.322 84	
618	0.895 390	0.423 153		0.679 08	0.320 92	
619	0.876 030	0.410 526		0.680 91	0.319 09	
620	0.856 297	0.398 057		0.682 66	0.317 34	
621	0.836 350	0.385 835		0.684 31	0.315 69	
622	0.816 290	0.373 951		0.685 82	0.314 18	
623	0.796 050	0.362 311		0.687 22	0.312 78	
624	0.775 610	0.350 863		0.688 53	0.311 47	
625	0.754 930	0.339 554		0.689 76	0.310 24	
626	0.733 990	0.328 309		0.690 94	0.309 06	
627	0.712 780	0.317 118		0.692 09	0.307 91	
628	0.691 290	0.305 936		0.693 21	0.306 79	
629	0.669 520	0.294 737		0.694 34	0.305 66	
630	0.647 467	0.283 493		0.695 48	0.304 52	
631	0.625 110	0.272 222		0.696 63	0.303 37	
632	0.602 520	0.260 990		0.697 76	0.302 24	
633	0.579 890	0.249 877		0.698 86	0.301 14	
634	0.557 370	0.238 946		0.699 94	0.300 06	
635	0.535 110	0.228 254		0.700 99	0.299 01	
636	0.513 240	0.217 853		0.702 02	0.297 98	
637	0.491 860	0.207 780		0.703 02	0.296 98	
638	0.471 080	0.198 072		0.704 00	0.296 00	
639	0.450 960	0.188 748		0.704 95	0.295 05	
640	0.431 567	0.179 828		0.705 87	0.294 13	
641	0.412 870	0.171 285		0.706 78	0.293 22	
642	0.394 750	0.163 059		0.707 68	0.292 32	
643	0.377 210	0.155 151		0.708 56	0.291 44	
644	0.360 190	0.147 535		0.709 42	0.290 58	
645	0.343 690	0.140 211		0.710 25	0.289 75	
646	0.327 690	0.133 170		0.711 04	0.288 96	
647	0.312 170	0.126 400		0.711 79	0.288 21	
648	0.297 110	0.119 892		0.712 49	0.287 51	
649	0.282 500	0.113 640		0.713 13	0.286 87	
650	0.268 329	0.107 633		0.713 71	0.286 29	
651	0.254 590	0.101 870		0.714 22	0.285 78	
652	0.241 300	0.096 347		0.714 65	0.285 35	
653	0.228 480	0.091 063		0.715 02	0.284 98	
654	0.216 140	0.086 010		0.715 34	0.284 66	
655	0.204 300	0.081 187		0.715 62	0.284 38	
656	0.192 950	0.076 583		0.715 87	0.284 13	
657	0.182 110	0.072 198		0.716 10	0.283 90	
658	0.171 770	0.068 024		0.716 32	0.283 68	
659	0.161 920	0.064 052		0.716 55	0.283 45	
660	0.152 568	0.060 281		0.716 79	0.283 21	

λ (nm)	$\bar{x}_{10}(\lambda)$	$\bar{y}_{10}(\lambda)$	$\bar{z}_{10}(\lambda)$	$x_{10}(\lambda)$	$y_{10}(\lambda)$	$z_{10}(\lambda)$
		Spectral Tristimulus Values			Chromaticity Coordinates	
661	0.143 670	0.056 697	0.000 000	0.717 03	0.282 97	0.000 00
662	0.135 200	0.053 292		0.717 27	0.282 73	
663	0.127 130	0.050 059		0.717 48	0.282 52	
664	0.119 480	0.046 998		0.717 69	0.282 31	
665	0.112 210	0.044 096		0.717 89	0.282 11	
666	0.105 310	0.041 345		0.718 08	0.281 92	
667	0.098 786 0	0.038 750 7		0.718 25	0.281 75	
668	0.092 610 0	0.036 297 8		0.718 42	0.281 58	
669	0.086 773 0	0.033 983 2		0.718 58	0.281 42	
670	0.081 260 6	0.031 800 4		0.718 73	0.281 27	
671	0.076 048 0	0.029 739 5		0.718 88	0.281 12	
672	0.071 114 0	0.027 791 8		0.719 01	0.280 99	
673	0.066 454 0	0.025 955 1		0.719 13	0.280 87	
674	0.062 062 0	0.024 226 3		0.719 24	0.280 76	
675	0.057 930 0	0.022 601 7		0.719 34	0.280 66	
676	0.054 050 0	0.021 077 9		0.719 44	0.280 56	
677	0.050 412 0	0.019 650 5		0.719 53	0.280 47	
678	0.047 006 0	0.018 315 3		0.719 61	0.280 39	
679	0.043 823 0	0.017 068 6		0.719 69	0.280 31	
680	0.040 850 8	0.015 905 1		0.719 76	0.280 24	
681	0.038 072 0	0.014 818 3		0.719 83	0.280 17	
682	0.035 468 0	0.013 800 8		0.719 89	0.280 11	
683	0.033 031 0	0.012 849 5		0.719 94	0.280 06	
684	0.030 753 0	0.011 960 7		0.719 98	0.280 02	
685	0.028 623 0	0.011 130 3		0.720 02	0.279 98	
686	0.026 635 0	0.010 355 5		0.720 05	0.279 95	
687	0.024 781 0	0.009 633 2		0.720 08	0.279 92	
688	0.023 052 0	0.008 959 9		0.720 11	0.279 89	
689	0.021 441 0	0.008 332 4		0.720 14	0.279 86	
690	0.019 941 3	0.007 748 8		0.720 16	0.279 84	
691	0.018 544 0	0.007 204 6		0.720 19	0.279 81	
692	0.017 241 0	0.006 697 5		0.720 22	0.279 78	
693	0.016 027 0	0.006 225 1		0.720 25	0.279 75	
694	0.014 896 0	0.005 785 0		0.720 27	0.279 73	
695	0.013 842 0	0.005 375 1		0.720 30	0.279 70	
696	0.012 862 0	0.004 994 1		0.720 31	0.279 69	
697	0.011 949 0	0.004 639 2		0.720 33	0.279 67	
698	0.011 100 0	0.004 309 3		0.720 34	0.279 66	
699	0.010 311 0	0.004 002 8		0.720 35	0.279 65	
700	0.009 576 88	0.003 717 74		0.720 36	0.279 64	
701	0.008 894 00	0.003 452 62		0.720 36	0.279 64	
702	0.008 258 10	0.003 205 83		0.720 36	0.279 64	
703	0.007 666 40	0.002 976 23		0.720 35	0.279 65	
704	0.007 116 30	0.002 762 81		0.720 34	0.279 66	
705	0.006 605 20	0.002 564 56		0.720 32	0.279 68	
706	0.006 130 60	0.002 380 48		0.720 31	0.279 69	
707	0.005 690 30	0.002 209 71		0.720 29	0.279 71	
708	0.005 281 90	0.002 051 32		0.720 27	0.279 73	
709	0.004 903 30	0.001 904 49		0.720 25	0.279 75	
710	0.004 552 63	0.001 768 47		0.720 23	0.279 77	

TABLE II. *(Continued)*

λ (nm)	Spectral Tristimulus Values			Chromaticity Coordinates		
	$\bar{x}_{10}(\lambda)$	$\bar{y}_{10}(\lambda)$	$\bar{z}_{10}(\lambda)$	$x_{10}(\lambda)$	$y_{10}(\lambda)$	$z_{10}(\lambda)$
711	0.004 227 50	0.001 642 36	0.000 000	0.720 20	0.279 80	0.000 00
712	0.003 925 80	0.001 525 35		0.720 18	0.279 82	
713	0.003 645 70	0.001 416 72		0.720 15	0.279 85	
714	0.003 385 90	0.001 315 95		0.720 12	0.279 88	
715	0.003 144 70	0.001 222 39		0.720 09	0.279 91	
716	0.002 920 80	0.001 135 55		0.720 06	0.279 94	
717	0.002 713 00	0.001 054 94		0.720 02	0.279 98	
718	0.002 520 20	0.000 980 14		0.719 99	0.280 01	
719	0.002 341 10	0.000 910 66		0.719 95	0.280 05	
720	0.002 174 96	0.000 846 19		0.719 91	0.280 09	
721	0.002 020 60	0.000 786 29		0.719 87	0.280 13	
722	0.001 877 30	0.000 730 68		0.719 83	0.280 17	
723	0.001 744 10	0.000 678 99		0.719 78	0.280 22	
724	0.001 620 50	0.000 631 01		0.719 74	0.280 26	
725	0.001 505 70	0.000 586 44		0.719 69	0.280 31	
726	0.001 399 20	0.000 545 11		0.719 64	0.280 36	
727	0.001 300 40	0.000 506 72		0.719 60	0.280 40	
728	0.001 208 70	0.000 471 11		0.719 55	0.280 45	
729	0.001 123 60	0.000 438 05		0.719 50	0.280 50	
730	0.001 044 76	0.000 407 41		0.719 45	0.280 55	
731	0.000 971 560	0.000 378 962		0.719 40	0.280 60	
732	0.000 903 600	0.000 352 543		0.719 34	0.280 66	
733	0.000 840 480	0.000 328 001		0.719 29	0.280 71	
734	0.000 781 870	0.000 305 208		0.719 24	0.280 76	
735	0.000 727 450	0.000 284 041		0.719 19	0.280 81	
736	0.000 676 900	0.000 264 375		0.719 13	0.280 87	
737	0.000 629 960	0.000 246 109		0.719 08	0.280 92	
738	0.000 586 370	0.000 229 143		0.719 02	0.280 98	
739	0.000 545 870	0.000 213 376		0.718 96	0.281 04	
740	0.000 508 258	0.000 198 730		0.718 91	0.281 09	
741	0.000 473 300	0.000 185 115		0.718 85	0.281 15	
742	0.000 440 800	0.000 172 454		0.718 79	0.281 21	
743	0.000 410 580	0.000 160 678		0.718 73	0.281 27	
744	0.000 382 490	0.000 149 730		0.718 67	0.281 33	
745	0.000 356 380	0.000 139 550		0.718 61	0.281 39	
746	0.000 332 110	0.000 130 086		0.718 55	0.281 45	
747	0.000 309 550	0.000 121 290		0.718 48	0.281 52	
748	0.000 288 580	0.000 113 106		0.718 42	0.281 58	
749	0.000 269 090	0.000 105 501		0.718 36	0.281 64	
750	0.000 250 969	0.000 098 428		0.718 29	0.281 71	
751	0.000 234 130	0.000 091 853		0.718 23	0.281 77	
752	0.000 218 470	0.000 085 738		0.718 16	0.281 84	
753	0.000 203 910	0.000 080 048		0.718 10	0.281 90	
754	0.000 190 350	0.000 074 751		0.718 03	0.281 97	
755	0.000 177 730	0.000 069 819		0.717 96	0.282 04	
756	0.000 165 970	0.000 065 222		0.717 89	0.282 11	
757	0.000 155 020	0.000 060 939		0.717 82	0.282 18	
758	0.000 144 800	0.000 056 942		0.717 75	0.282 25	
759	0.000 135 280	0.000 053 217		0.717 68	0.282 32	
760	0.000 126 390	0.000 049 737		0.717 61	0.282 39	

λ (nm)		Spectral Tristimulus Values			Chromaticity Coordinates	
	$\bar{x}_{10}(\lambda)$	$\bar{y}_{10}(\lambda)$	$\bar{z}_{10}(\lambda)$	$x_{10}(\lambda)$	$y_{10}(\lambda)$	$z_{10}(\lambda)$
761	0.000 118 100	0.000 046 491	0.000 000	0.717 54	0.282 46	0.000 00
762	0.000 110 370	0.000 043 464		0.717 46	0.282 54	
763	0.000 103 150	0.000 040 635		0.717 39	0.282 61	
764	0.000 096 427 0	0.000 038 000 0		0.717 32	0.282 68	
765	0.000 090 151 0	0.000 035 540 5		0.717 24	0.282 76	
766	0.000 084 294 0	0.000 033 244 8		0.717 16	0.282 84	
767	0.000 078 830 0	0.000 031 100 6		0.717 09	0.282 91	
768	0.000 073 729 0	0.000 029 099 0		0.717 01	0.282 99	
769	0.000 068 969 0	0.000 027 230 7		0.716 94	0.283 06	
770	0.000 064 525 8	0.000 025 486 0		0.716 86	0.283 14	
771	0.000 060 376 0	0.000 023 856 1		0.716 78	0.283 22	
772	0.000 056 500 0	0.000 022 333 2		0.716 70	0.283 30	
773	0.000 052 880 0	0.000 020 910 4		0.716 62	0.283 38	
774	0.000 049 498 0	0.000 019 580 8		0.716 54	0.283 46	
775	0.000 046 339 0	0.000 018 338 4		0.716 46	0.283 54	
776	0.000 043 389 0	0.000 017 177 7		0.716 38	0.283 62	
777	0.000 040 634 0	0.000 016 093 4		0.716 30	0.283 70	
778	0.000 038 060 0	0.000 015 080 0		0.716 22	0.283 78	
779	0.000 035 657 0	0.000 014 133 6		0.716 14	0.283 86	
780	0.000 033 411 7	0.000 013 249 0		0.716 06	0.283 94	
781	0.000 031 315 0	0.000 012 422 6		0.715 97	0.284 03	
782	0.000 029 355 0	0.000 011 649 9		0.715 89	0.284 11	
783	0.000 027 524 0	0.000 010 927 7		0.715 81	0.284 19	
784	0.000 025 811 0	0.000 010 251 9		0.715 72	0.284 28	
785	0.000 024 209 0	0.000 009 619 6		0.715 64	0.284 36	
786	0.000 022 711 0	0.000 009 028 1		0.715 55	0.284 45	
787	0.000 021 308 0	0.000 008 474 0		0.715 47	0.284 53	
788	0.000 019 994 0	0.000 007 954 8		0.715 38	0.284 62	
789	0.000 018 764 0	0.000 007 468 6		0.715 29	0.284 71	
790	0.000 017 611 5	0.000 007 012 8		0.715 21	0.284 79	
791	0.000 016 532 0	0.000 006 585 8		0.715 12	0.284 88	
792	0.000 015 521 0	0.000 006 185 7		0.715 03	0.284 97	
793	0.000 014 574 0	0.000 005 810 7		0.714 95	0.285 05	
794	0.000 013 686 0	0.000 005 459 0		0.714 86	0.285 14	
795	0.000 012 855 0	0.000 005 129 8		0.714 77	0.285 23	
796	0.000 012 075 0	0.000 004 820 6		0.714 68	0.285 32	
797	0.000 011 345 0	0.000 004 531 2		0.714 59	0.285 41	
798	0.000 010 659 0	0.000 004 259 1		0.714 50	0.285 50	
799	0.000 010 017 0	0.000 004 004 2		0.714 42	0.285 58	
800	0.000 009 413 63	0.000 003 764 73		0.714 32	0.285 68	
801	0.000 008 847 90	0.000 003 539 95		0.714 24	0.285 76	
802	0.000 008 317 10	0.000 003 329 14		0.714 14	0.285 86	
803	0.000 007 819 00	0.000 003 131 15		0.714 05	0.285 95	
804	0.000 007 351 60	0.000 002 945 29		0.713 96	0.286 04	
805	0.000 006 913 00	0.000 002 770 81		0.713 87	0.286 13	
806	0.000 006 501 50	0.000 002 607 05		0.713 78	0.286 22	
807	0.000 006 115 30	0.000 002 453 29		0.713 69	0.286 31	
808	0.000 005 752 90	0.000 002 308 94		0.713 60	0.286 40	
809	0.000 005 412 70	0.000 002 173 38		0.713 50	0.286 50	
810	0.000 005 093 47	0.000 002 046 13		0.713 41	0.286 59	

TABLE II. (*Continued*)

λ (nm)	Spectral Tristimulus Values			Chromaticity Coordinates		
	$\bar{x}_{10}(\lambda)$	$\bar{y}_{10}(\lambda)$	$\bar{z}_{10}(\lambda)$	$x_{10}(\lambda)$	$y_{10}(\lambda)$	$z_{10}(\lambda)$
811	0.000 004 793 80	0.000 001 926 62	0.000 000	0.713 32	0.286 68	0.000 00
812	0.000 004 512 50	0.000 001 814 40		0.713 22	0.286 78	
813	0.000 004 248 30	0.000 001 708 95		0.713 13	0.286 87	
814	0.000 004 000 20	0.000 001 609 88		0.713 04	0.286 96	
815	0.000 003 767 10	0.000 001 516 77		0.712 94	0.287 06	
816	0.000 003 548 00	0.000 001 429 21		0.712 85	0.287 15	
817	0.000 003 342 10	0.000 001 346 86		0.712 76	0.287 24	
818	0.000 003 148 50	0.000 001 269 45		0.712 66	0.287 34	
819	0.000 002 966 50	0.000 001 196 62		0.712 57	0.287 43	
820	0.000 002 795 31	0.000 001 128 09		0.712 47	0.287 53	
821	0.000 002 634 50	0.000 001 063 68		0.712 38	0.287 62	
822	0.000 002 483 40	0.000 001 003 13		0.712 28	0.287 72	
823	0.000 002 341 40	0.000 000 946 22		0.712 19	0.287 81	
824	0.000 002 207 80	0.000 000 892 63		0.712 09	0.287 91	
825	0.000 002 082 00	0.000 000 842 16		0.712 00	0.288 00	
826	0.000 001 963 60	0.000 000 794 64		0.711 90	0.288 10	
827	0.000 001 851 90	0.000 000 749 78		0.711 81	0.288 19	
828	0.000 001 746 50	0.000 000 707 44		0.711 71	0.288 29	
829	0.000 001 647 10	0.000 000 667 48		0.711 62	0.288 38	
830	0.000 001 553 14	0.000 000 629 70		0.711 52	0.288 48	

$^a \sum \bar{x}_{10}(\lambda) = 116.648\,519\,508\,908, \sum \bar{y}_{10}(\lambda) = 116.661\,877\,102\,312, \sum \bar{z}_{10}(\lambda) = 116.673\,980\,514\,647.$

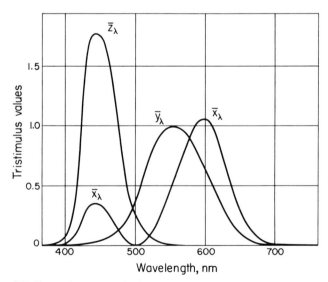

Fig. 2. CIE 2° standard spectral tristimulus values $\bar{x}_\lambda, \bar{y}_\lambda, \bar{z}_\lambda$ for color matching with imaginary primaries.

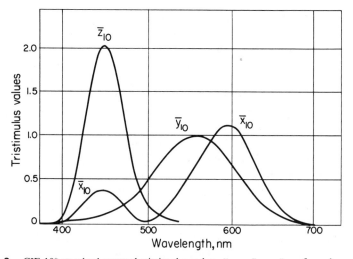

Fig. 3. CIE 10° standard spectral tristimulus values $\bar{x}_{10,\lambda}$, $\bar{y}_{10,\lambda}$, $\bar{z}_{10,\lambda}$ for color matching with imaginary primaries.

except that they refer to color mixture in larger fields (thus including effects of rod receptors) and have been determined as statistically reliable averages for observers with normal color vision. Since the $\bar{y}_{10}(\lambda)$ function is not identical with the $V(\lambda)$ function, the CIE 1964 Supplementary Colorimetric Observer data are not recommended for determining luminances; a convention that was adopted merely to avoid ambiguity of luminous values.

Graphical illustrations of the $\bar{x}(\lambda)$, $\bar{y}(\lambda)$, $\bar{z}(\lambda)$ 1931 spectral tristimulus values and the $\bar{x}_{10}(\lambda)$, $\bar{y}_{10}(\lambda)$, $\bar{z}_{10}(\lambda)$ 1964 spectral tristimulus values are shown in Figs. 2 and 3, respectively.

IV. CIE STANDARD VIEWING AND ILLUMINATING CONDITIONS

As noted earlier, the 1931 and 1964 standard observer data relate to two different sizes of stimuli, 2° and 10° ideally. In addition to size, measurements of samples that reflect or transmit light also depend in part on the geometry of illumination and viewing. Accordingly, the CIE has recommended four conditions of illumination and viewing for opaque reflecting samples. These conditions are referred to as 45/0, 0/45, d/0, and 0/d. The first pair comprises a reversed geometry relation, as does the second pair.

With the 45/0 geometry, the sample is illuminated by one or more beams whose axes are at an angle of $45° \pm 5°$ from the normal to the sample surface. The angle between the axes and any ray of the beam should not exceed 5°. Viewing should be normal to the sample surface or within 10° of the normal. As with illumination, viewing cones should not exceed 5°.

The 0/45 geometry is just the reverse of 45/0. The sample should be illuminated by a beam whose axis is at an angle not exceeding 10° from the normal to the sample surface. The sample should be viewed at an angle of 45° ± 5°. The angle between the axis and any ray of either the illuminating or viewing cone should not exceed 5°.

The d/0, which is an abbreviation for diffuse/normal, is a geometry in which the sample is illuminated with an integrating sphere. It is viewed through a port in the sphere such that the receptor receives light directly from the sample and at an angle that is normal or nearly so to the sample surface. The angle between the specimen and the axis of the viewing beam should not exceed 10° and the angle between the axis and any ray of the viewing beam should not exceed 5°. The integrating sphere may be of any diameter, provided the total area of the ports does not exceed 10% of the internal reflecting area of the sphere.

The 0/d geometry is one in which the sample is illuminated by a beam whose axis is at an angle not exceeding 10° from the normal to the sample surface. The reflected flux is collected by an integrating sphere, and viewing is accomplished by a receiver being directed toward an inner wall of the sphere. The angle between the axis and any ray of the illuminating beam should not exceed 5°. Again, the integrating sphere may be of any size, provided that the total area of its ports does not exceed 10% of the area of the inner surface of the sphere.

These viewing and illuminating conditions are illustrated in Fig. 4. Provision is made to reduce the influence of specular reflections in the d/0 and 0/d geometries for samples that are not completely diffuse reflectors. As shown in Fig. 4, a gloss trap may be incorporated in the sphere design. However, the CIE recommends that if a gloss trap is used, details of its size, shape, and location should be given as part of the measurement specification. Also because of the problem of specular reflections, particularly inter-reflections between sample and illumination optics, it is recommended that in the 0/45 and 0/d geometries the axis of the illumination beam not be at exactly 0° (normal) to the sample surface. Finally, it has been suggested that a baffle be placed within the sphere for the 0/d and d/0 geometries to prevent direct reflections from the reflected illumination beam to interfere with the viewing beam. That is, a baffle between the areas illuminated and viewed helps to eliminate partially integrated flux from confounding the diffuse measurements.

All the foregoing geometries relate to measurement of opaque reflecting samples. The CIE Colorimetry Committee has not made specific recom-

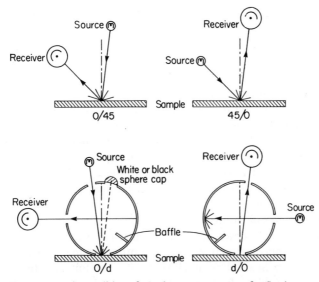

Fig. 4. Four geometric conditions for color measurement of reflecting samples recommended by the CIE; each identified by angle of illumination and viewing.

mendations for measuring translucent and transparent materials or for measuring radiant sources. However, analogous conditions to those used for reflecting samples are generally applied to such other measurements and recommendations have been made by the CIE Committee on Photometric Characteristics of Materials (CIE, 1978b) that deal with measurement problems for a wide variety of materials. It is always helpful for a clear understanding of specifications, to describe completely the conditions of viewing and illumination that were used or assumed in producing any colorimetric specification.

V. CIE STANDARD ILLUMINANTS AND SOURCES

There are many different sources of illumination that may be used to irradiate a sample whose colorimetric specification we may wish to make. Equation (12) showed that the tristimulus values of a colorimetric specification intended to represent a color match consisted of the integral, with respect to wavelength, of the spectral tristimulus function of the standard observer with the wavelength-by-wavelength product of the spectral power distribution of the source and the reflectance, radiance, or transmittance function of the sample. It will be clear from this that a colorimetric specification depends in part on the spectral power function of the source. To avoid use of a bewilderingly large number of potential sources of illumination, the CIE has recommended a limited number of illuminants and, where possible, actual sources. The distinction is made between "illuminant" and "source" in that an illuminant represents an aim spectral power distribution and a source represents an actual light whose characteristics are specified and whose spectral power distribution matches that of a specified illuminant. Originally, the CIE recommended only real sources (known as Sources A, B, and C), but since 1963 there have also been illuminants (D and E) recommended for which no real source exactly corresponds. In order that colorimetric specifications may be directly comparable, one of the CIE illuminants should be used or assumed in making measurements that form part of the specification, and the particular illuminant used should always be noted as a part of the specification.

CIE Illuminant A represents light from a full radiator at an absolute temperature of approximately 2856 K* according to the International Practical Temperature Scale, 1968 (CIPM, 1968). The full radiator (which has also been called a black-body or a Planckian radiator) is an incandescent or thermal radiator. It may be defined according to its absorptance characteristics as one in which absorption is complete for all incident radiation, whatever the wavelength, the direction of incidence, or the polarization. Such a radiator has maximum spectral concentration of radiant excitance at a given temperature for any wavelength. Its radiant excitance at a given wavelength, $M_{e,\lambda}$, is defined according to Planck's radiation law

$$M_{e,\lambda}(\lambda, T) = c_1 \lambda^{-5}(e^{c2/\lambda T} - 1)^{-1} \text{ W m}^{-3} \tag{23}$$

where the radiation constants are taken as

$$c_1 = 3.74150 \cdot 10^{-16} \text{ W m}^2$$

$$c_2 = 1.4388 \cdot 10^{-2} \text{ m K}$$

and the temperature T is set equal to

$$T = (1.4388/1.4350)2848 \text{ K} \cong 2856 \text{ K}.$$

The choice of temperature $T \cong 2856$ K assures that the spectral power distribution is identical to that of the CIE Illuminant A originally adopted in 1931 when the second radiation constant c_2 had a value of 1.4350×10^{-2} m K and T was therefore 2848 K in accordance with the International Practical Temperature Scale, 1927. The value of c_2 has changed a number

---

* Absolute temperature is measured in units of kelvins, symbolized K. The relationships of kelvins to Celsius and Fahrenheit degrees are $°C = K - 273.15$ and $°F = (9K/5) - 459.67$.

of times over the years since Illuminant A was first adopted by the CIE in 1931. Therefore, the value of T will be found to vary throughout the literature. However, the actual spectral power distribution has always been the same. The relative values of $S(\lambda)$ for CIE Illuminant A are listed in Table III. They were derived by computing $S(\lambda) = 100 M_{e,\lambda}/M_{e,560}$; that is, $S(\lambda) = 100$ for a wavelength of 560 nm.

CIE Source A is realized as a gas-filled coiled-coil tungsten filament lamp operating at an absolute temperature such as to produce the spectral power distribution tabulated in Table III. A lamp with a fused quartz

TABLE III. Relative Spectral Power Distribution of CIE Illuminant A

λ (nm)	(A) $S(\lambda)$	λ (nm)	(A) $S(\lambda)$	λ (nm)	(A) $S(\lambda)$	λ (nm)	(A) $S(\lambda)$
300	0.93	450	33.09	600	129.04	750	227.00
05	1.13	55	35.41	05	132.70	55	229.59
10	1.36	60	37.81	10	136.35	60	232.12
15	1.62	65	40.30	15	139.99	65	234.59
20	1.93	70	42.87	20	143.62	70	237.01
325	2.27	475	45.52	625	147.24	775	239.37
30	2.66	80	48.24	30	150.84	80	241.68
35	3.10	85	51.04	35	154.42	85	243.92
40	3.59	90	53.91	40	157.98	90	246.12
45	4.14	95	56.85	45	161.52	95	248.25
350	4.74	500	59.86	650	165.03	800	250.33
55	5.41	05	62.93	55	168.51	05	252.35
60	6.14	10	66.06	60	171.96	10	254.31
65	6.95	15	69.25	65	175.38	15	256.22
70	7.82	20	72.50	70	178.77	20	258.07
375	8.77	525	75.79	675	182.12	825	259.86
80	9.80	30	79.13	80	185.43	30	261.60
85	10.90	35	82.52	85	188.70		
90	12.09	40	85.95	90	191.93		
95	13.35	45	89.41	95	195.12		
400	14.71	550	92.91	700	198.26		
05	16.15	55	96.44	05	201.36		
10	17.68	60	100.00	10	204.41		
15	19.29	65	103.58	15	207.41		
20	20.99	70	107.18	20	210.36		
425	22.79	575	110.80	725	213.27		
30	24.67	80	114.44	30	216.12		
35	26.64	85	118.08	35	218.92		
40	28.70	90	121.73	40	221.67		
45	30.85	95	125.39	45	224.36		

envelope or window is recommended if the ultraviolet portion of the spectral power distribution of Illuminant A is to be realized with reasonable accuracy.

CIE Illuminant B was intended to represent direct sunlight with a correlated color temperature (to be defined shortly) of 4874 K. The relative spectral power distribution of this illuminant is listed in Table IV.

CIE Source B may be realized by using CIE Source A in conjunction with two liquid filters. The liquid filter solutions are given in Table V. The filter consists of a layer 1 cm thick of each of the two solutions B_1 and B_2 shown in that table.

TABLE IV.   Relative Spectral Power Distribution of CIE Illuminant B

λ (nm)	(B) $S(\lambda)$	λ (nm)	(B) $S(\lambda)$	λ (nm)	(B) $S(\lambda)$	λ (nm)	(B) $S(\lambda)$
300		450	85.40	600	98.00	750	85.20
05		55	86.88	05	98.08	55	84.80
10		60	88.30	10	98.50	60	84.70
15		65	90.08	15	99.06	65	84.90
20	0.02	70	92.00	20	99.70	70	85.40
325	0.26	475	93.75	625	100.36	775	
30	0.50	80	95.20	30	101.00	80	
35	1.45	85	96.23	35	101.56	85	
40	2.40	90	96.50	40	102.20	90	
45	4.00	95	95.71	45	103.05	95	
350	5.60	500	94.20	650	103.90	800	
55	7.60	05	92.37	55	104.59	05	
60	9.60	10	90.70	60	105.00	10	
65	12.40	15	89.65	65	105.08	15	
70	15.20	20	89.50	70	104.90	20	
375	18.80	525	90.43	675	104.55	825	
80	22.40	30	92.20	80	103.90	30	
85	26.85	35	94.46	85	102.84		
90	31.30	40	96.90	90	101.60		
95	36.18	45	99.16	95	100.38		
400	41.30	550	101.00	700	99.10		
05	46.62	55	102.20	05	97.70		
10	52.10	60	102.80	10	96.20		
15	57.70	65	102.92	15	94.60		
20	63.20	70	102.60	20	92.90		
425	68.37	575	101.90	725	91.10		
30	73.10	80	101.00	30	89.40		
35	77.31	85	100.07	35	88.00		
40	80.80	90	99.20	40	86.90		
45	83.44	95	98.44	45	85.90		

TABLE V. Liquid Filters to Be Used with CIE Source A
to Provide CIE Source B

Solution B_1:

Copper sulfate ($CuSO_4 \cdot 5H_2O$)	2.452 g
Mannite [$C_6H_8(OH)_6$]	2.452 g
Pyridine (C_5H_5N)	30.0 mliter
Distilled water to make	1000.0 mliter

Solution B_2:

Cobalt ammonium sulfate	21.71 g
[$CoSO_4 (NH_4)_2 SO_4 \cdot 6H_2O$]	
Copper sulfate ($CuSO_4 \cdot 5H_2O$)	16.11 g
Sulfuric acid (density 1.835 g mliter^{-1})	10.0 mliter
Distilled water to make	1000.0 mliter

CIE Illuminant C was intended to represent a phase of daylight (sunlight plus skylight) having a correlated color temperature of 6774 K. The relative spectral power distribution of CIE Illuminant C is tabulated in Table VI.

CIE Source C may be realized by CIE Source A in combination with a filter consisting of a layer 1 cm thick of each of two solutions C_1 and C_2 made up according to the formulae of Table VII.

CIE Illuminants D consist of a series of mathematical simulations of the relative spectral power distributions of various phases of natural daylight. These illuminants are based on various combinations of irradiation from sun and sky. Analysis of many measurements of the various combinations found in natural daylight indicates that there is a simple relationship between the correlated color temperature T_c of daylight and its relative spectral power distribution $S(\lambda)$ (Judd et al., 1964). On the basis of that work the CIE has recommended an operational definition of daylight illuminants. Of the many daylight illuminants thus defined, the CIE has selected that one representing a phase of daylight having a correlated color temperature of 6504 K as the preferred illuminant for colorimetry when daylight is of interest. That illuminant is designated D_{65}. It supplements CIE Illuminants A, B, and C but its use is to be preferred over B and C.

All CIE Daylight Illuminants are defined such that their CIE 1931 chromaticity coordinates satisfy the relation

$$y = -3.000x^2 + 2.870x - 0.275, \tag{24}$$

with x within the range of 0.250–0.380. The value of x may be determined for any such daylight illuminant having correlated color temperatures between about 4000 and 7000 K as

$$x = -4.6070(10^9/T_c^3) + 2.9678(10^6/T_c^2) + 0.09911(10^3/T_c) + 0.244063. \tag{25}$$

C. J. Bartleson

TABLE VI. Relative Spectral Power Distribution of CIE Illuminant C

λ (nm)	(C) $S(\lambda)$	λ (nm)	(C) $S(\lambda)$	λ (nm)	(C) $S(\lambda)$	λ (nm)	(C) $S(\lambda)$
300		450	124.00	600	89.70	750	59.20
05		55	123.60	05	88.83	55	58.50
10		60	123.10	10	88.40	60	58.10
15		65	123.30	15	88.19	65	58.00
20	0.01	70	123.80	20	88.10	70	58.20
325	0.20	475	124.09	625	88.06	775	
30	0.40	80	123.90	30	88.00	80	
35	1.55	85	122.92	35	87.86	85	
40	2.70	90	120.70	40	87.80	90	
45	4.85	95	116.90	45	87.99	95	
350	7.00	500	112.10	650	88.20	800	
55	9.95	05	106.98	55	88.20	05	
60	12.90	10	102.30	60	87.90	10	
65	17.20	15	98.81	65	87.22	15	
70	21.40	20	96.90	70	86.30	20	
375	27.50	525	96.78	675	85.30	825	
80	33.00	30	98.00	80	84.00	30	
85	39.92	35	99.94	85	82.21		
90	47.40	40	102.10	90	80.20		
95	55.17	45	103.95	95	78.24		
400	63.30	550	105.20	700	76.30		
05	71.81	55	105.67	05	74.36		
10	80.60	60	105.30	10	72.40		
15	89.53	65	104.11	15	70.40		
20	98.10	70	102.30	20	68.30		
425	105.80	575	100.15	725	66.30		
30	112.40	80	97.80	30	64.40		
35	117.75	85	95.43	35	62.80		
40	121.50	90	93.20	40	61.50		
45	123.45	95	91.22	45	60.20		

The value of x may be determined for daylight illuminants having correlated color temperatures from 7000 to approximately 25,000 K as

$$x = -2.0064(10^9/T_c^3) + 1.9018(10^6/T_c^2) + 0.24748(10^3/T_c) + 0.237040. \quad (26)$$

The relative spectral power distribution $S(\lambda)$ of a CIE Daylight Illuminant is defined as follows:

$$S(\lambda) = S_0(\lambda) + M_1 S_1(\lambda) + M_2 S_2(\lambda), \quad (27)$$

TABLE VII. Liquid Filters to Be Used with CIE Source A to Provide CIE Source C

Solution C_1:	
Copper sulfate ($CuSO_4 \cdot 5H_2O$)	3.412 g
Mannite [$C_6H_8(OH)_6$]	3.412 g
Pyridine (C_5H_5N)	30.0 mliter
Distilled water to make	1000.0 mliter
Solution C_2:	
Cobalt ammonium sulfate	30.58 g
[$CoSO_4 \cdot (NH_4)_2SO_4 \cdot 6H_2O$)]	
Copper sulfate ($CuSO_4 \cdot 5H_2O$)	22.52 g
Sulfuric acid (density 1.835 g $\cdot$ mliter^{-1})	10.0 mliter
Distilled water to make	1000.0 mliter

where $S_0(\lambda)$, $S_1(\lambda)$, $S_2(\lambda)$ are functions of wavelength listed in Table VIII and M_1 and M_2 are factors whose values are related to the chromaticity coordinates x, y of the illuminant as follows:

$$M_1 = \frac{-1.3515 - 1.7703x + 5.9114y}{0.0241 + 0.2562x - 0.7341y},$$

$$M_2 = \frac{0.0300 - 31.4424x + 30.0717y}{0.0241 + 0.2562x - 0.7341y}. \tag{28}$$

Although Eqs. (24)–(28) define CIE Daylight Illuminants over a wide range of correlated color temperatures, the CIE recommends that D_{65} be used as the preferred illuminant. When D_{65} cannot be used, it is recommended that one of the two illuminants D_{55} or D_{75}, having correlated color temperatures of 5503 and 7504, be used whenever possible. The reason for stating such preferences by the CIE is to reduce the number of different illuminants used in colorimetry and thereby to enhance the precision of communication. In practice, CIE Daylight Illuminants D_{75}, D_{65}, D_{55}, and D_{50} are commonly used and incorporated in various standards. Table IX lists the relative spectral power distributions of these four daylight illuminants.

At present there are no actual sources that are recommended by the CIE to realize any of these daylight illuminants. The CIE Colorimetry Committee is actively engaged in a search for such sources. Work along these lines is to be found in a status report (Wyszecki, 1970) and summary of the committee's activities through 1975 (Judd and Wyszecki, 1975).

CIE Illuminant E may be defined as a stimulus whose spectral radiant power is constant at all wavelengths; it has been called the equienergy

TABLE VIII. Values of Three Characteristic Vectors, $S_0(\lambda)$, $S_1(\lambda)$, $S_2(\lambda)$, used to define CIE, Illuminants D

λ (nm)	$S_0(\lambda)$	$S_1(\lambda)$	$S_2(\lambda)$
300	0.04	0.02	0.0
310	6.0	4.5	2.0
320	29.6	22.4	4.0
330	55.3	42.0	8.5
340	57.3	40.6	7.8
350	61.8	41.6	6.7
360	61.5	38.0	5.3
370	68.8	42.4	6.1
380	63.4	38.5	3.0
390	65.8	35.0	1.2
400	94.8	43.4	−1.1
410	104.8	46.3	−0.5
420	105.9	43.9	−0.7
430	96.8	37.1	−1.2
440	113.9	36.7	−2.6
450	125.6	35.9	−2.9
460	125.5	32.6	−2.8
470	121.3	27.9	−2.6
480	121.3	24.3	−2.6
490	113.5	20.1	−1.8
500	113.1	16.2	−1.5
510	110.8	13.2	−1.3
520	106.5	8.6	−1.2
530	108.8	6.1	−1.0
540	105.3	4.2	−0.5
550	104.4	1.9	−0.3
560	100.0	0.0	0.0
570	96.0	−1.6	0.2
580	95.1	−3.5	0.5
590	89.1	−3.5	2.1
600	90.5	−5.8	3.2
610	90.3	−7.2	4.1
620	88.4	−8.6	4.7
630	84.0	−9.5	5.1
640	85.1	−10.9	6.7
650	81.9	−10.7	7.3
660	82.6	−12.0	8.6
670	84.9	−14.0	9.8
680	81.3	−13.6	10.2
690	71.9	−12.0	8.3

λ (nm)	$S_0(\lambda)$	$S_1(\lambda)$	$S_2(\lambda)$
700	74.3	−13.3	9.6
710	76.4	−12.9	8.5
720	63.3	−10.6	7.0
730	71.7	−11.6	7.6
740	77.0	−12.2	8.0
750	65.2	−10.2	6.7
760	47.7	−7.8	5.2
770	68.6	−11.2	7.4
780	65.0	−10.4	6.8
790	66.0	−10.6	7.0
800	61.0	−9.7	6.4
810	53.3	−8.3	5.5
820	58.9	−9.3	6.1
830	61.9	−9.8	6.5

TABLE IX. Relative Spectral Power Distributions of CIE Illuminants D_{75}, D_{65}, D_{55}, and D_{50}

Wavelength (nm)	D_{50}	D_{55}	D_{65}	D_{75}	Wavelength (nm)	D_{50}	D_{55}	D_{65}	D_{75}
300	0.02	0.02	0.03	0.04	540	100.8	102.1	104.4	106.3
310	2.0	2.1	3.3	5.2	550	102.3	103.0	104.0	104.9
320	7.8	11.2	20.2	29.8	560	100.0	100.0	100.0	100.0
330	14.8	20.7	37.1	55.0	570	97.7	97.3	96.4	95.6
340	17.9	24.0	40.0	57.3	580	98.9	97.7	95.7	94.2
350	21.0	27.9	45.0	62.7	590	93.5	91.4	88.6	87.0
360	23.9	30.7	46.7	63.0	600	97.7	94.4	90.0	87.3
370	26.9	34.4	52.2	70.3	610	99.3	95.1	89.6	86.2
380	24.5	32.6	50.0	66.8	620	99.0	94.2	87.6	83.6
390	29.8	38.2	54.7	70.0	630	95.7	90.4	83.3	78.7
400	49.3	61.0	82.8	101.9	640	98.8	92.3	83.7	78.5
410	56.5	68.6	91.6	111.9	650	95.7	88.9	80.0	74.8
420	60.0	71.6	93.5	112.8	660	98.2	90.3	80.2	74.5
430	57.8	67.9	86.8	103.3	670	103.0	94.0	82.2	75.5
440	74.8	85.6	104.9	121.1	680	99.1	90.0	78.3	71.7
450	87.2	98.1	117.1	133.0	690	87.4	79.7	69.7	64.0
460	90.6	100.4	117.8	132.3	700	91.6	82.9	71.6	65.2
470	91.4	99.9	114.9	127.2	710	92.6	84.9	74.3	68.1
480	95.2	102.6	115.9	126.9	720	76.8	70.2	61.6	56.5
490	92.0	98.0	108.8	117.7	730	86.6	79.3	69.9	64.3
500	95.7	100.7	109.4	116.5	740	92.6	85.0	75.1	69.2
510	96.6	100.8	107.8	113.7	750	78.2	71.9	63.6	58.7
520	97.1	100.0	104.9	108.6	760	57.7	52.8	46.4	42.7
530	102.1	104.2	107.7	110.5	770	82.9	75.9	66.8	61.4

spectrum or equal energy source. It represents a concept that has been imbedded in CIE computations since 1931 but is only indirectly recommended by the CIE as an illuminant. There is no source that meets the requirements for equal power at all wavelengths over a sufficiently broad range (e.g., from about 300 to 800 nm), but its central role in normalizing spectral tristimulus values and derivative colorimetric characterizations justifies its being considered a standard CIE illuminant condition.

VI.   CIE 1931 TRISTIMULUS VALUES AND CHROMATICITY COORDINATES

The tristimulus values of a CIE colorimetric specification are, as noted before, defined integrals over wavelength of the spectral tristimulus values and the stimulus function. The tristimulus values X, Y, Z are expressed as

$$X = k \int_\lambda \varphi(\lambda)\bar{x}(\lambda)\, d\lambda,$$

$$Y = k \int_\lambda \varphi(\lambda)\bar{y}(\lambda)\, d\lambda, \qquad (29)$$

$$Z = k \int_\lambda \varphi(\lambda)\bar{z}(\lambda)\, d\lambda,$$

where $\varphi(\lambda)$ is the stimulus function. It is a product of $S(\lambda)$ and either $\rho(\lambda)$, $\beta(\lambda)$, or $\tau(\lambda)$; k is a constant.

For all practical purposes involving calculation, Eq. (29) may be approximated by summation; for example,

$$X = k \sum_\lambda \varphi(\lambda)\bar{x}(\lambda)\, \Delta\lambda,$$

$$Y = k \sum_\lambda \varphi(\lambda)\bar{y}(\lambda)\, \Delta\lambda, \qquad (30)$$

$$Z = k \sum_\lambda \varphi(\lambda)\bar{z}(\lambda)\, \Delta\lambda.$$

The constant k of either Eq. (29) or (30) may be chosen such as to provide tristimulus values X, Y, Z in units of luminance. When the value and dimensions of k are set at $683\ \mathrm{lm\ W^{-1}}$, the tristimulus values are called "absolute" and the tristimulus value Y is a measure of luminance if $\varphi(\lambda)$ is measured as spectral concentration of radiance in $\mathrm{W\ m^{-2}\ sr^{-1}\ nm^{-1}}$.

Alternatively, it is sometimes convenient to choose a value of k such that $Y = 100$ for samples having either reflectances, radiances, or transmittances of unity at all wavelengths. For example,

$$k = 100/\sum_{\lambda} S(\lambda)\bar{y}(\lambda)\,\Delta\lambda. \tag{31}$$

In this case the tristimulus value Y is equal to luminous reflectance, or luminance factor, or luminous transmittance. Tristimulus values derived in this way are relative in the sense that they are invariant with changes in illuminance.

The wavelength interval $\Delta\lambda$ of Eq. (30) may be either 1, 5, or 10 nm. Any increase in the size of the wavelength interval from that of the infinitesimally small one implied by integration will introduce error of some magnitude. However, in practice the spectral tristimulus functions are defined with a resolution of 1 nm. Accordingly, we may take the wavelength interval of 1 nm as reference in order to define the operationally correct values for Eq. (30). Errors generated by using wavelength intervals of 5 and 10 nm vary throughout the tristimulus space X, Y, Z, depending on the position of the correct tristimulus point. It is common practice to use wavelength intervals of 10 nm when greatest accuracy is not required or when the spectral selectivity of the stimulus function is low and to use 5 nm intervals when greater accuracy is required or the spectral selectivity of the stimulus function is moderate. Whenever greatest accuracy is required or spectral selectivity is high, wavelength intervals of 1 nm should be used. It is customary to note the wavelength interval of summation as a subscript to the summation sign—for example, $\sum_{10}$ where it is known that the range of wavelength summation is from 360 to 830 nm. In practice, the method of summation represented by Eq. (30) is used whenever the stimulus function and spectral power distribution of the reference illuminant is known, as, for example, when spectrophotometric and spectroradiometric data are available. The integration method of Eq. (29) is associated with colorimetric instruments in which the spectral response of the receiver is modified by optical filtering so that integration may be accomplished by optical means.

The chromaticity coordinates x, y, z are derived from the tristimulus values X, Y, Z as follows:

$$x = X/(X + Y + Z), \quad y = Y/(X + Y + Z), \quad z = Z/(X + Y + Z). \tag{32}$$

Because of the relationship $x + y + z = 1$, only two chromaticity coordinates suffice for a specification of chromaticity. The coordinates x and y are used for this purpose. Note that chromaticity expresses only the ratios of X or Y or Z to the sum of all three tristimulus values. Chromaticity provides no information about luminance or relative luminance. That information is contained in the tristimulus value Y. Accordingly, the triad of numbers x, y, Y form a complete specification consisting of chromaticity and luminance. They convey as much information as the triad of tristimulus values. The tristimulus values can be recovered from chromaticity and luminance specifications through the following relationship:

$$X = (x/y)Y, \qquad Y = Y, \qquad Z = (z/y)Y. \tag{33}$$

VII. CIE 1964 TRISTIMULUS VALUES AND CHROMATICITY COORDINATES

The tristimulus values and chromaticity coordinates for the CIE 1964 Supplementary Standard Colorimetric Observer are determined according to the same relationships as those for the CIE 1931 Standard Colorimetric Observer described earlier. The difference, of course, is that the spectral tristimulus values $\bar{x}_{10}$, $\bar{y}_{10}$, $\bar{z}_{10}$ of the 1964 observer are different from those of the 1931 observer's $\bar{x}$, $\bar{y}$, $\bar{z}$ functions. This distinction is noted by including a subscript numeral 10 to each of the symbols. Thus, we have as analogies to Eqs. (29)–(33), the following:

$$X_{10} = k \int_{\lambda} \varphi(\lambda)\bar{x}_{10}(\lambda)\, d\lambda,$$

$$Y_{10} = k \int_{\lambda} \varphi(\lambda)\bar{y}_{10}(\lambda)\, d\lambda, \tag{34}$$

$$Z_{10} = k \int_{\lambda} \varphi(\lambda)\bar{z}_{10}(\lambda)\, d\lambda,$$

and

$$X_{10} = k \sum_{\lambda} \varphi(\lambda)\bar{x}_{10}(\lambda)\, \Delta\lambda,$$

$$Y_{10} = k \sum_{\lambda} \varphi(\lambda)\bar{y}_{10}(\lambda)\, \Delta\lambda, \tag{35}$$

$$Z_{10} = k \sum_{\lambda} \varphi(\lambda)\bar{z}_{10}(\lambda)\, \Delta\lambda;$$

and

$$x_{10} = X_{10}/(X_{10} + Y_{10} + Z_{10}),$$

$$y_{10} = Y_{10}/(X_{10} + Y_{10} + Z_{10}), \tag{36}$$

$$z_{10} = X_{10}/(X_{10} + Y_{10} + Z_{10});$$

and

$$X_{10} = (x_{10}/y_{10})Y_{10}, \qquad Y_{10} = Y_{10}, \qquad Z_{10} = (z_{10}/y_{10})Y_{10}. \tag{37}$$

Since $\bar{y}_{10}(\lambda)$ is not identical to the $V(\lambda)$ function, tristimulus values Y_{10} in Eqs. (34)–(37) do not represent luminance.

VIII. CIE 1931 AND 1964 CHROMATICITY DIAGRAMS

The CIE 1931 Chromaticity Diagram is shown in Fig. 5. The CIE 1964 Chromaticity Diagram is shown in Fig. 6. They are both similar in appearance because they are derived from similar principles. The 1964 diagram is somewhat smaller in area than the 1931 diagram and represents chromaticity coordinates x_{10}, y_{10} rather than x, y. In other respects they are essentially the same. Accordingly, we will discuss only the 1931 chromaticity diagram in detail here and in the following sections that deal with relationships determined from the chromaticity diagram. Wherever there are differences, these will be pointed out.

The chromaticity diagram is the unit plane of the tristimulus space; $X + Y + Z = 1$ for the chromaticity plane. The primary vectors intersect

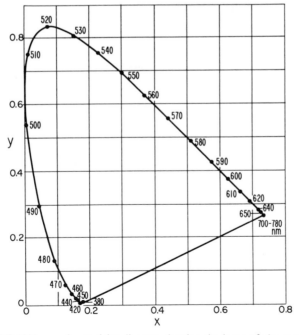

Fig. 5. CIE 1931 x, y chromaticity diagram showing the locus of chromaticities corresponding to spectral stimuli when color-matched by the CIE 2° standard observer.

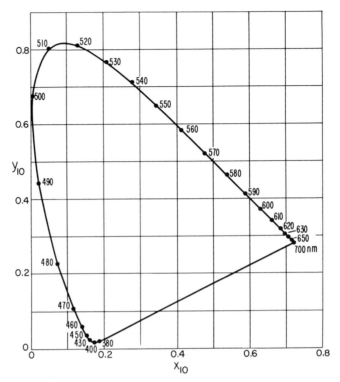

Fig. 6. CIE 1964 x_{10}, y_{10} chromaticity diagram showing the locus of chromaticities corresponding to spectral stimuli when color-matched by the CIE 10° standard observer.

the unit plane at the apices of the triangle formed by the chromaticity diagram. These apices have the following coordinates:

$$\mathfrak{X}: x = 1, y = 0; \qquad \mathfrak{Y}: x = 0, y = 1; \qquad \mathfrak{Z}: x = 0, y = 0.$$

The chromaticity triangle may be envisioned as the axonometric projection of the primary vector space where the conjunction of those vectors and all cases of equal tristimulus values are represented as the center of gravity of the triangle, as illustrated in Fig. 7.

The center of gravity (where $X = Y = Z$) corresponds to chromaticity coordinates of $x = y = \frac{1}{3}$. These are the chromaticity coordinates for the theoretical CIE illuminant E representing equal power at all wavelengths over the range of 360–830 nm. In fact, the original 1931 chromaticity coordinates for CIE Illuminant E were very close to $x = y = \frac{1}{3}$ to the sixth decimal place. However, smoothing and extrapolation of the spectral tristimulus data carried out in the intervening years has modified these values

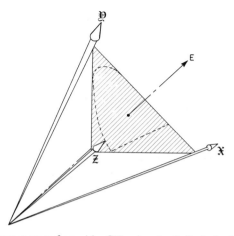

Fig. 7. Trichromatic vector space formed by CIE primaries $\mathfrak{X}$, $\mathfrak{Y}$, $\mathfrak{Z}$. A plane of constant luminance is represented as the triangular area. Dashed curves on that plane correspond to the spectrum locus. A vector representing equal proportions of the three primaries is shown as E.

slightly. Values for x, y and x_{10}, y_{10} determined by summation at 1 nm wavelength intervals are now as follows:

$$x = 0.333314, \qquad x_{10} = 0.333296$$

$$y = 0.333288, \qquad y_{10} = 0.333335.$$

The values for summation at 5 nm intervals are

$$x = 0.333334, \qquad x_{10} = 0.333296$$

$$y = 0.333330, \qquad y_{10} = 0.333339.$$

The values for summation at 10 nm intervals are

$$x = 0.333381, \qquad x_{10} = 0.333336$$

$$y = 0.333444, \qquad y_{10} = 0.333330.$$

Chromaticity coordinates of spectral stimuli are computed from the spectral tristimulus values according to Eqs. (32) and (36). Those chromaticity coordinates are listed in Tables I and II together with the spectral tristimulus values for the 2° and 10° observers. When they are plotted on the x, y chromaticity diagrams of Figs. 5 and 6, the result is the roughly horseshoe-shaped curve. That curve is the locus of chromaticities for spectral stimuli and is called the "spectrum locus," in CIE parlance.

A line has been drawn in each of Figs. 5 and 6 to connect the extremes of the spectrum locus. The line passes through all chromaticities corresponding to additive color mixtures of various proportions of spectral light of wavelengths 360 and 830 nm. It is called the "purple boundary" because such mixtures tend to appear purplish in hue under many ordinary viewing conditions.

The enclosed area defined by the spectrum locus and the purple boundary represents a maximum chromaticity gamut within which may be found the chromaticities of all real (physical) stimuli. Points that lie outside this gamut represent imaginary chromaticities, ones that cannot be produced by real stimuli. The chromaticities of the $\mathfrak{X}$, $\mathfrak{Y}$, $\mathfrak{Z}$ primaries, for example, lie outside the gamut because they are mathematical constructions that do not correspond to real stimuli. The $\mathfrak{R}$, $\mathfrak{G}$, $\mathfrak{B}$ primaries, however, do have chromaticities that are just contained within the gamut; their chromaticities lie on the spectrum locus at points corresponding to 700.0, 546.1, and 435.8 nm.

As with the purple boundary, the chromaticities of two-component stimulus mixtures lie along a line joining the chromaticities of the two components in the chromaticity diagram. Suppose that we draw a line between the chromaticities for each of the three $\mathfrak{R}$, $\mathfrak{G}$, $\mathfrak{B}$ primaries, whose chromaticities are, in fact,

$$\mathfrak{R}: x = 0.73469, y = 0.26531;$$

$$\mathfrak{G}: x = 0.27368, y = 0.71743;$$

$$\mathfrak{B}: x = 0.16654, y = 0.00888.$$

This has been done in Fig. 8. It may be seen that the gamut of two- and three-component mixtures does not include the chromaticities of all spectral stimuli. That is, a chromaticity match cannot be made for all real stimuli when using these real primaries. In fact, it can be shown that no real set of primaries will provide a gamut large enough to contain all possible chromaticities of real stimuli. This is equivalent to the statement that it is impossible to make color matches for all spectral stimuli without some negative values of spectral tristimulus values (as indicated in Fig. 1).

We see from Fig. 8, however, that lines joining the chromaticities of the imaginary primaries $\mathfrak{X}$, $\mathfrak{Y}$, $\mathfrak{Z}$ (which have chromaticities of $\mathfrak{X}: x = 1, y = 0$; $\mathfrak{Y}: x = 0, y = 1$; $\mathfrak{Z}: x = 0, y = 0$) describes a triangular gamut that does include chromaticities of all real stimuli (e.g., chromaticities for all spectral stimuli are found within the triangle). It may be seen from this that the chromaticity diagram provides a geometric representation of the mathematical conventions of tristimulus specification. In general it is true that the chromaticity diagram provides a representation of all mathematical relations expressed by the equations of tristimulus specification; this is true because

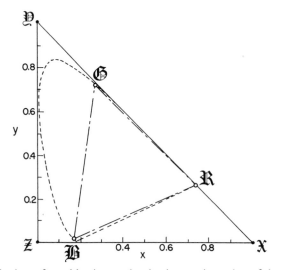

Fig. 8. Unit plane of equal luminance showing intersection points of the CIE imaginary primaries 𝕏, 𝕐, 𝖟, and real spectral primaries ℜ, 𝔊, 𝔅. The gamut of chromaticities defined by the real primaries is smaller than that defined by the imaginary primaries.

the chromaticity diagram is merely a construction based on those mathematical expressions. In short, the chromaticity diagram is a useful analog alternative to digital specification.

IX.   CIE DOMINANT WAVELENGTH AND PURITY

The chromaticity diagram provides certain other derived methods for specifying a tristimulus color match. As indicated earlier, these derived specifications represent mathematical relations among tristimulus values.

The first such specification to be discussed is called "dominant wavelength." It is based on the principle that the chromaticities of all additive stimulus mixtures of a two-component mixture lie along a straight line in the chromaticity diagram. Dominant wavelength of a stimulus represents the spectral stimulus that may be mixed with a reference illuminant (or "neutral" stimulus) to match the given stimulus. Figure 9 helps in understanding what is meant by the concept of dominant wavelength. Suppose that we have a stimulus whose chromaticity with respect to CIE Illuminant D_{65} is represented by point S_1 in the CIE 1931 chromaticity diagram of Fig. 9. The chromaticity of the D_{65} neutral reference is labeled D_{65}. If a straight line is drawn between points S_1 and D_{65}, we know that all additive

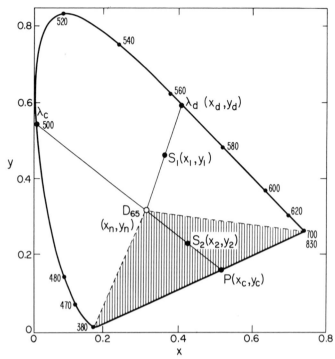

Fig. 9. CIE 1931 x, y chromaticity diagram illustrating definitions of dominant and complementary wavelength and excitation purity. See text for explanation.

mixtures of these two stimuli will have chromaticities that lie along that line. However, let us extend that line until it intersects the spectrum locus at the point labeled λ_d. We also know that the chromaticities of all additive mixtures of the neutral stimulus D_{65} and the spectral stimulus λ_d will lie along the line joining their chromaticities. That line passes through point S_1. Accordingly, S_1 may be matched by a mixture of the stimuli corresponding to D_{65} and λ_d. In fact, no other spectral stimulus may be mixed with D_{65} to provide the same chromaticity as that of S_1. Therefore, the spectral stimulus of point λ_d bears a unique relation to S_1 with respect to the neutral reference D_{65}. We say, then, that the wavelength of the spectral stimulus at λ_d (approximately 565 nm in this example) is the dominant wavelength of the stimulus at S_1 *with respect to CIE Illuminant D_{65}*. Note that a dominant wavelength specification for any stimulus must always be accompanied by a notation of the reference illuminant for which it was determined. Since different illuminants have different chromaticity coordinates, it is plain that S_1 will have different dominant wavelengths for all illuminants of chromaticity different from that of D_{65}.

Determination of dominant wavelength for stimuli whose chromaticities lie within the shaded area of Fig. 9 is impossible. Point S_2, for example, cannot be matched by a mixture of the D_{65} neutral reference and any spectral stimulus. S_2 has no dominant wavelength with respect to D_{65}. However, the concept of additive mixture may be applied somewhat differently in this case to determine what is called "complementary wavelength," symbolized as λ_c. Again, we know that a mixture of D_{65} and S_2 will yield chromaticities that lie along the line joining those two points in the chromaticity diagram. If we extend that line until it intersects the spectrum locus we will identify the spectral stimulus that can be additively mixed with the stimulus corresponding to S_2 to produce the same chromaticity as D_{65}. That spectral stimulus has a wavelength that is colorimetrically complementary to S_2. The wavelength is called the complementary wavelength of S_2 with respect to D_{65}. As with dominant wavelength, every complementary wavelength must be with respect to a specified neutral reference. Complementary wavelength is then the wavelength of the spectral stimulus that, when additively mixed in suitable proportions with the stimulus in question, matches in chromaticity the reference illuminant. In the example of Fig. 9 the complementary wavelength of S_2 with respect to D_{65} is located at λ_c, a wavelength of 500 nm.

Dominant and complementary wavelengths are most easily determined by graphical construction in a chromaticity diagram. They may, however, be determined by calculation and table lookup. This involves determining the slopes (g) of all the lines connecting spectral stimuli and a given reference neutral in a particular chromaticity diagram. Then the slope of the line joining the reference neutral chromaticity and that for the sample is computed

$$g = (x - x_n)/(y - y_n) \quad \text{or} \quad 1/g = (y - y_n)/(x - x_n).$$

To distinguish λ_c from λ_d, both slopes and values of y must be evaluated. For daylight or full radiators having correlated color temperatures between about 1500 and 15,000 K, the following relationships will identify λ_c:

$$\left(\frac{y_{360} - y_n}{x_{360} - x_n}\right) < \left(\frac{y - y_n}{x - x_n}\right) < \left(\frac{y_{830} - y_n}{x_{830} - x_n}\right)$$

and

$$y < y_n,$$

where the subscripts 360 and 830 refer to the chromaticity coordinates for spectral stimuli of 360 and 830 nm, respectively' and the subscript n refers to the reference illuminant used.

Dominant wavelength (and complementary wavelength) is sometimes taken as a rough approximation of relative hue, although there is no simple relationship between dominant wavelength and perceived hue that varies with adaptation and viewing conditions as well as with λ_d. The measure was developed for use with older visual colorimeters, called "vector colorimeters," that permitted color matching by varying wavelength and proportions of spectral to neutral stimuli. Used in conjunction with a measure of purity, a complete colorimetric specification may be obtained for chromaticity.

The CIE defines purity as a measure of the proportions of the amounts of a spectral stimulus and a specified neutral stimulus that, when additively mixed, provide a color match to a given stimulus in question. The proportions can be measured in various ways, resulting in quantities that are called "colorimetric purity," "excitation purity," and "psychometric purity." We will discuss only the first two at this point. In the case of stimuli having complementary wavelengths, a suitably proportioned mixture of two stimuli from opposite extremes of the spectrum locus (e.g., 360 and 830 nm) are used instead of a spectral stimulus.

CIE Colorimetric Purity is symbolized p_c and is defined by the equation

$$p_c = L_d/(L_n + L_d), \tag{38}$$

where L stands for luminance and the subscripts d and n refer, respectively, to the spectral stimulus and the reference neutral stimulus. In the case of stimuli having complementary rather than dominant wavelengths, the expression may be written

$$p_c = (L_{d1} + L_{d2})/[L_n + (L_{d1} + L_{d2})], \tag{39}$$

where the subscripts 1 and 2 refer to the spectral extreme components of the mixture that is substituted for a single spectral stimulus.

Referring to Fig. 9 again, we may see that Eq. (38) means that colorimetric purity is the proportion of luminance of the stimulus λ_d to the sum of the luminances of λ_d and D_{65} in that additive mixture that provides a color match for S_1. As with dominant wavelength, colorimetric purity must be specified with respect to a particular reference illuminant to have a unique meaning. The same is true of colorimetric purity specifications for stimuli having complementary rather than dominant wavelengths. The colorimetric purity of S_2 in Fig. 9 is determined as the ratio of luminances of the aspectral mixture labeled P to those of the sum of P plus D_{65} for that additive mixture of P and D_{65} that matches the color of S_2.

Both dominant wavelength and colorimetric purity are measures that were developed for use in vector colorimetry. A second purity quantity that derives from chromaticity relationships, without regard to luminance, is one called excitation purity.

CIE excitation purity is defined as the following chromaticity relationship:

$$p_e = \frac{x - x_n}{x_d - x_n}$$

or (40)

$$p_e = \frac{y - y_n}{y_d - y_n}.$$

The subscript n again refers to the reference illuminant. The subscript d indicates that the chromaticity to be used is that of the spectral stimulus whose wavelength is the dominant wavelength of the sample in question. In the case of a sample having a complementary wavelength, the subscript d would refer to the mixture along the purple boundary that is capable of being additively mixed with the neutral reference to create a color match for the sample in question. The form of Eq. (40) to be used is governed by whichever has the larger numerator.

We see from Eq. (40) that the excitation purity for sample S_1 in Fig. 9 is simply the ratio of the distance from D_{65} to S_1 to the distance from D_{65} to λ_d. Similarly, p_e for sample S_2 is the ratio of the distance from D_{65} to S_2 to the distance from D_{65} to P. In the chromaticity diagram of Fig. 9, excitation purity and colorimetric purity are related in the following manner:

$$p_c = p_e\left(\frac{y_d}{y}\right).$$ (41)

Colorimetric purity p_c is determined from luminance ratios and is therefore independent of the particular chromaticity diagram used to determine p_e. That is, p_e is a ratio of distances in a chromaticity diagram and, hence, its value depends on the metric used to express chromaticity. Accordingly, its use is limited to the CIE 1931 chromaticity diagram (in which case it is symbolized p_e) and to the CIE 1964 diagram (for which it is symbolized $p_{e,10}$).

Dominant or complementary wavelength together with excitation purity and luminance comprise a complete colorimetric specification. They contain the same information as the original tristimulus values and are derived in such a way that the tristimulus values can be recovered from them. We have also seen that chromaticity coordinates and luminance comprise a complete specification from which the original tristimulus values may be recovered. There are, then, three forms that a CIE 1931 or 1964 colorimetric specification may take: (1) tristimulus values, (2) chromaticity coordinates and luminance, and (3) dominant wavelength and excitation purity together with luminance. In each case luminance is included (the tristimulus value Y is equal to luminance or relative luminance). Although luminance is discussed in some detail elsewhere in this book, it is well to summarize here the colorimetric conventions concerning luminance.

X. CIE LUMINANCE

Luminance is defined as the integral over wavelength of spectral concentration of radiance $L_e(\lambda)$ and spectral luminous efficiency $V(\lambda)$. It is expressed as follows:

$$L = K_m \int_\lambda L_e(\lambda)V(\lambda)\, d\lambda \tag{42}$$

or may be approximated by summation as

$$L = K_m \sum_\lambda L_e(\lambda)V(\lambda)\, \Delta\lambda. \tag{43}$$

The luminous efficiency function of Eqs. (42) and (43), $V(\lambda)$, is a measure of the conditions for a restricted form of color matching. It relates to brightness matching without regard to hue and colorfulness. Such matches are called "heterochromatic brightness matches" because although the brightness attribute of color is matched, the hues and colorfulness of the two stimuli being compared may be different in any way. Heterochromatic brightness matches of spectral stimuli are of particular interest because they may be used to define the luminous efficiency of monochromatic radiations (i.e., the relative efficiency of different spectral stimuli to produce an intensive visual response). Although there are three accepted experimental techniques for making such matches, only two of them figured significantly in establishing the $V(\lambda)$ function. The direct comparison method involves comparing and adjusting a test spectral stimulus with a reference stimulus until a brightness match is achieved. There are several variations of this technique, but even the best of them is difficult to do because the color match is not complete. Only the brightness is matched. When that is accomplished, there is a residual difference in color between the stimuli being matched.

There is, however, another way in which the brightnesses of two spectral lights may be matched by indirect means. This method relies on what is called the "critical flicker frequency" of the perception of alternating lights. This critical frequency varies with luminance and with the size of the stimulus. There is also a difference in critical flicker frequency between chromatic and achromatic vision and this is the foundation stone on which "flicker photometry" is based. When two lights (subtending the same visual angle) are alternated in time, their colors will be apparent to the observer until the alternation rate rises above a certain cyclic frequency, the value of which depends on size and luminance, at which point the observer is no longer able to distinguish differences in hue. In other words, if the lights are alternated rapidly enough, a chromatic additive mixture is achieved. There will still be flicker apparent because complete fusion has not occurred. The observer will still see differences in brightness between the two stimuli, and this is manifest as a continuing flicker. When the alternation is increased still further, the remaining flicker will diminish and disappear, at which point the alternating stimuli appear as one; there is complete temporal additive color mixture. In the flicker region where chromatic blending has occurred, but not yet complete fusion, an adjustment to the luminance of one member of the stimulus pair will affect the amount of remaining flicker. If one of the two stimuli remains fixed in luminance as a standard of reference, increasing the luminance of the other member by a sufficient amount will cause the flicker to disappear completely. The alternation rate may then be reduced just enough to make the flicker reappear. The variable luminance is then adjusted until flicker again disappears. The process is repeated until a minimum alternation rate that provides fusion is found. Under that condition a luminance setting of the variable stimulus that minimizes or just eliminates the residual flicker may be determined with considerable precision. The luminance so determined is taken as that which will provide a heterochromatic brightness match. Ives (1912) reported very little difference between results by direct heterochromatic brightness matching and such flicker photometry for stimuli subtending 2° visual angle. Accordingly, the CIE in 1924 adopted a $V(\lambda)$ function that was based on a weighted average of both direct matching and flicker photometry data then available. This size restriction (2° subtense) was, incidentally, the primary reason for using 2° in establishing the CIE 1931 Standard Colorimetric Observer (Wright, 1969); it was necessary to invoke the 1924 luminance method in order to convert the experimentally determined chromaticity data of Guild and Wright to form spectral tristimulus functions [cf. Eq. (21)]. We now recognize a systematic difference between direct heterochromatic results and those from flicker photometry and this is why Judd (1951) proposed corrections to the $V(\lambda)$ function that would yield somewhat different spectral tristimulus values according to the method of Wright and Guild.

The CIE $V(\lambda)$ function is then identical to the spectral tristimulus values $\bar{y}(\lambda)$ listed in Table I. If $\bar{y}(\lambda)$ is substituted for $V(\lambda)$ in Eqs. (42) and (43), and

$L_e(\lambda)$ is taken to be the stimulus function $\varphi(\lambda)$, it may be seen that these two equations are the same as the formulae for computing the Y tristimulus value in Eqs. (29) and (30).

The constant K_m of Eqs. (42) and (43) is referred to as the maximum spectral luminous efficiency because it represents the maximum of the $V(\lambda)$ function. That maximum occurs at a frequency of 540.0154×10^{12} Hz, which corresponds in air to a wavelength of 555 nm. The value of K_m was chosen such that L is 600,000 cd m^{-2} when $dL_e/d\lambda$ is the spectral radiance of the full radiator at the temperature of freezing platinum on the thermo-dynamic temperature scale, approximately 2045 K. Luminances* determined according to these values are known as "photopic luminances." They relate to an integrated measure of flux according to the spectral weighting of the $V(\lambda)$ function as determined for the standard observer adapted to a level of illumination providing not less than 1 and preferrably more than 3 cd m^{-2}. There is a second CIE standard for lower ("scotopic") light levels, from about 10^{-6} to 10^{-3} cd m^{-2} corresponding to "night vision." It uses a different spectral efficiency function $V'(\lambda)$ and a different constant K'_m. However, the scotopic luminance figures are not applicable to colorimetry. Color vision normally takes place in photopic levels of illumination providing luminances from about 10^0 to 10^6 cd m^{-2}. Under those conditions the value of K_m chosen by the CIE is 683 L W^{-1}. When that value is used to determine the tristimulus value Y for the CIE 1931 Standard Colorimetric Observer, the specification represents luminance in candelas per square meter.

It is often convenient to express luminance in relative units. For this purpose the CIE has recommended that luminance factor β† be taken as the ratio of luminance of an object to that of a perfect reflecting or transmitting diffuser identically illuminated. The perfect diffuser (or what has often been called the "Lambert reflector") reflects light equally in all directions and has a reflectance equal to unity; an analogous definition applies to the case of transmission. Before 1969 smoked magnesium oxide was used as a reflectance standard and many references state β with respect to that standard of reference. Such values will tend to be about 2% higher than more recent values of β.

---

* The luminance of a stimulus (in a given direction and at a point on the surface of the stimulus or at a point on the path of a beam) is defined by the CIE as the quotient of the luminous flux leaving, arriving at, or passing through an element of surface at this point and propagated in directions defined by an elementary cone containing the given direction, to the product of solid angle of the cone and the area of the orthogonal projection of the element of surface on a plane perpendicular to the given direction.

† The term β is used here as an abbreviation for β_v. That symbol refers to luminance factor as distinguished from radiance factor, symbolized β_e. When the context is clear, either factor may be represented by β alone.

Finally, it should be noted that luminous reflectance factors are often used to provide a measure of the luminance of an object with respect to the luminance of some other object of particular interest. The reference object is typically one that appears white. In such cases the term "relative luminance" is to be preferred.

XI. RELATIVE COLORIMETRY

Just as it is sometimes convenient to deal with relative luminance, so it is also often convenient to make relative colorimetric specifications. These are sometimes referred to as a process of "colorimetry by difference." It involves making measurements with respect to an arbitrary but operationally useful reference. Such measurements can be made quickly and with high precision but they need not require the accuracy of what is called, by contrast, "absolute colorimetry." In fact, very few applications require the accuracy of absolute colorimetry. By far the overwhelming majority of colorimetry is practiced as relative colorimetry.

We may examine the difference between absolute and relative colorimetry by detailing the operations of both according to the methods set forth in Eqs. (29) and (30). First consider what is done to solve Eq. (30), the summation method. Spectrophotometric or spectroradiometric measurements are made of the stimulus function $\varphi(\lambda)$. If the sample is a reflecting one, for example, these measurements consist of a series of spectral reflectance values over the wavelength range of 360–830 nm at some wavelength interval $\Delta\lambda$, let us say 10 nm. The $\rho(\lambda)$ spectral reflectance function is then approximated by 48 spectral reflectance values. A like number of values represent the relative spectral power function $S(\lambda)$ of the source illuminating the sample. Those values are either measured by spectroradiometry or assumed from a table of standard illuminants. Similarly, the standard observer's spectral tristimulus functions $\bar{x}(\lambda)$, $\bar{y}(\lambda)$, $\bar{z}(\lambda)$ are approximated by 48 values each as obtained from appropriate tables. To determine the tristimulus value X, the values $\bar{x}_\lambda$, ρ_λ, S_λ are multiplied together, wavelength by wavelength Then the 48 products are summed and divided by 10, the value of $\Delta\lambda$. Finally, k is used to multiply the result so that we have the tristimulus value X. Exactly the same procedure is used to determine tristimulus values Y and Z, except, of course, that $\bar{y}_\lambda$ and $\bar{z}_\lambda$ values are used.

If the value of k is equal to K_m, our tristimulus values will be in luminance units and will vary with illuminance, although their ratios, as in the chromaticity coordinates, will remain constant with lighting level. If k is equal to 100 over the sum of the wavelength by wavelength products of S_λ and $\bar{y}_\lambda$ divided by 10, then our tristimulus specification will be in terms of relative luminance. In the first case our specification is with respect to the standard illuminant. In the second case the specification is with respect to a perfect reflector illuminated by the standard illuminant. Both specifications can be traced directly to the basic standards of colorimetry.

Such measurements are often called "absolute" because the method involves reference to basic standards and assumes that there are no errors in any of the component functions. In the example used, all values except those of the sample's spectral reflectances could be determined from standard tables. They are the kinds of measurements that may be made with a single-beam spectrophotometer. It is assumed that such instruments measure spectral reflectance with no error. If the wavelength calibration of the instrument is correct, the bandwidth sufficiently narrow, and its response linear with intensity, this may not be a bad assumption.

Suppose, however, that these requirements are not met exactly. Then the assumption of zero error is false and the specifications will be incorrect by some amount. Double-beam spectrophotometers are designed to try to minimize such problems. They compare the sample with a reference in rapid alternation so that it may be assumed that the instrument is responding in the same way to both sample and reference over a very short period of time. The reference is usually a white-appearing plaque for reflectance measurements or an appropriate thickness of air or solvent for transmittance measurements. The reference is carefully calibrated with respect to a basic standard such as the perfect diffuser. The difference between the working standard of reference and the basic standard at each wavelength is then used to, adjust the measured values of reflectance for the sample. In this way more accurate measurements can be made with the same instrument when it is used as a double-beam device than when it is operated in the single-beam mode. Again this is true because the wavelength errors and photometric scale errors have negligible effect on the difference signal; the error made in measuring the sample will have the same direction and essentially the same magnitude as the error made in measuring the working standard. Thus the errors tend to cancel out. If we designate the errors as e, we may state

$$\Delta\rho(\lambda) = [\rho_1(\lambda) - e_1(\lambda)] - [\rho_2(\lambda) - e_2(\lambda)]$$

$$\Delta\rho(\lambda) = \rho_1(\lambda) - \rho_2(\lambda) \tag{44}$$

where the subscript 1 refers to the calibrated standard and the subscript 2 identifies the sample.

When ρ_1 and ρ_2 are very nearly the same, $\Delta\rho$ will be essentially zero. It will be obvious from this that accuracy can be maximum whenever the working standard is a near-replica of the samples to be measured. In fact, the overwhelming majority of routine spectrophotometric measurements may be made in just this way. Recall the example of a plastics tableware manufacturer used in the introduction to this chapter. If he wishes to make routine colorimetric measurements of samples from his production of yellow plates, he may use one of the plates that he knows to be on aim as his working standard. Henceforth, his spectrophotometric measurements are taken with respect to that reference. He thereby has some assurance that the errors of his instrument are not significantly influencing his measurements. The measurements are now *relative* colorimetric measurements. The computed values of tristimulus may not be exactly correct in terms of the basic standards (i.e., their *absolute* values may not be correct), but the *differences* in tristimulus values that he measures should be correct (i.e., their *relative* values will be correct), even though his instrument may be somewhat in error.

Most manufacturers will not rely on spectrophotometry for routine measurements. Instead, they will use less expensive and more easily controlled instruments called "tristimulus colorimeters." The tristimulus colorimeter is a device that incorporates optically filtered photodetectors. The spectral sensitivity of the photodetector combined with the spectral transmittances of the filters are designed to match the standard observer's spectral tristimulus functions $\bar{x}(\lambda)$, $\bar{y}(\lambda)$, $\bar{z}(\lambda)$. When the photometric response of the device is linear with intensity, tristimulus values are determined according to Eq. (29), the integration method. Optical rather than mathematical integration is used in such instruments. However, no such instrument has ever been built in which its spectral sensitivities as filtered exactly match the $\bar{x}(\lambda)$, $\bar{y}(\lambda)$, $\bar{z}(\lambda)$ functions. The poorest tristimulus colorimeters differ greatly from those functions and even the best do not match closely enough for all purposes.

Manufacturers of tristimulus colorimeters usually specify the "accuracy" of their instruments as a percentage deviation from the standard observer's response. What they imply by this is the mean error integrated over wavelength. That is,

$$e_x = 100 \int \left[\bar{x}(\lambda) - s_x(\lambda) \right] d\lambda,$$

$$e_y = 100 \int \left[\bar{y}(\lambda) - s_y(\lambda) \right] d\lambda, \qquad (45)$$

$$e_z = 100 \int \left[\bar{z}(\lambda) - s_z(\lambda) \right] d\lambda,$$

where s refers to the filtered sensitivity of the colorimeter. This figure, however, will only be correct when the instrument is used to measure the theoretical CIE Illuminant E. Errors are larger than indicated by Eq. (45) whenever the spectral selectivity of the source and sample is greater than that of CIE Illuminant E, as it is, in fact, with all real situations. This may be seen from Eq. (46):

$$e_x = 100 \int \{[S(\lambda)\rho(\lambda)\bar{x}(\lambda)] - [S(\lambda)\rho(\lambda)s_x(\lambda)]\} \, d\lambda,$$

$$e_y = 100 \int \{[S(\lambda)\rho(\lambda)\bar{y}(\lambda)] - [S(\lambda)\rho(\lambda)s_y(\lambda)]\} \, d\lambda, \qquad (46)$$

$$e_z = 100 \int \{[S(\lambda)\rho(\lambda)\bar{z}(\lambda)] - [S(\lambda)\rho(\lambda)s_z(\lambda)]\} \, d\lambda.$$

The problem becomes more acute when highly spectrally selective light sources are to be measured. For example, some light-emitting diodes and television phosphors (e.g., europium yttrium vanadate phosphors now commonly used on television receiver screens) have very narrowband emission spectra with at least one emission peak around 698 nm. Now a typical colorimeter spectral response function might have a value of about 0.009 for s_y at this wavelength rather than a value of 0.004676404 of the $\bar{y}(\lambda)$ function. A typical integrated error value according to Eq. (45) might be 2% for such an instrument. However, the error at 698 nm is actually 108% as seen from the figures just given. In the case of a light-emitting diode, that might be the only value of importance, and Eq. (46) then differs from the results of Eq. (45) by a factor of 2.08. The nominally "2% error" instrument would then tell us that the luminance of the light-emitting diode is 62.4 cd m^{-2}, when, in fact, it is only 30 cd m^{-2}.

The point is simply that the best tristimulus colorimeters may produce erroneous results by factors much larger than the colorimeter manufacturer's specifications under unfavorable conditions. Actually, in all real situations the error will tend to be intermediate between the integrated figure of Eq. (45) and the maximum error of the mismatch between s_x, s_y, s_z and $\bar{x}$, $\bar{y}$, $\bar{z}$ at a single wavelength where the mismatch is greatest. Accordingly, relative colorimetry procedures are generally even more useful when using tristimulus colorimeters than when spectrophotometers are used. This is merely a statement of fact. It is not a condemnation of tristimulus colorimeters. Such instruments are used for routine colorimetry far more often than spectrophotometers, and they are generally found to yield satisfactory results, particularly if used as *relative* colorimeters.

The sensitivity of a tristimulus colorimeter for the $\bar{x}(\lambda)$ channel [which will be used in the following examples with the understanding that exactly analogous equations may be written for $\bar{y}(\lambda)$ and $\bar{z}(\lambda)$] may be expressed as

$$s_x(\lambda) = \bar{x}(\lambda) + e_x(\lambda). \tag{47}$$

The response q of the instrument is then written as

$$q_x = \int \rho(\lambda)S(\lambda)s_x(\lambda)\,d\lambda. \tag{48}$$

From Eq. (47) we may expand the integral of Eq. (48) to read

$$q_x \simeq \int \rho(\lambda)S(\lambda)\bar{x}(\lambda)\,d\lambda + \int \rho(\lambda)S(\lambda)e_x(\lambda)\,d\lambda. \tag{49}$$

Suppose that we now have a working standard that has been carefully calibrated by a national standardizing laboratory or some other agency where absolute colorimetry is practiced.* We can place this calibrated standard in our tristimulus instrument and adjust the instrumental controls so that $q_x = X$, $q_y = Y$, $q_z = Z$, where X, Y, Z are the true tristimulus values from the calibration. This operation is the equivalent of setting

$$0 = \int \rho(\lambda)S(\lambda)e_x(\lambda)\,d\lambda$$

and so on, for e_y and e_z.

The adjustment operation eliminates the error function for the standard. It will also be zero for any other sample with the *same* spectral reflectance distribution as the standard. It will not be zero for samples with different spectral reflectance distributions *unless* their spectral reflectances are *proportional* at all wavelengths to those of the working standard. In that case we have

$$q_1 = aq_2,$$

where a is a proportionality constant and the subscript 1 refers to the standard and 2 to the sample. Our hypothetical manufacturer of yellow plates may then use one sample that he knows to have the aim tristimulus values, adjust his tristimulus colorimeter controls to match those values, and then measure all other production samples with respect to that condition. As with his practice of relative colorimetry on the spectrophotometer, his tristimulus colorimeter measurements may not be accurate in an absolute sense, but their relative accuracy may be high.

In practice, relative colorimetry is often conducted in situations where

---

* Calibrated transmitting glass filters and reflecting ceramic tiles are available from some instrument manufacturers and from the National Bureau of Standards.

the spectral distributions of samples and working standards are neither the same nor proportional. Spectral reflectance distributions of most objects are both spectrally broad and not highly selective. If this is also true of the distribution for the working standard, then "on average" the error integrals may approach zero. The process may be represented as

$$(q_{x,1} - q_{x,2}) \simeq (X_1 - X_2) + \int (\rho_1(\lambda) - \rho_2(\lambda))S(\lambda)e_x(\lambda)\, d\lambda. \qquad (50)$$

Generally, if $\rho_1(\lambda)$ and $\rho_2(\lambda)$ intersect at three or more wavelengths, the error integrals for most slightly selective samples will be reasonably small.

Relative colorimetry is, then, a method for practicing colorimetric measurement under less than ideal conditions. Only in a few cases will conditions ever be found to approach closely the ideal. Almost all colorimetry is carried out with instruments that are inaccurate to some significant degree. Relative colorimetry helps to minimize the influence of these instrumental inaccuracies on the validity of the tristimulus measurements made with them.

XII. DISTRIBUTION TEMPERATURE, COLOR TEMPERATURE, AND CORRELATED COLOR TEMPERATURE

The ratio of emittance of a radiator to that of a full radiator is called "emissivity." When emissivity does not vary with wavelength, the source in question is said to be nonselective. The radiance of the source is proportional by the same factor at all wavelengths to the radiance of a full radiator. A measure called "distribution temperature" is used to characterize sources with radiances proportional, or nearly so, to that of a full radiator at some absolute temperature. That is, nonselective or only slightly selective sources are specified according to their distribution temperatures. The CIE has defined distribution temperature as the absolute temperature of a full radiator for which the ordinates of the spectral distribution curve of its radiance are proportional (or approximately so), in the region of wavelengths from about 360 to 830 nm, to those of the distribution curve of the source under consideration. In such a case, both radiators will have the same or nearly the same chromaticity. Using Planck's radiation formula [cf. Eq. (23)], we may express the solution for distribution temperature as follows:

$$\int_{\lambda} \left(1 - \frac{M'_{e,\lambda}}{aM_{e,\lambda}(\lambda, T)}\right)^2 d\lambda \rightarrow \text{minimum}, \qquad (51)$$

where $M'_{e,\lambda}$ is the exitance of the test source, $M_{e,\lambda}(\lambda, T)$ is given by the Planckian equation, and a is an arbitrary constant. The values of a and T, the Planckian temperature, are adjusted simultaneously until the integral is minimized. When this has been done, the final value of T is the distribution temperature.

It is customary to restrict the use of the term "distribution temperature" to cases where the ratio $M'_{e,\lambda}/M_{e,\lambda}(\lambda, T)$ does not change by more than $\pm 5\%$.

To a close first-order approximation, a specification of distribution temperature defines both the spectral power distribution and the chromaticity coordinates of a source. However, many sources are selective radiators in the sense that their spectral emissivities vary significantly with wavelength. That is, their spectral distributions bear no simple (proportionate) relationship to that of any full radiator. Yet such a source may have the same chromaticity coordinates as a full radiator. When this is the case, a specification of "color temperature" is used to characterize the color-matching appearance of the source.

The CIE has defined color temperature as the absolute temperature of the full radiator that emits radiation of the same chromaticity as that of the source under consideration. Color temperature may be determined from any chromaticity diagram since the chromaticity coordinates of a test source must match those of one of the series of full radiators defined by the Planckian radiation formula. The chromaticities of full radiators lie along a curve in the CIE 1931 chromaticity diagram, as illustrated in Fig. 10.

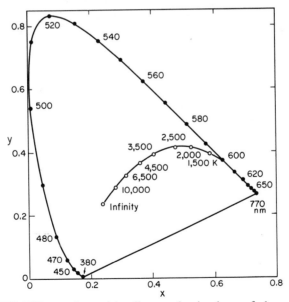

Fig. 10. CIE 1931 x, y chromaticity diagram showing locus of chromaticities corresponding to full radiators. Numbers along that locus indicate color temperatures of the full radiators.

When a selective radiator has chromaticity coordinates that are not the same as any full radiator but that, nevertheless, are not greatly different from those of a full radiator at some temperature, neither distribution temperature nor color temperature is an appropriate specification of the source characteristics. Instead, a specification called "correlated color temperature" is used.

Correlated color temperature values were originally determined by Judd (1936). He subjected the CIE 1931 chromaticity diagram to a mathematical transformation that was intended to array differences in chromaticity as a constant proportion of the corresponding visual difference in color. In that new diagram, Judd constructed "isotemperature lines," lines that passed through a point on the locus of full radiator chromaticities and along which were arrayed the chromaticities of stimuli that were minimally different in color appearance from the color of the full radiator whose chromaticity was intersected by the isotemperature line. He found these lines to have smooth, systematic, variation in angle as a function of the full radiator temperature. However, the slopes of the lines corresponding to the tangents of those angles are obviously dependent on the particular chromaticity diagram for which they are determined. Since Judd's uniform-chromaticity diagram was not a standard one, no standard existed for determining correlated color temperature.

Most sources used in practice cannot be properly specified by either distribution temperature or color temperature. Consequently, some method of specifying selective sources is desirable. The concept of correlated color temperature allows people to express some characteristic of such sources as the temperature of the full radiator whose color appearance most closely resembles that of the selective radiator in question. For this reason the CIE has adopted a measure of correlated color temperature. Correlated color temperature is defined as the color temperature corresponding to the point on the locus of full radiator chromaticities nearest to the chromaticity point of the source in question when chromaticities are arrayed on an agreed uniform-chromaticity-scale diagram. The presently agreed uniform-chromaticity-scale diagram is the CIE 1960-UCS diagram, which will be discussed in the following sections.

There are also other CIE uniform-chromaticity-scale diagrams (e.g., the CIE 1976 u', v' diagram, *vide infra*). This has raised some question about which diagram should be used to determine the isotemperature lines, since, as noted earlier, the slopes of those lines vary from one diagram to another (e.g., MacAdam, 1977). For example, the slope of the 5000 K isotemperature line in the CIE 1931 x, y diagram is approximately $+12.5$ when isotemperature is determined in the CIE 1960 u, v diagram, but it is about -2.5 if determined in the CIE 1976 u', v' diagram. Those slopes differ in angular extent by about 15% and represent chromaticities that do not match. Clearly there is some question about which, if either, is correct. However, measurements have been made recently (Grum *et al.*, 1978) that suggest the question may not be answered with certainty. In that work, nine observers adjusted the chromaticities of full radiators to minimize the color differences between the full radiator and one of a series of test sources having chromaticities that lay near, but not on, the locus of full radiator chromaticities. The results indicate that there is considerable disagreement among observers. For example, a source with a nominal correlated color temperature of 6500 K was "matched" by full radiators whose color temperatures ranged from about 6337 to 6702 K, a variation of 58% in angle of isotemperature lines. This is roughly an imprecision of 50–200 K over the scale of commonly encountered illuminants. Since the interobserver differences are nearly four times as large as that resulting from determination of isotemperature lines in the CIE 1960 or 1976 diagrams, it has been decided that the CIE 1960-UCS diagram will continue to be the reference for isotemperature lines. In that diagram they are defined as lines perpendicular to the locus of chromaticities corresponding to full radiators, also called the "Planckian locus." No standard has been set for the distance along such a line that is considered to be too large a deviation in chromaticity for the specification of correlated color temperature to be appropriate. However, it is generally understood that the distance should be "small." One will not be wrong by too much if "small" is interpreted to mean about twice the distance represented by the separation of chromaticities corresponding to phases of natural daylight from the Planckian locus (Kelly, 1963; Krochmann *et al.*, 1978).

In addition to the locus of chromaticities corresponding to full radiators, there is also a locus of chromaticities corresponding to daylight illumination, and it is defined by Eq. (24) in an earlier section of this chapter. Figure 11 illustrates both the Planckian and daylight loci in the CIE 1931 x, y diagram. Isotemperature lines, determined in the CIE 1960 u, v diagram, are shown for five correlated color temperatures. In addition, color temperatures are marked off on the Planckian locus.

It is sometimes convenient to express both color temperature and correlated color temperature as reciprocal values: $10^6/T_c$. The value $10^6/T_c$ is called "reciprocal megakelvins" because of its relation to the unit of color temperature, kelvin. In some of the older literature the same quantity may be referred to as "mireds," which was coined from "micro reciprocal degrees." The reciprocal megakelvin has the useful property that a given sum of such units corresponds to approximately the same color difference anywhere on the scale from about 100 to 1000 (corresponding to color temperatures from 10,000 to 1000 K). Reciprocal megakelvins are often used to specify

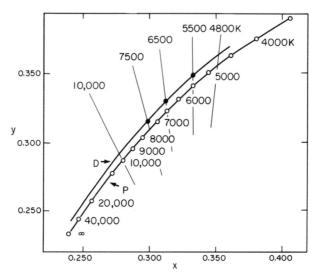

Fig.. 11. A portion of the CIE 1931 x, y chromaticity diagram showing the loci of chromaticities corresponding to full or Planckian radiators P and to CIE daylight radiators D. Isotemperature lines for correlated color temperatures of 4800, 5500, 6500, 7500, and 10,000 K are also indicated. Numbers along the Planckian locus indicate color temperatures of full radiators.

the color-temperature-altering power of optical filters. Such conversion filters will alter the color temperature by approximately the same number of reciprocal megakelvins anywhere on the scale and are, therefore, identified by the reciprocal megakelvin shift they introduce.

Reciprocal megakelvins (expressed in absolute temperature: $10^6/T$) have also been used to estimate correlated color temperature by calculational methods suitable for computers (Robertson, 1968) and "desktop" calculators (Schanda *et al.*, 1978). The underlying rationale of these computational approximations involves the assumption that the chromaticity coordinates of the Planckian radiator x_p, y_p nearest in color appearance to that of a test source x_i, y_i are such that x_p is a polynomial function of reciprocal megakelvins:

$$x_p = a_j(10^6/T)^j + \cdots + a_2(10^6/T)^2 + a_1(10^6/T)^1 + a_0,$$

and, in turn, y_p is a polynomial of x_p:

$$y_p = b_k x^k + \cdots + b_2 x^2 + b_1 x + b_0.$$

To find the correlated color temperature of a source with chromaticity coordinates x_i, y_i, the following algorithm is solved:

$$\{[u(x_i, y_i) - u(x_p, y_p)]^2 + [v(x_i, y_i) - v(x_p, y_p)]^2\}T \rightarrow \text{minimum},$$

where $u(x, y)$ and $v(x, y)$ are the functions transforming x and y to CIE 1960 UCS coordinates [cf. Eq. (67), *vide infra*]. The minimum solution to this expression yields x_p and y_p chromaticity coordinates that correspond to the Planckian or full radiator whose chromaticity is approximately closest to that of the source in question in the CIE 1960 UCS diagram. Derivation of x_p and y_p by a fourth-order polynomial generally yields results with errors of only about 0.1%. For details of the computations, the reader is referred to the cited publications.

XIII. METAMERISM

Grassman's laws indicate that we may have two lights or two stimuli that match in color appearance even though their relative spectral radiance distributions differ. This is also clear from the discussion of color temperature, where we see that two lights may have the same chromaticity although one may be a full radiator and the other may have a highly selective emissivity distribution. This kind of condition was referred to as "metamerism." We may represent a metameric match in the following way:

$$\int \varphi_1(\lambda)\bar{x}(\lambda)\,d\lambda = \int \varphi_2(\lambda)\bar{x}(\lambda)\,d\lambda,$$

$$\int \varphi_1(\lambda)\bar{y}(\lambda)\,d\lambda = \int \varphi_2(\lambda)\bar{y}(\lambda)\,d\lambda, \qquad (52)$$

$$\int \varphi_1(\lambda)\bar{z}(\lambda)\,d\lambda = \int \varphi_2(\lambda)\bar{z}(\lambda)\,d\lambda,$$

where $\varphi_1(\lambda)$ and $\varphi_2(\lambda)$ are the two different stimulus functions. They may differ for various reasons. If they represent different illuminants

$$\varphi_1(\lambda) = S_1(\lambda), \qquad \varphi_2(\lambda) = S_2(\lambda),$$

or they may represent different objects illuminated by the same illuminant

$$\varphi_1(\lambda) = \beta_1(\lambda)S(\lambda), \qquad \varphi_2(\lambda) = \beta_2(\lambda)S(\lambda),$$

or different objects illuminated by different illuminants

$$\varphi_1(\lambda) = \beta_1(\lambda)S_1(\lambda), \qquad \varphi_2(\lambda) = \beta_2(\lambda)S_2(\lambda).$$

Occasionally reference has been made to still another kind of difference: the difference or disagreement between two observers as to whether a color match obtains. By some form of inverted logic this situation has been called "observer metamerism," but it is nothing more than a difference among observers. To be properly called metamerism, the situation should imply different fundamental sensitivities among observers but the *same* color match. The biochemical processes that give rise to such fundamental sensitivities do vary somewhat, but it seems doubtful that two observers with fully transitive identical match responses would have different underlying fundamental sensitivities. There are some *in*transitive cases of this kind. A person with anomalous color vision will generally be satisfied with all the matches that are made by a normal trichromat, but the trichromat will not be satisfied with all the matches made by the person with anomalous color vision. Except for this unusual case, we will assume that observers with different spectral sensitivities will not be satisfied with the same color match. Accordingly, the term "metamerism" will be reserved for matching colors where the spectral distributions of the stimuli differ, as in Eq. (52).

When two light sources are metameric, they will appear to be of the same color when the standard observer looks directly at them. Similarly, when they each illuminate a spectrally nonselective object, such as the perfect diffuser, that object will also have the same color appearance (hence chromaticity coordinates) under both sources. However, when two metameric sources are used each to illuminate a spectrally selective object, the object will not necessarily appear the same or have the same chromaticity coordinates with respect to each source.

Typically, a metameric match is specific to one observer and one illuminant. When either illuminant or observer is changed, it is most common to find that the metameric match breaks down. There are, however, some instances in which the metameric match may hold for a second illuminant (e.g., from daylight to incandescent light). Usually this will only be true where, for example, reflectance values of two samples that are metameric are equal at three or more wavelengths. That is, the stimulus functions intersect in at least three wavelength regions of the spectrum. Such samples will tend to be metameric under a particular source, and if the location of intersections is appropriate, they may continue to provide a color match for a second illuminant. It is possible to imagine other samples that provide metameric matches for three or more illuminants. However, the number of intersections must increase with the number of illuminants for which they are metameric matches. This is simply another way of saying that the two reflectance distributions must be increasingly similar if they are to match under a larger number of illuminants. In the limiting case, where they must match under all illuminants, the spectral reflectance functions must be identical.

Accordingly, some measure of the difference in spectral reflectance functions between two samples that provide a metameric color match under some illuminant may be taken as an indication of the probability that they will also be a metameric match under some different illuminant. The term "degree of metamerism" has been coined to refer to the inverse of that probability; that is, two metameric samples with greatly different spectral reflectance functions, and hence a high *im*probability of maintaining the metameric match for a second illuminant, are said to have a high degree of metamerism. Such highly metameric samples are likely to be troublesome when the source is changed (they will no longer match in color) or when a different observer or different tristimulus colorimeter is used to measure them. On the other hand, two samples with a low degree of metamerism are not as likely to be troublesome when illuminants or observers vary.

One obvious measure of the difference between a pair of metameric samples is simply the difference in stimulus functions; for example,

$$D = \{\textstyle\sum_{\lambda} [\varphi_1(\lambda) - \varphi_2(\lambda)]^2\}^{1/2}. \tag{53}$$

This expression provides a singular index of the difference between $\varphi_1(\lambda)$ and $\varphi_2(\lambda)$. However, it weights all wavelengths equally and we know from the spectral tristimulus functions of the standard observer that this is not necessarily desirable. It would be better to give different wavelengths different weights in the summation. Further, Eq. (53) does not address the question of what may happen when the illuminant is changed, the situation that is generally of most interest when dealing with metameric samples.

For these reasons the CIE has adopted a somewhat different approach to specifying degree of metamerism. The degree of metamerism M_t of two samples illuminated by a reference light source $\beta_1(\lambda)S_r(\lambda)$ and $\beta_2(\lambda)S_r(\lambda)$ is measured as the color difference ΔE found between the samples when they are illuminated by a test source $S_t(\lambda)$. That is,

$$M_t = \Delta E = f[\beta_1(\lambda):\beta_2(\lambda);S_r(\lambda):S_t(\lambda)]. \tag{54}$$

This is known as the CIE Special Metamerism Index: Change in Illuminant (CIE, 1971b). The preferred reference illuminant is CIE D_{65} and the preferred test illuminant is either CIE Illuminant A or one of three spectral power distributions that represent commonly encountered fluorescent lamps. Relative spectral power distributions for these four illuminants are tabulated at 10 nm intervals from 380 to 780 nm in Table X. The higher

TABLE X. Relative Spectral Power Distributions of
Illuminants Used as References in
Determining the CIE Special Metamerism
Index: Change in Illuminant

λ (nm)	A	F1	F2	F3
380	9.8	5.4	10.7	23.0
90	12.1	5.6	12.0	27.5
400	14.7	5.8	13.9	33.4
10	17.7	6.1	16.8	43.6
20	21.0	7.4	20.8	55.0
30	24.7	10.6	28.0	67.7
40	28.7	17.2	37.9	81.0
450	33.1	26.5	48.8	94.2
60	37.8	33.6	58.5	104.6
70	42.9	38.3	64.4	111.1
80	48.2	39.7	66.5	114.3
90	53.9	39.8	67.0	115.5
500	59.9	40.7	66.6	114.2
10	66.1	43.6	67.7	111.4
20	72.5	49.9	69.9	107.6
30	79.1	57.4	73.2	103.6
40	85.9	67.8	78.7	101.0
550	92.9	82.3	88.4	99.8
60	100.0	100.0	100.0	100.0
70	107.2	113.2	110.4	101.1
80	114.4	125.7	116.0	102.7
90	121.7	112.9	115.3	102.7
600	129.0	103.2	111.2	101.2
10	136.3	93.3	104.6	99.5
20	143.6	109.8	104.0	98.9
30	150.8	145.7	104.9	97.4
40	158.0	143.6	103.6	92.7
650	165.0	272.2	116.9	96.5
60	172.0	296.5	147.7	96.0
70	178.8	86.9	62.3	63.6
80	185.4	35.6	40.5	47.2
90	191.9	21.2	30.2	38.1
700	198.3	12.4	23.6	31.4
10	204.4	8.0	18.0	25.3
20	210.4	5.2	14.0	20.5
30	216.1	3.5	10.8	16.7
40	221.7	2.0	9.3	13.5
750	227.0	1.0	6.6	11.0
60	232.1	0.2	5.2	9.0
70	237.0	0.0	4.0	7.3
80	241.7	0.0	3.1	6.0

λ (nm)	A	$F1$	$F2$	$F3$
404.7[a]		27.2	42.3	77.7
435.8		84.0	112.1	182.4
546.1		77.7	77.7	100.8
577.8		23.7	23.0	29.1

[a] Power at principal mercury lines taken above continuum and spread evenly over a 10 nm band.

the value of $\Delta E = M_t$, the greater is the degree of metamerism of the match with respect to the reference illuminant.

Of course, other illuminants than those recommended may be used in the functional notation of Eq. (54), but they should be specified so that the conditions under which ΔE is determined may be understood. In any case, it is necessary to have a measure of ΔE that is meaningful in visual terms. This can only be derived from some metric space where extents of ΔE correspond to visually equal differences or ratios. We need, in other words, a uniform color space in which to determine ΔE.

XIV. COLOR DIFFERENCES

A metric for measuring and specifying differences in color should ideally be one in which all distances of a given value correspond to equal-sized differences in perceived color regardless of direction and location within the space. It would be called a "uniform color space." The problem of specifying color differences by colorimetry is, then, essentially one of determining an appropriate mathematical transformation from the three-dimensional tristimulus space to an n-dimensional metric space having the properties of color uniformity. It will be recalled from the introduction to this chapter that such a space may be uniform in distance (the relationship $\Delta\theta/\Delta\theta = a$ constant) or in magnitude ratios (the relationship $\Pi/\Pi = a$ constant) but usually not both at the same time. The metric distance relationship has been the subject of greatest study, particularly during past years. That relationship is very important in setting and interpreting production tolerances. Much of the data that have been produced and studied in an attempt to derive a uniform color distance space relate to thresholds of sensation. Weber's and Fechner's laws have played a large role in extending threshold data to form scales of color differences, particularly small color differences.

The earliest quantitative studies of this kind addressed thresholds for single attributes of color (e.g., lightness or brightness, hue, and colorfulness). In studies of lightness thresholds a common experimental technique involves

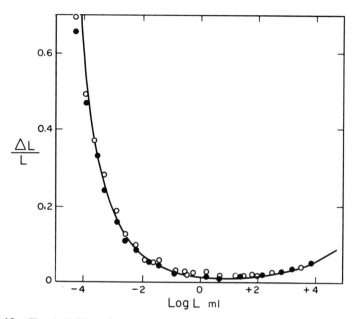

$\dfrac{\Delta L}{L}$

Fig. 12. Threshold Weber fractions $\Delta L/L$ as a function of test luminance in millilamberts (mL) measured by König and Brodhun (1889).

measuring the Weber fraction $[=\delta\Phi/\Phi$, from Eq. (2)] as the luminance of the test stimulus is increased. Results of an early and classical determination of this kind are illustrated in Fig. 12. The data were collected by König and Brodhun (1889). They are generally representative of subsequent data, although the shape and location of the curves vary with illuminance, wavelength, purity, size, and surround of the test stimulus. Those results show that the Weber fraction is not constant over the entire range of luminances studied. A horizontal line would represent a constant Weber fraction. The experimental data follow that line only throughout the middle range of stimulus luminances. The Weber fraction is higher at both lower and higher stimulus luminances.* However, since the abscissa of Fig. 12 is logarithmic, we see that the Weber fraction is very nearly constant over a range of more than 10,000:1 in luminance, a range that normally includes most luminances of practical interest.

Similar determinations have been made of the threshold difference in wavelength necessary to see a hue difference. Figure 13 illustrates the results of such a determination made by König and Dieterici (1884), which

* Some doubt has been cast on the increase in the Weber fraction at high luminances by later investigations, but it seems clear that there is some increase for complex fields even though there appears to be little or none when the eye is adapted to a uniform field (e.g., Alpern et al., 1970).

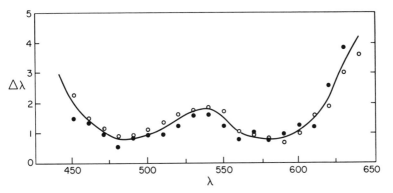

Fig. 13. Wavelength difference $\Delta\lambda$ required to perceive a just noticeable difference in hue shown as a function of wavelength λ measured by König and Dieterici (1884).

is again an early and classic experiment of this kind. These results are also generally representative of subsequent data. Other studies indicate somewhat more change in the threshold curve, particularly in the wavelength region from 400 to 500 nm, and the shapes of the wavelength discrimination threshold curves do depend on stimulus intensity, size, surround, and observer characteristics.

Figure 14 illustrates data from a classic determination of purity thresholds (Wright and Pitt, 1937). Again we see that the threshold varies considerably with wavelength. The data on which Fig. 14 is based were derived by measuring the first perceptible decrement in colorfulness when adding neutral-appearing light to that of a spectral stimulus. When, instead, spectral light is added to that of a neutral stimulus (e.g., Priest and Brickwedde, 1938),

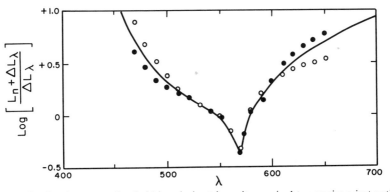

Fig. 14. Log increment threshold in colorimetric purity required to perceive a just noticeable difference in colorfulness shown as a function of wavelength λ, measured by Wright and Pitt (1937).

the purity threshold curve tends to be more steep, but these differences are ones of degree and not of kind (Kaiser *et al.*, 1976) and are related to the colorfulness of the stimulus (Bartleson, 1979a).

A color difference metric may, in theory, be constructed according to Fechner's law [cf. Eq. (5)] for each of the attributes represented by the data of Figs. 12–14. It is usually assumed that such metric will be most useful in the region of the thresholds (i.e., for small color differences). The metric may represent one, two, or three of the attributes involved. For example, Fechner's law was applied to the luminance versus lightness relationship for many years. That is, a scale of lightnesses was built up by the addition of luminance JND's. However, such scales have been found not to provide very useful predictions of perceived lightness and generally have been abandoned in favor of lightness scales based on suprathreshold data obtained by direct scaling. Threshold data in two dimensions (wavelength and purity) were collected from available reports by Judd (1935) and used to derive a transformation of the CIE 1931 chromaticity diagram to make a uniform-chromaticity-scale diagram in which, as we have seen, lines of constant correlated color temperature were determined. Judd's UCS diagram was arrayed as an equilateral triangle. Within four years of its publication another transformation was proposed that would simply render Judd's diagram in rectangular coordinates so that it could be more easily compared with the CIE 1931 chromaticity diagram (Breckenridge and Schaub, 1939). By that time, however, some workers were beginning to make experimental determinations of color thresholds in two dimensions, and these data are now more often the basis of threshold-derived scales of color difference.

Wright (1941, 1943) reported results of extensive studies of chromaticity thresholds determined by the method of limits. He had observers adjust additive stimulus mixtures until a just noticeable difference in color was detected from each of a large number of reference mixtures, all having the same luminance. These visually equal chromaticity differences are illustrated in Fig. 15. The lines plotted there represent a difference about three times as large as a JND for Wright's experimental conditions. Data such as these show the manner in which the CIE 1931 chromaticity diagram deviates from a condition of visual uniformity for chromaticity.

In 1942 MacAdam published results of his first study of chromaticity thresholds based on the method of average error (MacAdam, 1942). A single observer made many replicate color matches of each of a large number of stimuli. Assuming bivariate normality of the distribution of match points, MacAdam constructed ellipses that represent the variability about each of the stimulus chromaticities examined. He determined that these ellipses

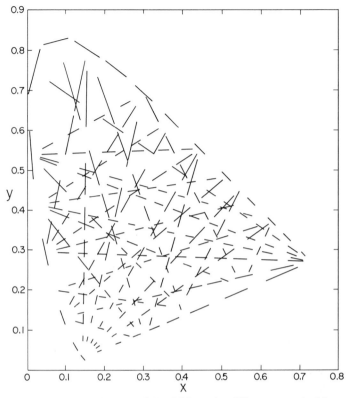

Fig. 15. Chromaticity differences of 2° additive color differences required to produce a perceptually just noticeable difference in color as measured by Wright (1941). Each line is approximately three times enlarged.

represented about $\frac{1}{3}$ of a JND. They are plotted in Fig. 16 on a scale of 10 times the radial size found by MacAdam.

Subsequently, color matching ellipsoids were determined for variability in both chromaticity and luminance of matches made by two observers (Brown and MacAdam, 1949) and average ellipsoids were determined for color matching by 12 observers (Brown, 1957). Cross sections of those ellipsoids, enlarged 10 times for clarity, are illustrated in Fig. 17. Most recently, Wyszecki and Fielder (1971) have determined such ellipsoids for three observers. Cross sections of those ellipsoids are shown in Fig. 18, where the ellipse axes are five times actual size.

The data illustrated in Figs. 15–18 represent virtually the entire extent of chromaticity threshold information that has been gathered to date. The

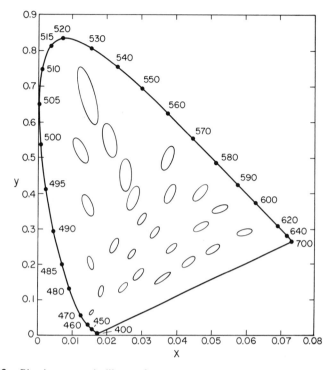

Fig. 16. Bivariate normal ellipses of chromaticity matching determined by MacAdam (1942). Each ellipse axis is approximately $\frac{10}{3}$ the size of a just noticeable difference.

directions and extents of differences in chromaticity that represent equal visual color differences, according to the Fechnerian assumption that JNDs are visually equal, do not differ greatly among the data shown in these four figures. Pointer (1974) has shown that the lines of threshold chromaticity difference remain largely unchanged for variations in color temperature of illumination to which an observer is adapted. However, Wyszecki and Fielder (1971) published results of careful experimentation which indicated that the sizes and orientations of the color-matching ellipses vary significantly among observers and also for an individual over repeated sessions. Those experimenters also determined ellipsoids for direct matching of color *differences*. Although the correlation between orientations of the color-matching and color-difference matching elliptical sections was reasonable on average, the axes of the color-matching ellipses did not sum to suitable predictions of the small color differences found by direct matching (Wyszecki, 1972).

The point, then, is that there are not many data on which to base scales of chromaticity thresholds and there is some question about the reliability

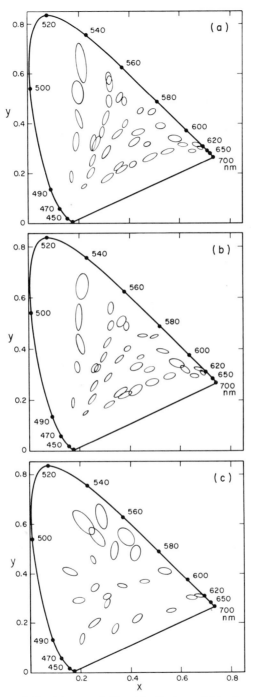

Fig. 17. Sections of trivariate normal ellipsoids of color matching for observers (a) W. R. J. B. and (b) D. L. M. determined by Brown and MacAdam (1949) and (c) averages for 12 observers determined by Brown (1957). Each section is expanded linearly by a factor of ten.

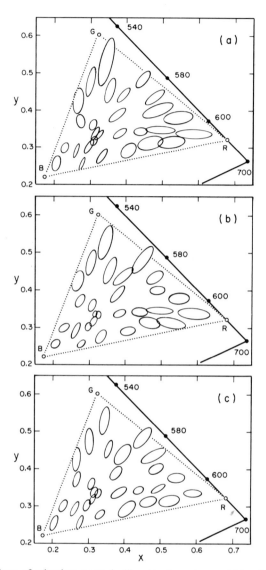

Fig. 18. Sections of trivariate normal ellipsoids for color matching for observers (a) G. W., (b) G. F., and (c) A. R. determined by Wyszecki and Fielder (1971). Each section is expanded linearly by a factor of 5.

of those data that are at hand in the sense of their being general and accurate predicters of small, equisized, color differences. Nonetheless, the data have been used both to determine and to evaluate transformations intended to yield uniform chromaticity-scale diagrams. Generally two kinds of approaches have been used: theoretical and empirical. The theoretical approach assumes certain things about color vision and represents those assumptions and their consequences with a mathematical model. The empirical approach simply attempts to determine a mathematical transformation that renders the color differences or ellipses as uniform circles. Both approaches often invoke suprathreshold scaling data as a check on the effectiveness of the transformations. Before discussing these approaches, it will be well to summarize briefly the suprathreshold data on color differences that have been determined by exhaustive scaling experiments.

The most often used scales of color difference are those associated with the Munsell Book of Color. The Munsell system actually represents an atlas of color. It was first introduced in 1905 (Munsell, 1905) and has undergone two modifications since that time. The most extensive modification, and the one presently used, is referred to as the "Munsell renotation" (Newhall *et al.*, 1943). The renotation work was carried out by the Optical Society of America subcommittee on the spacing of Munsell colors. It involved over three million experimental judgments of color relations and is generally considered to be the most reliable set of estimates of suprathreshold differences in color appearance for opaque object colors. The Munsell renotation system is organized along three dimensions called "value," "hue," and "chroma." These correspond, respectively, to the perceptual dimensions or attributes of "lightness," "hue," and "relative colorfulness" (viz., "chroma"). Munsell hues and chromas at any one level of value may be expressed as CIE 1931 chromaticity coordinates. When this is done for value 5 (which corresponds to a luminance factor of roughly 0.2), the hue and chroma contours of Fig. 19 result. All points on that diagram lie on the plane of luminance factor equal to 0.1977 and all are assigned Munsell value 5. Other such contours correspond to different value levels. In general, the shapes and orientations of the contours remain the same for different luminance factor planes, but the radial extent of the chroma contours expands at lower luminance factors and contracts for higher ones.

A uniform-chromaticity-scale transformation that renders distances between chromaticities of fixed pairs of Munsell hue and chroma locations equal is usually considered to be successful in its attempt to provide a uniform chromaticity metric. Small differences of this kind are sometimes used to develop metric systems for specifying small- to moderate-sized color differences.

Two other color spaces determined by direct scaling should also be

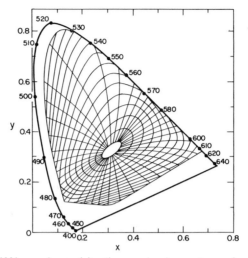

Fig. 19. CIE 1931 x, y chromaticity diagram showing contours of constant Munsell hue (roughly radial lines) and constant Munsell chroma (roughly concentric circles) for the plane of Munsell value 5 ($\beta_v = 0.1977$). All chromaticities are with respect to CIE Illuminant C.

mentioned. The first is the Natural Colour System developed in Sweden (Hård, 1970; Steen, 1970). This color space was developed by scaling color appearances directly and may not necessarily relate to color differences in a simple way. However, it has been compared quantitatively with both the Munsell system (Judd and Nickerson, 1975) and the CIE color spaces (Tonnquist, 1975). Both the Munsell and Natural Colour Systems are related to CIE specifications in complex ways. However, a uniform color difference space developed by a committee of the Optical Society of America is related to CIE specifications in an explicit manner. That uniform space was determined from scaling comparisons of the color differences exhibited by object colors forming polyhedra of 12 points about each of a large number of central stimulus points throughout color space. All the scaled color differences were analyzed and a three-dimensional space was defined as a transformation of CIE 1931 x, y, Y specifications such that the new space represents equality of suprathreshold color difference in all directions about any centroid point for the same numerical metric distance. The coordinates of the OSA uniform color space are $\mathscr{L}, g, j$. $\mathscr{L}$ represents lightness and is defined as

$$\mathscr{L} = 5.9(Y_0^{1/3} - \tfrac{2}{3}) + 0.042(Y_0 - 30)^{1/3} \tag{55}$$

where Y_0 is determined from CIE 1931 x, y, Y coordinates as

$$Y_0 = Y(4.4934x^2 + 4.3034y^2 - 4.276xy - 1.3744x$$
$$- 2.5643y + 1.8103) \qquad (56)$$

The chromaticness coordinates g, j (representing greenness and yellowness, with j standing for the French *jaune*, to avoid confusion with the chromaticity coordinate y) are defined in terms of a set of fundamental primaries as follows:

$$g = C(-13.7R^{1/3} + 17.7G^{1/3} - 4.0B^{1/3})$$
$$j = C(1.7R^{1/3} + 8.0G^{1/3} - 9.7B^{1/3}) \qquad (57)$$

where

$$C = \frac{\mathscr{L}}{5.9(Y_0^{1/3} - \frac{2}{3})},$$
$$R = 0.799X + 0.4194Y - 0.1648Z,$$
$$G = -0.4493X + 1.3265Y + 0.0927Z,$$
$$B = -0.1149X + 0.3394Y + 0.717Z.$$

Equations (55)–(57) are as set forth by MacAdam (1974). The OSA uniform color space has been compared with both CIE (Wyszecki, 1975) and Munsell (Nickerson, 1975) spaces.

Since the Munsell spacing is the oldest of these three sets of scaled data, it has been used in the overwhelming majority of cases where suprathreshold color difference data are invoked to derive or evaluate uniform chromaticity transforms.

Theoretical approaches to specifying color differences generally have tended to follow the example set down by Hermann Ludwig Ferdinand von Helmholtz (1891, 1892, 1896). Helmholtz was a firm adherent to the concept of trichromacy of color vision as proposed by Sir Thomas Young (1820a,b). In fact, Helmholtz's contributions were so extensive that the trichromatic theory has come to be known as the Young–Helmholtz theory of color vision. The specific question that concerned Helmholtz was whether or not hue could be discriminated on the basis of gradations in intensity of three fundamental sensitivity processes (R, G, B). If so, each of the three processes could be treated in the manner of Weber. That is, a Weber fraction could be assumed to represent each of the processes at threshold. Distance in an n-dimensional space could then be represented by combinations of JNDs of each of the processes. He did not assume that n was necessarily equal to 3.

Instead, he adopted the general metric for distance that had been proposed by Riemann for multiply extended manifolds. If the threshold difference is represented by ds and minimal sensation steps of each of three chromatic processes are represented as ds_1, ds_2, and ds_3, then the JND (ds) is represented in a Riemannian metric as the following definite and positive quadratic expression:

$$ds = [g_{11}(ds_1)^2 + g_{22}(ds_2)^2 + g_{33}(ds_3)^2$$
$$+ 2g_{12}ds_1 ds_2 + 2g_{23}ds_2 ds_3 + 2g_{31}ds_3 ds_1]^{1/2} \qquad (58)$$

If, as Helmholtz assumed, the processes are independent, then $g_{11} = g_{22} = g_{33} = 1$ and $g_{12} = g_{23} = g_{31} = 0$. This special case represents a Euclidean three-dimensional space in which ds is represented simply as

$$ds = [(ds_1)^2 + (ds_2)^2 + (ds_3)^2]^{1/2} \qquad (59)$$

Equations (58) and (59) are usually called "line elements." Since ds_i represent Weber fractions for each of the trichromatic processes, we may write Helmholtz's line element for normally high intensities (where the Weber fraction is, in fact, essentially constant) as follows:

$$3(ds)^2 = \left(\frac{dR}{R}\right)^2 + \left(\frac{dG}{G}\right)^2 + \left(\frac{dB}{B}\right)^2 \qquad (60)$$

Helmholtz tested his line element by predicting JND's for wavelength discrimination and comparing them with the experimental data of König and Dieterici (1884), which are illustrated in Fig. 13. He did get reasonably satisfactory agreement, but to do so he had to assume a set of fundamental primaries that were unreasonably different from those based on physiological considerations of his time (König and Dieterici, 1892). Schrödinger (1920) later pointed out that Helmholtz's line element would also predict a luminous efficiency function, what is now known as the $V(\lambda)$ curve, with two maxima rather than one. He then proposed a modification to Helmholtz's line element that would correct many of the original deficiencies. He introduced the assumption that the minimum integral of JNDs ($\int ds \rightarrow$ minimum) between two samples of different chromaticity defines equality of brightness when any change in luminance increases the value of the integral. Using the fundamental sensitivity derived by König and Dieterici, Schrödinger's brightness is then proportional to $a_1 R + a_2 G + a_3 B$, and his line element at normally high intensities becomes

$$(ds)^2 = \left(\frac{1}{a_1 R + a_2 G + a_3 B}\right)\left(\frac{a_1\, dR^2}{R} + \frac{a_2\, dG^2}{G} + \frac{a_3\, dB^2}{B}\right). \qquad (61)$$

Since Schrödinger's line element also fails to predict some relevant forms of experimental data (e.g., Bouma and Heller, 1935), others have continued to develop improved line elements. Perhaps the most significant modification was made by Stiles (1946); on the basis of extensive experimentation with two-color thresholds, he determined that the Weber fractions for each of the three processes were not equal, as Helmholtz had assumed, but stood in the ratio of about 0.8:1:4.5 for R, G, and B, respectively. Thus, at normally high levels of intensity, Stiles's line element reduces to

$$(ds)^2 = \left(\frac{dR}{0.8R}\right)^2 + \left(\frac{dG}{G}\right)^2 + \left(\frac{dB}{4.5B}\right)^2. \tag{62}$$

Stiles' line element has been tested in a number of ways and found to be generally useful in predicting experimental threshold data. By way of example, Fig. 20 shows the ellipses (three times enlarged) predicted by Stiles'

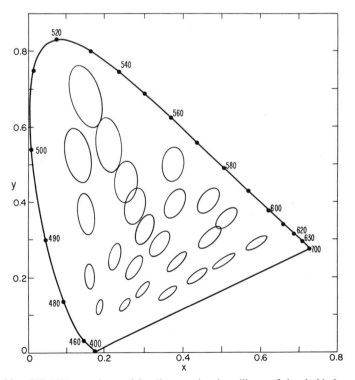

Fig. 20.   CIE 1931 x, y chromaticity diagram showing ellipses of threshold chromaticity differences as computed from Stiles' line element equation (Stiles, 1946) for the same centers as MacAdam's (1942) color-matching data. Each ellipse is enlarged by a linear factor of 3.

line element for the same centers as MacAdam's 1942 data, which are illustrated in Fig. 16. The high degree of similarity between shapes and orientations of the experimental threshold ellipses and those predicted from the line element suggest that such theoretical treatments may have considerable utility.

Each of Eqs. (60)–(62) assumes a kind of independence of the three chromatic processes. That is, none of the line element equations includes cross products as shown in the definite positive quadratic of Eq. (58). This assumption is a key one and one that has been tested by Stiles (Boynton *et al.*, 1964; Stiles, 1967). The details of this test and implications of the results need not concern us here, but the outcome indicates that the processes are *not* independent as assumed. Despite this finding, others have continued to explore and develop line element approaches to predicting threshold color differences. Notable among those who have are Vos and Walraven (1971, 1972a,b); they have paid particular attention to the critical choice of fundamental primaries and color vision model used in their line element. Others have developed formulae for predicting perceptual sizes of color differences and these formulae have had significant impact on the industrial practice of colorimetry.

Friele (1961, 1965, 1966a,b, 1967) developed a line element that assumes that the R, G, B processes do not have equal Weber fractions and are followed by a neural (or "zone") stage of antagonistic chromatic response not unlike that of the Müller zone theory of color vision (Müller, 1930). Friele attempted to make his line element optimum for predicting the ellipses of MacAdam and Brown-MacAdam discussed earlier, but he also included data from commercial color tolerance ("acceptability") ellipses (Davidson and Friede, 1953). It is questionable that acceptability data are unbiased in their relation to perceptibility. In part for this reason, MacAdam (1964a,b, 1965a,b, 1966) determined the metric coefficients of Friele's line element such as to make only the prediction of the color-matching ellipses optimum. Chickering (1967) further refined the coefficients by making the predictions optimum for only the MacAdam 1942 ellipses. The resultant formula is referred to as the Friele–MacAdam–Chickering formula No. 1, abbreviated FMC-1. It was one of a number of color difference formulae proposed for field testing by the CIE between 1967 and 1971. However, as experience was gained with its use in industry, it became obvious to Chickering that additional factors were required to alter the sizes of ellipses as a function of luminance factor. Accordingly, still another modification was made (Chickering, 1971), and this formula became known as the FMC-2 metric.

A color difference ΔE is computed from the FMC-2 metric as follows:

$$\Delta E_{\text{FMC-2}} = [(\Delta C)^2 + (\Delta L)^2]^{1/2}, \tag{63}$$

where

$$\Delta C = K_1 \Delta C_1$$

$$\Delta C_1 = \left[\left(\frac{\Delta C_{rg}}{a} \right)^2 + \left(\frac{\Delta C_{yb}}{b} \right)^2 \right]^{1/2},$$

$$\Delta C_{rg} = \frac{Q \Delta P - P \Delta Q}{(P^2 + Q^2)^{1/2}},$$

$$\Delta C_{yb} = \left[\frac{S \Delta L_1}{(P^2 + Q^2)^{1/2}} \right] - S,$$

$$\Delta L = K_2 \Delta L_2,$$

$$\Delta L_1 = \frac{(P \Delta P + Q \Delta Q)}{(P^2 + Q^2)^{1/2}},$$

$$\Delta L_2 = \frac{0.279 \Delta L_1}{a},$$

$$K_1 = 0.55669 + 0.049434Y - 0.82575 \times 10^{-3}Y^2$$
$$+ 0.79172 \times 10^{-5}Y^3 - 0.30087 \times 10^{-7}Y^4,$$

$$K_2 = 0.17548 + 0.027556Y - 0.57262 \times 10^{-3}Y^2$$
$$+ 0.63893 \times 10^{-5}Y^3 - 0.26731 \times 10^{-7}Y^4,$$

$$a = \left(\frac{17.3 \times 10^{-6}(P^2 + Q^2)}{[1 + 2.73P^2Q^2/(P^4 + Q^4)]} \right)^{1/2},$$

$$b = [3.098 \times 10^{-4}(S^2 + 0.2015Y^2)]^{1/2},$$

$$P = 0.724X + 0.383Y - 0.098Z,$$

$$Q = -0.48X + 1.37Y + 0.1276Z,$$

$$S = 0.686Z,$$

and X, Y, Z are the CIE 1931 tristimulus values.

This equation serves to demonstrate the lengths of complexity to which color difference metrics have been stretched. It is doubtful, from work done by Wyszecki and Fielder (1971), that such precision of metric coefficients is justified. Nonetheless, Eq. (63) has been used widely in industry, as, in fact, have a number of other color difference formulae approaching the complexity of that one. Work along the lines of developing new line element based color difference equations still continues (e.g., Friele, 1978).

However, a parallel path of development has been that of the empirical approach. Rather than attempt to formulate a model of color vision that adequately predicts threshold color differences, this approach simply tries to "fit" equations to threshold and suprathreshold data in order to provide an operationally useful measure of color difference. As we have seen, Judd (1935, 1936) was among the first to use this approach. Using color threshold data then available, Judd derived a projective transformation of the type

$$a = \frac{c_{11}x + c_{12}y + c_{13}}{c_{31}x + c_{32}y + c_{33}},$$

$$b = \frac{c_{21}x + c_{22}y + c_{23}}{c_{31}x + c_{32}y + c_{33}}.$$

The coefficients c_{ij} were chosen to fit color-difference data arrayed on a two-dimensional chromaticity diagram; hence a and b become transform chromaticity coordinates. Judd's work in this determination is important not only for his use of the transform diagram to derive isotemperature lines of correlated color temperature but also because his transformation, after modification by Hunter (1942), became the NBS (National Bureau of Standards) recommended method of designating color differences.

The NBS method was used for quite a few years, particularly in American industry, and is expressed as follows:

$$\Delta E_{\text{NBS}} = G([22(\overline{Y})^{1/4}\{(\Delta a)^2 + (\Delta b)^2\}^{1/2}]^2 + \{k(\Delta Y^{1/2})\}^2)^{1/2} \qquad (64)$$

where

$$\overline{Y} = (Y_1 + Y_2)/2, \qquad \Delta Y^{1/2} = Y_1^{1/2} - Y_2^{1/2},$$

$$\Delta a = a_1 - a_2, \qquad \Delta b = b_1 - b_2,$$

$$a = \frac{2.4266x - 1.3631y - 0.3214}{x + 2.2633y + 1.1054},$$

$$b = \frac{0.5710x + 1.2447y - 0.5708}{x + 2.2633y + 1.1054}.$$

G represents a "gloss factor" and is taken as $Y/(Y + a)$, where a varies with ambient illumination conditions. A representative value of a is 2.5.

Another color difference formula that has been used frequently is referred to as the "Adams–Nickerson" equation. Nickerson and Stultz (1944) developed the equation for measuring amounts of dye fading. They used distances in the Adams (1942) "chromatic-value" diagram, one of the many uniform-chromaticity diagrams that have been proposed. Adams's transformation involved essentially treating CIE tristimulus values X, Y, Z by the same function used to determine Munsell value (V) from the tristimulus value Y. A number of constants have been proposed for the Adams–Nickerson formula but those most frequently used in recent years are included in Eq. (65):

$$\Delta E_{AN40} = [(\Delta L_{AN})^2 + (\Delta A)^2 + (\Delta B)^2]^{1/2}, \qquad (65)$$

where

$$L_{AN} = 9.2V_y,$$

$$A = 40(V_x - V_y), \qquad B = 16(V_y - V_z),$$

and

$$100X/X_n = 1.2219V_x - 0.23111(V_x)^2 + 0.23951(V_x)^3$$
$$- 0.021009(V_x)^4 + 0.0008404(V_x)^5,$$

and $100Y/Y_n$, $100Z/Z_n$ are identical functions of V_y and V_z.

Equation (65) has been recommended by the International Standards Organization (ISO) for measurement of color differences among textile products (McLaren and Coates, 1972). This and many other color difference equations have been used over the years. During the 1950s over a dozen such formulae were in common use. Since each formula implies a different color space, it is impossible to compare them with precision. Obviously, there has been a need for standardization. However, it has been difficult to obtain agreement on which of the many color difference formulae should be recommended by the CIE; preferences for different formulae have been strong among several industries. An attempt was made by the CIE Committee on Colorimetry to standardize on a single formula in 1963, but that recommendation did not meet with general acceptance. Accordingly, two additional color difference formulae were recommended in 1976. As with all CIE recommendations, these are subject to change at some later date, should conditions warrant such action. These recommendations will be summarized in the next section.

XV.   CIE UNIFORM SPACES AND COLOR DIFFERENCE FORMULAE

The first uniform-chromaticity-scale diagram to be recommended by the CIE was based on a diagram originally proposed by MacAdam (1937). This diagram was soon extended to three dimensions according to a proposal of Wyszecki (1963). The UCS diagram is referred to as the CIE 1960 u, v diagram and the color space as the CIE 1964 $U^*V^*W^*$ uniform color space. U^*, V^*, W^* are defined as follows:

$$U^* = 13W^*(u - u_n),$$
$$V^* = 13W^*(v - v_n),$$
$$W^* = 25(Y)^{1/3} - 17,$$

(66)

where W^* simulates the Munsell value function and applies to relative luminances ($100Y/Y_n$) between about 1 and 100. The UCS coordinates u, v are defined as

$$u = \frac{4X}{(X + 15Y + 3Z)},$$
$$v = \frac{6Y}{(X + 15Y + 3Z)},$$

(67)

and the subscript n refers to the reference illuminant or object.

Figure 21 illustrates the CIE 1960 u, v diagram. Stiles' ellipses, from Fig. 20, have been plotted in Fig. 21 as well so that they may be compared with the shapes in the 1931 x, y diagram. The ellipses become more nearly circular in shape, but they are still not arrayed as circles of equal radial extent.

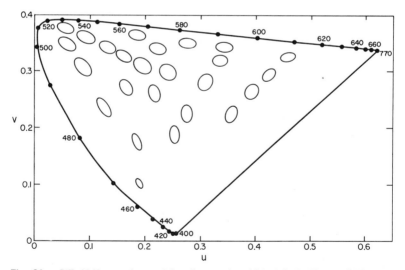

Fig. 21. CIE 1960 u, v chromaticity diagram in which Stiles' ellipses of Fig. 20 have been plotted. The CIE UCS diagram renders the ellipses as more nearly circles than does the CIE 1931 x, y chromaticity diagram.

The CIE 1964 color difference equation consists of Euclidean distances in the U^*, V^*, W^* color space:

$$\Delta E_{\text{CIE}1964} = [(\Delta U^*)^2 + (\Delta V^*)^2 + (\Delta W^*)^2]^{1/2}, \tag{68}$$

where

$$\Delta U^* = U_1^* - U_2^*, \qquad \Delta V^* = V_1^* - V_2^*, \qquad \Delta W^* = W_1^* - W_2^*.$$

This measure is intended to apply to color differences between samples of the same size and shape viewed against identical white- to gray-appearing backgrounds illuminated with a source having chromaticity coordinates not too different from that of CIE illuminant D_{65}.

Equation (68) is still part of the CIE method for assessing the color-rendering properties of illuminants. However, it did not meet with general acceptance in commerce and industry. A number of workers pointed out that the recommended color difference formula did not bring about the desired improvements in predicting the perceptual sizes of color differences in many practical applications. Accordingly, the CIE continued its search for a useful metric for expressing color differences. Other formulae were tested, and in 1976 two color difference equations and underlying color spaces were recommended.

The first of these is a modification to the CIE 1964 uniform color space. It is identified as the CIE 1976 L^*, u^*, v^* space and is produced by plotting in rectangular coordinates the values of L^*, u^*, v^* defined as

$$L^* = 116(Y/Y_n)^{1/3} - 16, \quad u^* = 13L^*(u' - u_n'), \quad v^* = 13L^*(v' - v_n'), \tag{69}$$

where

$$u' = \frac{4X}{X + 15Y + 3Z}, \qquad v' = \frac{9Y}{X + 15Y + 3Z}.$$

It will be seen from this that the CIE 1976 UCS diagram, represented by u', v', is the same as the CIE 1960 UCS diagram except that $v' = 1.5v$. Figure 22 illustrates the u', v' diagram.

Color differences are then calculated in the L^*, u^*, v^* space as follows:

$$\Delta E_{uv} = [(\Delta L^*)^2 + (\Delta u^*)^2 + (\Delta v^*)^2]^{1/2} \tag{70}$$

The abbreviation "CIELUV" is recommended to identify the color space of Eq. (69) and the color difference formula of Eq. (70). For constant values of L^*, the u', v' UCS diagram is a projective transformation of the CIE 1931 x, y diagram. Straight lines in the x, y diagram remain straight in the u', v' diagram. This feature is considered important in cases where additive mixtures of lights are involved. L^* is only a minor modification of the 1964

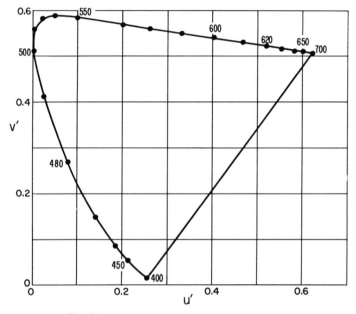

Fig. 22.   CIE 1976 u', v' chromaticity diagram.

W^* function. The 1964 coefficient $25 \times 100^{1/3} = 116.04$, or approximately 116 as used in the 1976 formula of Eq. (69). The constant -16 is used (rather than -17) so that the L^* value of a white reference object is exactly equal to 100.

It is sometimes desirable to identify the components of a color difference as correlates of perceived hue, colorfulness, and lightness. L^* is used for the latter. Relative colorfulness or chroma is then defined from Eq. (70) as

$$C_{uv}^* = (u^{*2} + v^{*2})^{1/2}, \tag{71}$$

and the relative colorfulness attribute called saturation is defined as

$$s_{uv} = 13[(u^*)^2 + (v^*)^2]^{1/2}, \tag{72}$$

or

$$s_{uv} = C_{uv}^*/L^*. \tag{73}$$

Hue angle is

$$h_{uv} = \arctan(v^*/u^*), \tag{74}$$

and hue difference is

$$\Delta H_{uv}^* = [(\Delta E_{uv}^*)^2 - (\Delta L^*)^2 - (\Delta C^*)^2]^{1/2}. \tag{75}$$

When calculating L^* values for Y/Y_n greater than 0.008856, the expression of Eq. (69) should be used:

$$L^* = 116(Y/Y_n)^{1/3} - 16, \tag{76}$$

but L^* for Y/Y_n less than 0.008856 may be calculated as

$$L^* = 903.3(Y/Y_n). \tag{77}$$

The second CIE 1976 color space is based on the Adams–Nickerson equations. Its coordinates L^*, a^*, b^* are

$$L^* = 116(Y'/Y_n)^{1/3} - 16,$$
$$a^* = 500[(X/X_n)^{1/3} - (Y/Y_n)^{1/3}], \tag{78}$$
$$b^* = 200[(Y/Y_n)^{1/3} - (Z/Z_n)^{1/3}],$$

where X/X_n, Y/Y_n, Z/Z_n are greater than 0.01. The abbreviation "CIELAB" is recommended to identify this space. Unlike the CIELUV space, there is no chromaticity diagram associated with CIELAB because values of a^* and b^* depend on L^*. If L^* is constant, straight lines in the CIE 1931 x, y diagram become curved lines, in general, in the a^*, b^* coordinate system.

As with the CIELUV relations, there is a color difference formula for the CIELAB space. It is defined as

$$\Delta E_{ab}^* = [(\Delta L^*)^2 + (\Delta a^*)^2 + (\Delta b^*)^2]^{1/2}. \tag{79}$$

And the corresponding components of color difference are analogous to those of the CIELUV space.

Chroma is defined as

$$C_{ab}^* = (a^{*2} + b^{*2})^{1/2}. \tag{80}$$

Saturation cannot be given for CIELAB since there is no constant luminance diagram of chromaticities. However, hue angle and hue difference may be calculated according to Eqs. (81) and (82):

$$h_{ab} = \arctan(b^*/a^*), \tag{81}$$
$$\Delta H_{ab}^* = [(\Delta E_{ab}^*)^2 - (\Delta L^*)^2 - (\Delta C_{ab}^*)^2]^{1/2}. \tag{82}$$

Hue difference (ΔH_{ab}^* or ΔH_{uv}^*) is defined such that ΔL^*, ΔC^*, ΔH^* squares sum to the square of ΔE^*. Differences in h_{ab} or h_{uv} do not have this property. However, for small color differences sufficiently removed from the neutral axis, ΔH^* may be reasonably estimated from Δh as $\Delta H^* \cong C^* \Delta h(\pi/180)$.

Despite the fact that the CIE has recommended these color difference equations and associated color spaces, the problem of deriving suitably uniform metrics for perceived small extents of color is not considered solved. The CIE continues to carry out coordinated research on color difference evaluation (Robertson, 1978). In addition, the CIE continues to explore methods for expressing large color differences or ratios and for accounting for the influence on color appearance of variations in chromatic adaptation. Brief mention should be made of the present status of work in these two areas, and that will be done in the following two sections.

XVI. CHROMATIC ADAPTATION

The CIE has defined "adaptation" as the process by which the properties of the visual mechanism are modified by observation of stimuli having various spectral power distributions, and "chromatic adaptation" refers to such adaptation when emphasis is on the variation in relative spectral power. Essentially, adaptation is a process whereby the sensitivities of fundamental visual response mechanisms are altered by exposure to light. The study of chromatic adaptation consists of attempts to determine the manner in which these alterations in sensititivities take place. Many such studies have been conducted over the years and they are reviewed in a number of publications (Jameson and Hurvich, 1972; Wyszecki and Stiles, 1967; Terstiege, 1972; Bartleson, 1978).

The first to propose a method for characterizing chromatic adaptation was Johannes Adolph von Kries (1877, 1904, 1905, 1911). He postulated three independent fundamental sensitivity mechanisms that remained invariant in relative spectral distribution but that could be altered in absolute level by adaptation to light flux. This leads to a simple coefficient expression of change in sensitivity with variation in spectral power:

$$R' = a_r R, \qquad G' = a_g G, \qquad B' = a_b B, \tag{83}$$

where R', G', B' represent altered sensitivities and R, G, B are the original or rest-state sensitivities. Two basic questions arise from Eq. (83): (1) "Is the adaptation process linearly proportional to variations in light flux as manifest in the fundamental sensitivities?" and (2) "What is the nature of these fundamental sensitivities?"

We know from the laws of color matching that a minimum of three fundamental sensitivities will suffice to represent the visual color mechanism. However, there are theoretically an infinite number of choices for these fundamentals. That is, many different sets of sensitivities will satisfy the requirements for trichromacy and color matching. Accordingly, the exact nature of the spectral sensitivity distributions to be used as "fundamentals" has been a matter of conjecture and controversy. Some candidates have been determined by considerations of responses of "color blind" people (who are assumed to have only two of the three mechanisms operating) in comparison to normal trichromatic responses (e.g., König and Dieterici, 1892). Others have been deduced from two-color threshold data (e.g., Pugh and Sigel, 1978). Some have been derived from considerations of a number of facts, including color matching, line element theory, and microspectro-photometric measurements of cone receptor photopigments (e.g., Vos, 1978). Three such fundamental functions are illustrated in Fig. 23. Those of König-Dieterici and of Vos are linearly related to the CIE spectral tristimulus values $\bar{x}$, $\bar{y}$, $\bar{z}$. They are tabulated in Table XI. The König–Dieterici fundamentals are related to $\bar{x}$, $\bar{y}$, $\bar{z}$ as follows:

$$r = 0.0713\bar{x} + 0.9625\bar{y} - 0.0147\bar{z},$$

$$g = -0.3952\bar{x} + 1.1668\bar{y} + 0.0815\bar{z}, \tag{84}$$

$$b = 0.5610\bar{z},$$

and those proposed by Vos have the following relationship:

$$\bar{r} = 0.1551646\bar{x}' + 0.5430763\bar{y}' - 0.0370161\bar{z}',$$

$$\bar{g} = -0.1551646\bar{x}' + 0.4569237\bar{y}' + 0.0296946\bar{z}', \tag{85}$$

$$\bar{b} = 0.0073215\bar{z}',$$

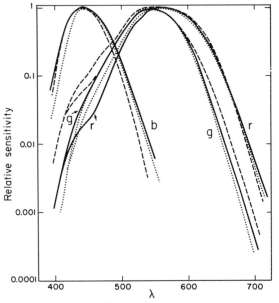

Fig. 23.   Relative spectral sensitivity of fundamental visual color mechanisms r, g, b as proposed by König and Dieterici $\cdots$, Pugh and Sigel $---$, and Vos———.

TABLE XI. Relative Spectral Sensitivity Distributions of Visual Fundamentals Proposed by König and by Vos

Wavelength (λ)	König			Vos		
	$\bar{r}_\lambda$	$\bar{g}_\lambda$	$\bar{b}_\lambda$	$\bar{r}_\lambda$	$\bar{g}_\lambda$	$\bar{b}_\lambda$
400	0.0004	0.0004	0.0381	0.0015	0.0014	0.1077
10	0.0012	0.0011	0.1164	0.0039	0.0040	0.2849
20	0.0039	0.0042	0.3622	0.0088	0.0104	0.6593
30	0.0110	0.0143	0.7773	0.0136	0.0200	0.9075
40	0.0213	0.0316	0.9801	0.0195	0.0349	1.0000
50	0.0345	0.0559	0.9941	0.0246	0.0519	0.9104
60	0.0539	0.0912	0.9364	0.0327	0.0758	0.7990
70	0.0826	0.1339	0.7223	0.0553	0.1214	0.6890
80	0.1286	0.1907	0.4561	0.0969	0.1830	0.4676
90	0.1956	0.2680	0.2610	0.1591	0.2638	0.2763
500	0.3072	0.3971	0.1526	0.2611	0.3947	0.1635
10	0.4825	0.5961	0.0888	0.4220	0.5944	0.0955
20	0.6867	0.8098	0.0439	0.6163	0.8082	0.0474
30	0.8408	0.9438	0.0237	0.7730	0.9419	0.0256
40	0.9386	1.0000	0.0114	0.8833	0.9977	0.0124
50	0.9884	0.9904	0.0049	0.9538	0.9877	0.0054
60	1.0000	0.9263	0.0022	0.9934	0.9236	0.0025
70	0.9706	0.8090	0.0012	0.9974	0.8074	0.0014
80	0.9027	0.6531	0.0010	0.9646	0.6515	0.0011
90	0.8018	0.4778	0.0006	0.8943	0.4776	0.0008
600	0.6830	0.3166	0.0004	0.7947	0.3180	0.0006
10	0.5556	0.1907	0.0002	0.6704	0.1935	0.0003
20	0.4276	0.1069	0.0001	0.5302	0.1103	0.0002
30	0.3009	0.0553		0.3800	0.0583	0.0001
40	0.2004	0.0272		0.2565	0.0296	
50	0.1232	0.0128		0.1591	0.0144	
60	0.0705	0.0060		0.0914	0.0071	
70	0.0370	0.0028		0.0482	0.0034	
80	0.0197	0.0013		0.0257	0.0017	
90	0.0095	0.0006		0.0124	0.0008	
700	0.0048	0.0003		0.0062	0.0004	
10	0.0024	0.0002		0.0032	0.0002	
20	0.0012			0.0016	0.0001	
30	0.0006			0.0008	0.0001	
40	0.0003			0.0004		
50	0.0001			0.0002		

where

$$\bar{x}' = \frac{0.39564\bar{x} - 0.12666\bar{y} + 0.08853}{-0.54397\bar{x} - 0.51866\bar{y} + 1},$$

$$\bar{y}' = \frac{-0.03833\bar{x} + 0.50787\bar{y} + 0.01056}{-0.54397\bar{x} - 0.51866\bar{y} + 1}.$$

When such fundamentals are used, von Kries' Eq. (83) may be written as follows:

$$R' = \left[\frac{a_{11}X'_n + a_{12}Y'_n + a_{13}Z'_n}{a_{11}X_n + a_{12}Y_n + a_{13}Z_n}\right](a_{11}X + a_{12}Y + a_{13}Z),$$

$$G' = \left[\frac{a_{21}X'_n + a_{22}Y'_n + a_{23}Z'_n}{a_{21}X_n + a_{22}Y_n + a_{23}Z_n}\right](a_{21}X + a_{22}Y + a_{23}Z), \qquad (86)$$

$$B' = \left[\frac{a_{31}X'_n + a_{32}Y'_n + a_{33}Z'_n}{a_{31}X_n + a_{32}Y_n + a_{33}Z_n}\right](a_{31}X + a_{32}Y + a_{33}Z),$$

where the superscript prime refers to the altered state and the subscript n identifies the illuminant chromaticity. In other words, the von Kries coefficients a_i of Eq. (83) may be simply determined as the ratio of fundamental tristimulus values for the two illuminants in question.

Equations (83) and (86) represent a linear proportionate alteration in fundamental sensitivities as the spectral power to which an observer is adapted changes. While such a rule may indicate first-order effects, it has been found generally inadequate to predict the influence of adaptation on color appearances of objects in a manner that has utility in practice. Consequently, a number of alternatives to the von Kries coefficient rule have been proposed (cf. Bartleson, 1978). Many of these substitute a nonlinear process for the linear one of Eq. (83). Bartleson (1979a,b) has proposed a prototypical form for such nonlinear adaptation equations:

$$R' = k_r(a_r R)^{p_r}, \qquad G' = k_g(a_g G)^{p_g}, \qquad B' = k_b(a_b B)^{p_b}. \qquad (87)$$

where

$$R = g_{11}(X) + g_{12}(Y) + g_{13}(Z),$$

$$G = g_{21}(X) + g_{22}(Y) + g_{23}(Z),$$

$$B = g_{31}(X) + g_{32}(Y) + g_{33}(Z);$$

$$a_r = R'_n/R_n,$$

$$a_g = G'_n/G_n,$$

$$a_b = B'_n/B_n;$$

$$p_r = c_{r1}(a_r)^{d_{r1}} + c_{g1}(a_g)^{d_{g1}} + c_{b1}(a_b)^{d_{b1}},$$

$$p_g = c_{r2}(a_r)^{d_{r2}} + c_{g2}(a_g)^{d_{g2}} + c_{b2}(a_b)^{d_{b2}},$$

$$p_b = c_{r3}(a_r)^{d_{r3}} + c_{g3}(a_g)^{d_{g3}} + c_{b3}(a_b)^{d_{b3}};$$

and

$$k_r = (a_r R_n)/(a_r R_n)^{p_r},$$

$$k_g = (a_g G_n)/(a_g G_n)^{p_g},$$

$$k_b = (a_b B_n)/(a_b B_n)^{p_b}.$$

When illuminants such as those recommended by the CIE (A, B, C, D_{55}, D_{65}, D_{75}) are used, Eq. (87) simplifies to linear (von Kries–type) equations for R and G and only B is significantly nonlinear. If the König-type fundamentals of Table XI are chosen, the only parameters beyond the a_i coefficients to be determined are

$$p_b = 0.326(a_r)^{27.45} + 0.325(a_g)^{-3.91} + 0.340(a_b)^{-0.45}, \tag{88}$$

and

$$k_b = (a_b B_n)/(a_b B_n)^{p_b}. \tag{89}$$

Equations (87)–(89) represent only one of many different approaches to characterizing the chromatic adaptation process for colorimetric purposes. It is included here as an alternative to the linear method of Eq. (83). Together, these two methods comprise a reasonable representation of present state-of-the-art techniques for predicting the chromaticities of a sample that, under a new condition of chromatic adaptation, will match in color appearance a sample represented by a given set of chromaticities under some reference adaptation condition. The CIE Committee on Colorimetry formed a Subcommittee on Chromatic Adaptation in 1971, and this subcommittee continues to study the problem. As yet, the CIE has made no direct recommendations for a chromatic adaptation transformation method. However, the CIE Committee on Color Rendering has chosen to use the von Kries linear method of Eq. (83) in making small adjustments to account for differences in illuminants to be compared for color-rendering properties.

XVII.   CIE COLOR-RENDERING METHOD

We have seen that two sources may be metameric. That is, their chromaticities may be the same but their spectral power distributions may differ significantly. Accordingly, two sources with the same color temperature or correlated color temperature will not necessarily yield the same chromaticities for reflecting or transmitting samples with spectrally selective relative radiance distributions. In short, we cannot assume that sources with equal chromaticities will render the same color appearances for a given sample illuminated by each. In an effort to cope with this problem the CIE has developed a method for assessing the color-rendering properties of sources (CIE, 1974). Indices of color-rendering determined by this method are useful supplements to specifications of color temperature or correlated color temperature of light sources.

The method is called the "CIE 1974 Method of Measuring and Specifying Colour Rendering Properties of Light Sources" (CIE, 1974). It is usually shortened to the "CIE 1974 Color-Rendering Method." It supercedes an earlier (1965) method. The purpose of the method is to provide quantitative means for rating the color-rendering properties of general-purpose illuminants. The method consists of a series of "special color-rendering indices" and an average "general color-rendering index." Each of the special color-rendering indices represents CIE 1964 color differences between the chromaticities for a specified sample under the test illuminant and a specified reference illuminant. Small differences between the chromaticities of the test and reference illuminants are normalized by a von Kries–type chromatic adaptation transform. Test illuminants with correlated color temperatures less than 5000 K are compared with a Planckian or full radiator at the nearest color temperature to within 100 K. Test illuminants with correlated color temperatures above 5000 K are compared with the nearest daylight radiator as defined by Eq. (27). A tolerance is set for maximum CIE 1960 chromaticity difference between the test and reference illuminant. It is defined as $\Delta C <$ 5.4×10^{-3}, where

$$\Delta C = [(u_k - u_r)^2 + (v_k - v_r)^2]^{1/2}, \tag{90}$$

where u, v are CIE 1960 chromaticity coordinates, k refers to the test source, and r represents the reference illuminant.

The chromaticities of 14 specified test samples are computed for both test and reference illuminants. Spectral reflectances of these 14 test samples are listed in Table XII. Each of these samples gives rise to a special index (R_i) and the average of the first eight special indices is called the general index (R_a). These indices are calculated from "resultant color shifts" between the two illumination conditions, after normalizing the "adaptive shifts" between illuminant chromaticities. The procedure may be summarized with the following step-by-step guide for computing color-rendering indices:

(1) Determine the CIE 1960 u, v chromaticity coordinates of the test source and the reference illuminant and each of the 14 test samples illuminated by each of the two sources.

TABLE XII. Spectral Reflectance Distributions of 14 Samples Used in Determing CIE Color Rendering Indices

λ	1	2	3	4	5	6	7	8
360	0.116	0.053	0.058	0.057	0.143	0.079	0.150	0.075
365	0.136	0.055	0.059	0.059	0.187	0.081	0.177	0.078
370	0.159	0.059	0.061	0.062	0.233	0.089	0.218	0.084
375	0.190	0.064	0.063	0.067	0.269	0.113	0.293	0.090
380	0.219	0.070	0.065	0.074	0.295	0.151	0.378	0.104
385	0.239	0.079	0.068	0.083	0.306	0.203	0.459	0.129
390	0.252	0.089	0.070	0.093	0.310	0.265	0.524	0.170
395	0.256	0.101	0.072	0.105	0.312	0.339	0.546	0.240
400	0.256	0.111	0.073	0.116	0.313	0.410	0.551	0.319
405	0.254	0.116	0.073	0.121	0.315	0.464	0.555	0.416
410	0.252	0.118	0.074	0.124	0.319	0.492	0.559	0.462
415	0.248	0.120	0.074	0.126	0.322	0.508	0.560	0.482
420	0.244	0.121	0.074	0.128	0.326	0.517	0.561	0.490
425	0.240	0.122	0.073	0.131	0.330	0.524	0.558	0.488
430	0.237	0.122	0.073	0.135	0.334	0.531	0.556	0.482
435	0.232	0.122	0.073	0.139	0.339	0.538	0.551	0.473
440	0.230	0.123	0.073	0.144	0.346	0.544	0.544	0.462
445	0.226	0.124	0.073	0.151	0.352	0.551	0.535	0.450
450	0.225	0.127	0.074	0.161	0.360	0.556	0.522	0.439
455	0.222	0.128	0.075	0.172	0.369	0.556	0.506	0.426
460	0.220	0.131	0.077	0.186	0.381	0.554	0.488	0.413
465	0.218	0.134	0.080	0.205	0.394	0.549	0.469	0.397
470	0.216	0.138	0.085	0.229	0.403	0.541	0.448	0.382
475	0.214	0.143	0.094	0.254	0.410	0.531	0.429	0.366
480	0.214	0.150	0.109	0.281	0.415	0.519	0.408	0.352
485	0.214	0.159	0.126	0.308	0.418	0.504	0.385	0.337
490	0.216	0.174	0.148	0.332	0.419	0.488	0.363	0.325
495	0.218	0.190	0.172	0.352	0.417	0.469	0.341	0.310
500	0.223	0.207	0.198	0.370	0.413	0.450	0.324	0.299
505	0.225	0.225	0.221	0.383	0.409	0.431	0.311	0.289
510	0.226	0.242	0.241	0.390	0.403	0.414	0.301	0.283
515	0.226	0.253	0.260	0.394	0.396	0.395	0.291	0.276
520	0.225	0.260	0.278	0.395	0.389	0.377	0.283	0.270
525	0.225	0.264	0.302	0.392	0.381	0.358	0.273	0.262
530	0.227	0.267	0.339	0.385	0.372	0.341	0.265	0'256
535	0.230	0.269	0.370	0.377	0.363	0.325	0.260	0.251
540	0.236	0.272	0.392	0.367	0.353	0.309	0.257	0.250
545	0.245	0.276	0.399	0.354	0.342	0.293	0.257	0.251
550	0.253	0.282	0.400	0.341	0.331	0.279	0.259	0.254
555	0.262	0.289	0.393	0.327	0.320	0.265	0.260	0.258
560	0.272	0.299	0.380	0.312	0.308	0.253	0.260	0.264
565	0.283	0.309	0.365	0.296	0.296	0.241	0.258	0.269
570	0.298	0.322	0.349	0.280	0.284	0.234	0.256	0.272
575	0.318	0.329	0.332	0.263	0.271	0.227	0.254	0.274

λ	9	10	11	12	13	14
360	0.069	0.042	0.074	0.189	0.071	0.036
365	0.072	0.043	0.079	0.175	0.076	0.036
370	0.073	0.045	0.086	0.158	0.082	0.036
375	0.070	0.047	0.098	0.139	0.090	0.036
380	0.066	0.050	0.111	0.120	0.104	0.036
385	0.062	0.054	0.121	0.103	0.127	0.036
390	0.058	0.059	0.127	0.090	0.161	0.037
395	0.055	0.063	0.129	0.082	0.211	0.038
400	0.052	0.066	0.127	0.076	0.264	0.039
405	0.052	0.067	0.121	0.068	0.313	0.039
410	0.051	0.068	0.116	0.064	0.341	0.040
415	0.050	0.069	0.112	0.065	0.352	0.041
420	0.050	0.069	0.108	0.075	0.359	0.042
425	0.049	0.070	0.105	0.093	0.361	0.042
430	0.048	0.072	0.104	0.123	0.364	0.043
435	0.047	0.073	0.104	0.160	0.365	0.044
440	0.046	0.076	0.105	0.207	0.367	0.044
445	0.044	0.078	0.106	0.256	0.369	0.045
450	0.042	0.083	0.110	0.300	0.372	0.045
455	0.041	0.088	0.115	0.331	0.374	0.046
460	0.038	0.095	0.123	0.346	0.376	0.047
465	0.035	0.103	0.134	0.347	0.379	0.048
470	0.033	0.113	0.148	0.341	0.384	0.050
475	0.031	0.125	0.167	0.328	0.389	0.052
480	0.030	0.142	0.192	0.307	0.397	0.055
485	0.029	0.162	0.219	0.282	0.405	0.057
490	0.028	0.189	0.252	0.257	0.416	0.062
495	0.028	0.219	0.291	0.230	0.429	0.067
500	0.028	0.262	0.325	0.204	0.443	0.075
505	0.029	0.305	0.347	0.178	0.454	0.083
510	0.030	0.365	0.356	0.154	0.461	0.092
515	0.030	0.416	0.353	0.129	0.466	0.100
520	0.031	0.465	0.346	0.109	0.469	0.108
525	0.031	0.509	0.333	0.090	0.471	0.121
530	0.032	0.546	0.314	0.075	0.474	0.133
535	0.032	0.581	0.294	0.062	0.476	0.142
540	0.033	0.610	0.271	0.051	0.483	0.150
545	0.034	0.634	0.248	0.041	0.490	0.154
550	0.035	0.653	0.227	0.035	0.506	0.155
555	0.037	0.666	0.206	0.029	0.526	0.152
560	0.041	0.678	0.188	0.025	0.553	0.147
565	0.044	0.687	0.170	0.022	0.582	0.140
570	0.048	0.693	0.153	0.019	0.618	0.133
575	0.052	0.698	0.138	0.017	0.651	0.125

TABLE XII. (*Continued*)

λ	1	2	3	4	5	6	7	8
580	0.341	0.335	0.315	0.247	0.260	0.225	0.254	0.278
585	0.367	0.339	0.299	0.229	0.247	0.222	0.259	0.284
590	0.390	0.341	0.285	0.214	0.232	0.221	0.270	0.295
595	0.409	0.341	0.272	0.198	0.220	0.220	0.284	0.316
600	0.424	0.342	0.264	0.185	0.210	0.220	0.302	0.348
605	0.435	0.342	0.257	0.175	0.200	0.220	0.324	0.384
610	0.442	0.342	0.252	0.169	0.194	0.220	0.344	0.434
615	0.448	0.341	0.247	0.164	0.189	0.220	0.362	0.482
620	0.450	0.341	0.241	0.160	0.185	0.223	0.377	0.528
625	0.451	0.339	0.235	0.156	0.183	0.227	0.389	0.568
630	0.451	0.339	0.229	0.154	0.180	0.233	0.400	0.604
635	0.451	0.338	0.224	0.152	0.177	0.239	0.410	0.629
640	0.451	0.338	0.220	0.151	0.176	0.244	0.420	0.648
645	0.451	0.337	0.217	0.149	0.175	0.251	0.429	0.663
650	0.450	0.336	0.216	0.148	0.175	0.258	0.438	0.676
655	0.450	0.335	0.216	0.148	0.175	0.263	0.445	0.685
660	0.451	0.334	0.219	0.148	0.175	0.268	0.452	0.693
665	0.451	0.332	0.224	0.149	0.177	0.273	0.457	0.700
670	0.453	0.332	0.230	0.151	0.180	0.278	0.462	0.705
675	0.454	0.331	0.238	0.154	0.183	0.281	0.466	0.709
680	0.455	0.331	0.251	0.158	0.186	0.283	0.468	0.712
685	0.457	0.330	0.269	0.162	0.189	0.286	0.470	0.715
690	0.458	0.329	0.288	0.165	0.192	0.291	0.473	0.717
695	0.460	0.328	0.312	0.168	0.195	0.296	0.477	0.719
700	0.462	0.328	0.340	0.170	0.199	0.302	0.483	0.721
705	0.463	0.327	0.366	0.171	0.200	0.313	0.489	0.720
710	0.464	0.326	0.390	0.170	0.199	0.325	0.496	0.719
715	0.465	0.325	0.412	0.168	0.198	0.338	0.503	0.722
720	0.466	0.324	0.431	0.166	0.196	0.351	0.511	0.725
725	0.466	0.324	0.447	0.164	0.195	0.364	0.518	0.727
730	0.466	0.324	0.460	0.164	0.195	0.376	0.525	0.729
735	0.466	0.323	0.472	0.165	0.196	0.389	0.532	0.730
740	0.467	0.322	0.481	0.168	0.197	0.401	0.539	0.730
745	0.467	0.321	0.488	0.172	0.200	0.413	0.546	0.730
750	0.467	0.320	0.493	0.177	0.203	0.425	0.553	0.730
755	0.467	0.318	0.497	0.181	0.205	0.436	0.559	0.730
760	0.467	0.316	0.500	0.185	0.208	0.447	0.565	0.730
765	0.467	0.315	0.502	0.189	0.212	0.458	0.570	0.730
770	0.467	0.315	0.505	0.192	0.215	0.469	0.575	0.730
775	0.467	0.314	0.510	0.194	0.217	0.477	0.578	0.730
780	0.467	0.314	0.516	0.197	0.219	0.485	0.581	0.730
785	0.467	0.313	0.520	0.200	0.222	0.493	0.583	0.730
790	0.467	0.313	0.524	0.204	0.226	0.500	0.585	0.731
795	0.466	0.312	0.527	0.210	0.231	0.506	0.587	0.731

λ	9	10	11	12	13	14
580	0.060	0.701	0.125	0.017	0.680	0.118
585	0.076	0.704	0.114	0.017	0.701	0.112
590	0.102	0.705	0.106	0.016	0.717	0.106
595	0.136	0.705	0.100	0.016	0.729	0.101
600	0.190	0.706	0.096	0.016	0.736	0.098
605	0.256	0.707	0.092	0.016	0.742	0.095
610	0.336	0.707	0.090	0.016	0.745	0.093
615	0.418	0.707	0.087	0.016	0.747	0.090
620	0.505	0.709	0.085	0.016	0.749	0.090
625	0.581	0.708	0.082	0.016	0.748	0.087
630	0.641	0.710	0.080	0.018	0.748	0.086
635	0.682	0.711	0.079	0.018	0.748	0.085
640	0.717	0.712	0.078	0.018	0.748	0.084
645	0.740	0.714	0.078	0.018	0.748	0.084
650	0.758	0.716	0.078	0.019	0.748	0.084
655	0.770	0.718	0.078	0.020	0.748	0.084
660	0.781	0.720	0.081	0.023	0.747	0.085
665	0.790	0.722	0.083	0.024	0.747	0.087
670	0.797	0.725	0.088	0.026	0.747	0.092
675	0.803	0.729	0.093	0.030	0.747	0.096
680	0.809	0.731	0.102	0.035	0.747	0.102
685	0.814	0.735	0.112	0.043	0.747	0.110
690	0.819	0.739	0.125	0.056	0.747	0.123
695	0.824	0.742	0.141	0.074	0.746	0.137
700	0.828	0.746	0.161	0.097	0.746	0.152
705	0.830	0.748	0.182	0.128	0.746	0.169
710	0.831	0.749	0.203	0.166	0.745	0.188
715	0.833	0.751	0.223	0.210	0.744	0.207
720	0.835	0.753	0.242	0.257	0.743	0.226
725	0.836	0.754	0.257	0.305	0.744	0.243
730	0.836	0.755	0.270	0.354	0.745	0.260
735	0.837	0.755	0.282	0.401	0.748	0.277
740	0.838	0.755	0.292	0.446	0.750	0.294
745	0.839	0.755	0.302	0.485	0.750	0.310
750	0.839	0.756	0.310	0.520	0.749	0.325
755	0.839	0.757	0.314	0.551	0.748	0.339
760	0.839	0.758	0.317	0.577	0.748	0.353
765	0.839	0.759	0.323	0.599	0.747	0.366
770	0.839	0.759	0.330	0.618	0.747	0.379
775	0.839	0.759	0.334	0.633	0.747	0.390
780	0.839	0.759	0.338	0.645	0.747	0.399
785	0.839	0.759	0.343	0.656	0.746	0.408
790	0.839	0.759	0.348	0.666	0.746	0.416
795	0.839	0.759	0.353	0.674	0.746	0.422

TABLE XII. (*Continued*)

λ	1	2	3	4	5	6	7	8
800	0.466	0.312	0.531	0.218	0.237	0.512	0.588	0.731
805	0.466	0.311	0.535	0.225	0.243	0.517	0.589	0.731
810	0.466	0.311	0.539	0.233	0.249	0.521	0.590	0.731
815	0.466	0.311	0.544	0.243	0.257	0.525	0.590	0.731
820	0.465	0.311	0.548	0.254	0.265	0.529	0.590	0.731
825	0.464	0.311	0.552	0.264	0.273	0.532	0.591	0.731
830	0.464	0.310	0.555	0.274	0.280	0.535	0.592	0.731
405	0.254	0.116	0.073	0.121	0.315	0.464	0.555	0.416
436	0.232	0.122	0.073	0.140	0.341	0.539	0.550	0.471
546	0.247	0.277	0.400	0.352	0.340	0.290	0.257	0.251
578	0.332	0.333	0.321	0.254	0.264	0.226	0.254	0.276
589	0.385	0.340	0.287	0.217	0.234	0.221	0.267	0.292

(2) Normalize the chromaticity differences between test and reference sources. This is done by converting the chromaticity coordinates u, v to u'', v'', representing chromaticities after adaptive color shift, assuming that the standard observer is completely adapted to each of the two sources in question. The u'', v'' coordinates are computed as follows:

$$u''_{k,i} = \frac{10.872 + 0.404(c_r/c_k)c_{k,i} - 4(d_r/d_k)d_{k,i}}{16.518 + 1.481(c_r/c_k)c_{k,i} - (d_r/d_k)d_{k,i}},$$

$$v''_{k,i} = \frac{5.52}{16.518 + 1.481(c_r/c_k)c_{k,i} - (d_r/d_k)d_{k,i}},$$

(91)

where k is the source, r is the reference source, i stands for samples 1 through 14, $c = (4 - u - 10v)/v$, and $d = (1.708v + 0.404 - 1.481u)/v$.

The values $u''_{k,i}$ and $v''_{k,i}$ are the chromaticity coordinates of a test sample i after adaptive shift to normalize any chromaticity differences between test and reference illuminants; for example, $u''_k = u_r$ and $v''_k = v_r$.

(3) Next determine the CIE 1964 tristimulus differences U^*, V^*, W^* for each sample under the two illuminants. These are computed as follows:

Reference Illuminant

$$W^*_{r,i} = 25(100 Y_{r,i})^{1/3} - 17,$$

$$U^*_{r,i} = 13 W^*_{r,i}(u_{r,i} - u_r),$$

$$V^*_{r,i} = 13 W^*_{r,i}(v_{r,i} - v_r);$$

(92)

λ	9	10	11	12	13	14
800	0.839	0.759	0.359	0.680	0.746	0.428
805	0.839	0.759	0.365	0.686	0.745	0.434
810	0.838	0.758	0.372	0.691	0.745	0.439
815	0.837	0.757	0.380	0.694	0.745	0.444
820	0.837	0.757	0.388	0.697	0.745	0.448
825	0.836	0.756	0.396	0.700	0.745	0.451
830	0.836	0.756	0.403	0.702	0.745	0.454
405	0.052	0.067	0.121	0.068	0.313	0.039
436	0.047	0.074	0.104	0.169	0.366	0.044
546	0.034	0.638	0.244	0.040	0.493	0.155
578	0.056	0.700	0.130	0.017	0.668	0.122
589	0.096	0.704	0.107	0.016	0.714	0.107

Test Illuminant

$$W_{k,i}^* = 25(100Y_{k,i})^{1/3} - 17,$$
$$U_{k,i}^* = 13W_{k,i}^*(u_{k,i}'' - u_k''), \qquad (93)$$
$$V_{k,i}^* = 13W_{k,i}^*(v_{k,i}'' - v_k'');$$

where the symbols are the same as in Eq. (91).

(4) The resultant color shift ΔE_i is determined for each of the test samples according to the following:

$$\Delta E_i = [(U_{r,i}^* - U_{k,i}^*)^2 + (V_{r,i}^* - V_{k,i}^*)^2 + (W_{r,i}^* - W_{k,i}^*)^2]^{1/2}. \qquad (94)$$

(5) The special color-rendering indices R_i are then determined for each of the 14 samples as follows:

$$R_i = 100 - 4.6\,\Delta E_i. \qquad (95)$$

R_i should be rounded to the nearest whole number. Values of R_i have been scaled so that 100 represents no difference in color-rendering properties. A value of about 50 typically represents the color-rendering properties of a standard warm white fluorescent lamp as compared with an incandescent lamp.

(6) If a single index is desired, the value of R_a may be calculated as

$$R_a = \frac{1}{8}\sum_{i=1}^{8} R_i. \qquad (96)$$

Note that only the first 8 test samples are used to compute R_a.

The scales of R_i and R_a have a resolution of one unit; that is, they are rounded to a scale precision of 1%. However, uncertainties of determination of R_i and R_a have been found to be up to three units. In addition, threshold color differences may correspond to five units on the scale. Accordingly, it is doubtful that distinctions in color-rendering properties of less than about 5% have very much practical meaning.

Color-rendering indices are not absolute quantities in the sense of implying an equality among sources with equal values of R_a or R_i. Two sources with equally high color-rendering indices but quite different correlated color temperatures will not have the same color-rendering properties. Moreover, their respective reference sources will not have the same color-rendering properties. The reference sources are merely arbitrarily chosen spectral power distributions that are said to have satisfactory color-rendering properties *for their correlated color temperatures*. In this sense, then, color-rendering indices are *relative* figures of merit.

It is recognized that the direction of color differences may be as important as any average index of color rendering. Neither R_i nor R_a account for direction of color difference. It is therefore helpful to plot each of the color differences from Eq. (94) to aid in evaluating the properties of a test source. The CIE continues to consider the question of how best to characterize such differences among light sources. At present, the method outlined here is recommended for use as a comparative method of evaluating color-rendering properties of light sources for general-purpose illumination.

XVIII. MEASURING COLOR APPEARANCES

It will likely have been inferred from the foregoing that there are substantial difficulties in developing methods for measuring and specifying color differences, chromatic adaptation, and measurement applications where those techniques must be used. It is not surprising, therefore, that little or no agreement has been achieved for ways in which to measure and specify the vastly more complicated question of color appearance. As yet the CIE has made no recommendations for these very important methods. Both the CIE Committee on Colorimetry and the Committee on Vision continue to pursue the question. A number of workers have determined that ratios of color sensations may be generally represented by variations on the power function of Brentano [cf. Eq. (7) in the introduction to this chapter]. However, the manner in which the relations should be expressed is far from being agreed on.

Perhaps the largest amount of work has thus far gone into the question of representing perceived brightness and lightness. A brief example will be given here merely to indicate the additional considerations that must attend the choice of such relationships as compared with the simple psychometric quantity L^*. The following is based on a single proposal (Bartleson, 1979b) and is not to be taken as a preferred method. At the present time, there are no preferred methods for measuring color appearances. Nevertheless, let us examine one method for predicting the magnitudes of perceived brightness and lightness of object colors.

Brightness is defined by the CIE as the attribute of a visual sensation according to which an area appears to exhibit more or less light. Brightness is a perceptually absolute quantity; it has an absolute zero but no upper limit. Brightness describes the intensive aspect of all objects, related and unrelated, chromatic and achromatic. Any measure that attempts to represent brightness must take such factors into account. One function that meets this requirement for the factors just alluded to may be expressed quite simply in a manner not unlike the definition of L^*:

$$Q^{**} = a(kL)^{1/3} - b, \tag{97}$$

where Q^{**} stands for a measure of brightness; L is luminance in candelas per square meter; and the constants a, b, and k are as defined in the following.

The constants a and b vary with viewing conditions and with illuminance. Both are power functions of illuminance and may be expressed as a dependence on the luminance of a reference white object:

$$a = q_a(L_n^{-0.164}), \tag{98}$$

$$b = q_b(L_n^{0.215}), \tag{99}$$

where L_n represents the luminance of a reference white. Values of q are determined as functions of induction I as follows:

$$q_a = 35.2069(I)^{-0.1303}, \tag{100}$$

$$q_b = 0.3885(I)^{0.3616} - 0.39. \tag{101}$$

The symbol I stands for an empirical index of the extent to which the brightness of a focal area is changed from what it would be with a dark surround. That is, I represents changes in Q^{**} resulting from response interactions among related colors. I, therefore, depends on spatial and luminance relationships within the field of view. It may be arbitrarily expressed as

$$I = (m_f L_f + m_s L_s)/(m_f L_f), \tag{102}$$

where m_f is the fractional area of the focal sample relative to a circular area of diameter 40°, m_s the fractional area of the surround relative to a circular are of diameter 40°, L_f the luminance of a white reference, and L_s the luminance of a uniform surround or average luminance of a nonuniform surround. The factor k in Eq. (97) is a chromatic adjustment term. It accounts for the fact that samples of equal luminance but different chromaticity are not usually equal in brightness; k has been defined as

$$k = \text{antilog}(gp_e^n) \tag{103}$$

where p_e is the CIE 1931 excitation purity of the focal sample. The constants g and n are functions of "chromatic purity" $\hat{P}$, according to the following relationships:

$$g = 0.2983(\log \hat{P}) - 0.1834, \tag{104}$$

$$n = 1.3268(\log \hat{P}) - 0.0304, \tag{105}$$

and

$$\hat{P} = \frac{(C_1^2 + C_2^2)^{1/2}}{A}, \tag{106}$$

where $C_1 = 1.66R - 2.23G + 0.37B$, $\quad C_2 = 0.34R + 0.06G - 0.71B$, and $A = 0.85R + 0.15G$. R, G, B in this expression are the fundamental tristimulus values of the sample when using König-type fundamentals. R, G, B may be determined from Eq. (84). They are subject to modification with adaptation according to Eq. (87).

Thus, Q^{**} is a function of luminance, illuminance, spatial relationships, chromaticity, induction, and chromatic adaptation. There are other perceptual factors that could be included as well, and some of them have been proposed from time to time. Those included here suffice to illustrate the point that a reasonable predictor of perceived brightness and lightness may be quite complex. Lightness is simply brightness relative to that of a white reference. Thus, we may state L^{**} (for a measure of perceived lightness) as a ratio of brightnesses:

$$L^{**} = Q^{**}/Q_n^{**}. \tag{107}$$

The steps necessary to solve Eq. (107) are considerably more than those required for computing L^* in Eq. (76); an indication of the degree of complexity required to proceed from psychometric measures to ones that correspond more nearly with color appearance.

It will be seen, then, that the CIE has progressed steadily from the most simple to the most difficult concepts of color measurement. Specification of the stimulus conditions necessary for a color match—the condition of equality—was the first method to be subject to CIE recommendation. Over 40 years after they were made, those recommendations are still used more than any others. By invoking concepts of Weber and Fechner, the CIE has tried to provide methods for measuring and specifying small color differences. But while many years of careful work and deliberation have gone into recommendations for specifying small color differences, we see that they have not been completely successful. Only provisional agreement may be said to exist for color difference methods. Perhaps the lack of complete acceptance for those methods stems in large part from the fact that we cannot define a completely uniform visual space over both large and small extents. We see that there is no agreement as yet about how to specify perceived suprathreshold extents of color. The problem of color appearance is still far from solved.

Thus, when we speak of color measurement we should be careful not to attach more significance to a particular measure than is justified. In colorimetry each form of color measurement has an explicit operational definition. When we use such measurements according to those definitions, they become useful tools. When we assume that they tell us more than is implied by the operational definition, they may mislead us.

REFERENCES

Adams, E. Q. (1942). *J. Opt. Soc. Am.* **32**, 168.
Alpern, M., Rushton, W. A. H., and Torii, S. (1970). *J. Physiol.* (*London*) **207**, 463.
Bartleson, C. J. (1978). *In* "Color 77" (F. W. Billmeyer, Jr. and G. Wyszecki, eds.), pp. 63–96. Hilger, London.
Bartleson, C. J. (1979a). *Color Res. Appl.* **4**, 167.
Bartleson, C. J. (1979b). *Color Res. Appl.* **4**, 191.
Bouma, P. J., and Heller, G. (1935). *Proc. K. Ned. Acad. Wet.* **38**, 258.
Boynton, R. M., Ikeda, M., and Stiles, W. S. (1964). *Vision Res.* **4**, 87.
Breckenridge, F. C., and Schaub, W. R. (1939). *J. Opt. Soc. Am.* **29**, 370.
Brentano, F. (1874). "Psychologie vom empirischen Standpunkte." Duncker & Humblot, Leipzig.
Brown, W. R. J., and MacAdam, D. L. (1949). *J. Opt. Soc. Am.* **39**, 808.
Chickering, K. D. (1967). *J. Opt. Soc. Am.* **57**, 537.
Chickering, K. D. (1971). *J. Opt. Soc. Am.* **61**, 118.
CIE (1970). "Principles of Light Measurements." Bureau Central CIE, Paris.
CIE (1971a). "Colorimetry." Bureau Central CIE, Paris.
CIE (1971b). "Special Metamerism Index: Change in Illuminant." Bureau Central CIE, Paris.
CIE (1974). "Method of Measuring and Specifying Colour Rendering Properties of Light Sources," 2nd Ed., Publ. No. 13.2. Bureau Central CIE, Paris.
CIE (1978a). "Recommendations on Uniform Color-Spaces—Color Difference Equations—Psychometric Color Terms." Bureau Central CIE, Paris.
CIE (1978b). "Radiometric and Photometric Characteristics of Materials and their Measurement." Bureau Central CIE, Paris.

CIPM (1968). "Annex II, Conférence Générale des Poids et Mesures, 13th Compte Rendu." Com. Int. Poids Mes., Paris.

Davidson, H. R., and Friede, E. (1953). *J. Opt. Soc. Am.* **43**, 581.

Fechner, G. T. (1860). "Elemente der Psychophysik." Breitkopf & Härtel, Leipzig.

Friele, L. F. C. (1961). *Farbe* **10**, 193.

Friele, L. F. C. (1965). *J. Opt. Soc. Am.* **55**, 1314.

Friele, L. F. C. (1966a). *Farbe* **14**, 192.

Friele, L. F. C. (1966b). *J. Opt. Soc. Am.* **56**, 259.

Friele, L. F. C. (1967). *Proc. Symp. Colour Meas. Ind.*, *Colour Group* (*G.B.*), pp 44–54, London

Friele, L. F. C. (1978). In "Color 77" (F. W. Billmeyer, Jr. and G. Wyszecki, eds.), pp. 529–532. Hilger, London.

Fullerton, G. S., and Cattell, J. McR. (1892). "On the Perception of Small Differences." Univ. Pennsylvania Press, Philadelphia, Pennsylvania.

Grassman, H. (1853). *Poggendorfs Ann. Phys.* **89**, 69.

Grassman, H. (1854). *Philos. Mag.* **7**, 254.

Grum, F., Saunders, S. B., and MacAdam, D. L. (1978). *Color Res. Appl.* **3**, 17.

Guild, J. (1931). *Philos. Trans. R. Soc. London, Ser. A* **230**, 149.

Guild, J. (1932). In "Discussion on Vision," pp. 1–26. Phys. Soc., London.

Hård, A. (1970). In "Color 69" (M. Richter, ed.), Vol. 1, pp. 351–368. Musterschmidt-Verlag, Göttingen.

Helmholtz, H. von (1891). *Z. Psychol. Physiol. Sinnesorg.* **2**, 1.

Helmholtz, H. von (1892). *Z. Psychol. Physiol. Sinnesorg.* **3**, 1.

Helmholtz, H. von (1896). "Handbuch der physiologischen Optik," Band 2. Voss, Hamburg.

Herbart, J. F. (1824). "Psychologie als Wissenschaft," Unger, Königsberg.

Hunter, R. S. (1942). Photoelectric Tristimulus Colorimetry with Three Filters. *Natl. Bur. Stand. (U.S.), Circ.* C429.

Ives, H. E. (1912). *Philos. Mag.* **24**, 845.

Jameson, D., and Hurvich, L. M. (1972). In "Handbook of Sensory Physiology. Vol. VII/4: Visual Psychophysics" (D. Jameson and L. M. Hurvich, eds.), pp. 568–581. Springer-Verlag, Berlin and New York.

Judd, D. B. (1935). *J. Opt. Soc. Am.* **25**, 24.

Judd, D. B. (1936). *J. Opt. Soc. Am.* **26**, 421.

Judd, D. B. (1951). *CIE Proc.* **1**, Part 7, p. 11.

Judd, D. B., and Nickerson, D. (1975). *J. Opt. Soc. Am.* **65**, 85.

Judd, D. B., and Wyszecki, G. (1975). "Color in Business, Science, and Industry," 3rd Ed. Wiley, New York.

Judd, D. B., MacAdam, D. L., and Wyszecki, G. (1964). *J. Opt. Soc. Am.* **54**, 1031.

Kaiser, P. K., Comerford, J. P., and Bodinger, D. B. (1976). *J. Opt. Soc. Am.* **66**, 818.

Kelly, K. L. (1963). *J. Opt. Soc. Am.* **53**, 999.

König, A., and Brodhun, E. (1889). *Zweite Mitt. Sitzungsber. Preuss. Akad. Wiss.* **27**, 641.

König, A., and Dieterici, C. (1884). *Ann. Phys. Chem.* **22**, 589.

König, A., and Dieterici, C. (1892). *Z. Psychol. Physiol. Sinnesorg.* **4**, 231.

Krochmann, J., Özver, Z, and Rattunde, R., (1978). In "Color 77" (F. W. Billmeyer, Jr. and G. Wyszecki, eds.), pp. 289–292. Hilger, London.

MacAdam, D. L. (1937). *J. Opt. Soc. Am.* **27**, 294.

MacAdam, D. L. (1942). *J. Opt. Soc. Am.* **32**, 247.

MacAdam, D. L. (1964a). *J. Opt. Soc. Am.* **54**, 249.

MacAdam, D. L. (1964b). *J. Opt. Soc. Am.* **54**, 1161.

MacAdam, D. L. (1965a). *J. Opt. Soc. Am.* **55**, 91.

MacAdam, D. L. (1965b). *Acta Chromat.* **1**, 147.

MacAdam, D. L. (1966). *J. Opt. Soc. Am.* **56**, 1784.

MacAdam, D. L. (1974). *J. Opt. Soc. Am.* **64**, 1691.

MacAdam, D. L. (1977). *J. Opt. Soc. Am.* **67**, 839.

McLaren, K., and Coates, E. (1972). *J. Soc. Dyers Colour.* **88**, 28.

Merkel, J. (1888). *Philos. Stud.* **4**, 541.

Müller, G. E. (1930). *Z. Psychol. Ergenzungsber.* **17/18**, 1–430, 435–647.

Munsell, A. H. (1905). "A Color Notation." Ellis, Boston, Massachusetts.

Newhall, S. M., Nickerson, D., and Judd, D. B. (1943). *J. Opt. Soc. Am.* **33**, 385.

Nickerson, D. (1975). *J. Opt. Soc. Am.* **65**, 205.

Nickerson, D., and Stultz, K. F. (1944). *J. Opt. Soc. Am.* **34**, 550.

Pointer, M. R. (1974). *J. Opt. Soc. Am.* **64**, 750.

Priest, I. G., and Brickwedde, F. G. (1938). *J. Opt. Soc. Am.* **28**, 133.

Pugh, E. N., Jr., and Sigel, C. (1978). *Vision Res.* **18**, 317.

Robertson, A. R. (1968). *J. Opt. Soc. Am.* **58**, 1528.

Robertson, A. R. (1978). *Color Res. Appl.* **3**, 149.

Schanda, J., Mészáros, M., and Czibula, G. (1978). *Color Res. Appl.* **3**, 65.

Schrödinger, E. (1920). *Ann. Phys. (Leipzig)* **63**, 397, 427, 481.

Speranskaya, N. I. (1959). *Opt. Spectrosc. (USSR)* **7**, 424.

Steen, P. (1970). *In* "Color 69" (M. Richter, ed.), pp. 369–376. Musterschmidt-Verlag, Göttingen.

Stevens, S. S. (1946). *Science* **103**, 677.

Stevens, S. S. (1975). "Psychophysics," Wiley, New York.

Stiles, W. S. (1946). *Proc. Phys. Soc., London* **58**, 41.

Stiles, W. S. (1967). *J. Colour Group (G.B.)* **11**, 106.

Stiles, W. S., and Burch, J. M. (1959). *Opt. Acta* **6**, 1.

Terstiege, H. (1972). *J. Color Appearance* **1**, 19.

Tonnquist, G. (1975). "Comparison Between CIE and NCS Colour Spaces," FOA Rep. C 30032-E1. Försvarets Forskningsanstalt, Stockholm.

Urban, F. M. (1909). *Arch. Gesamte. Psychol.* **15**, 261; **16**, 168.

von Kries, J. A. (1877). *Albrecht von Graefes Arch. Ophthalmol.* **23**, 17.

von Kries, J. A. (1904). *In* "Die Physiologie der Sinne III" (W. Nagel, ed.), pp. 109–282. Vieweg, Braunschweig.

von Kries, J. A. (1905). *In* "Handbuch der Physiologie der Menschen" (W. Nagel, ed.), pp. 109–282. Vieweg, Braunschweig.

von Kries, J. A. (1911). *In* "Handbuch der Physiologisches Optik," Vol. II, (W. Nagel, ed.) pp. 366–369. Leopold Voss, Hamburg.

Vos, J. J. (1978). *Color Res. Appl.* **3**, 125.

Vos, J. J., and Walraven, P. L. (1971). *In* "Color Metrics" (J. J. Vos, L. F. C. Friele, and P. L. Walraven, eds.), pp. 69–81. AIC/Holland, Soesterberg.

Vos, J. J., and Walraven, P. L. (1972a). *Vision Res.* **12**, 1327.

Vos, J. J., and Walraven, P. L. (1972b). *Vision Res.* **12**, 1345.

Weber, E. H. (1834). "De Pulsu, Resorptione, Auditu et Tactu: Annotationes Anatomicae et Physiologicae." Köhler, Leipzig.

Wright, W. D. (1928–1929). *Trans. Opt. Soc. London* **30**, 141.

Wright, W. D. (1941). *Proc. Phys. Soc., London* **53**, 93.

Wright, W. D. (1943). *J. Opt. Soc. Am.* **33**, 632.

Wright, W. D. (1969). "The Measurement of Colour," 4th Ed. Hilger, London.

Wright, W. D., and Pitt, F. H. G. (1937). *Proc. Phys. Soc., London* **49**, 329.

Wyszecki, G. (1963). *J. Opt. Soc. Am.* **53**, 1318.

Wyszecki, G. (1970). *Farbe* **19**, 43.

Wyszecki, G. (1972). *J. Opt. Soc. Am.* **62**, 117.

Wyszecki, G. (1975). *J. Opt. Soc. Am.* **65**, 456.

Wyszecki, G., and Fielder, G. H. (1971). *J. Opt. Soc. Am.* **61**, 1135.

Wyszecki, G., and Stiles, W. S. (1967). "Color Science: Concepts and Methods, Quantitative Data and Formulas." Wiley, New York.

Young, T. (1820a). *Philos. Trans., 1802* p. 12. (Reprinted from an 1807 private printing.)

Young, T. (1820b). *Philos. Trans., 1802* p. 387. (Reprinted from an 1807 private printing.)

4

Modern Illuminants

FREDERICK T. SIMON

Clemson University
Clemson, South Carolina

I. INTRODUCTION

Light is one of the three fundamental components of color, the other two being the spectral characteristics of the object and the visual mechanism or, loosely, the eye of the observer. All these parameters influence the way in which radiant energy stimulates vision. As we have seen in the chapter on colorimetry, the efficiency of radiant energy to evoke a visual response varies considerably with wavelength. For normal light sources at normal intensity levels, the range of wavelengths that is commonly considered to be effective in stimulating vision is from about 360 to 830 nm. Radiant energy having wavelengths within that range is called "light." Therefore, consideration of radiation beyond this wavelength range is of limited interest when color phenomena are involved. The spectral characteristics of fluorescent whitening agents are a special case since they are excited by near-ultraviolet energy to produce light in the visible spectrum range; consequently, the wavelength region between 300 and 400 nm is also important for fluorescent objects.

This chapter will describe sources of light and will discuss some of the conventions by which standard illuminants are defined. The distinction that has been made between an illuminant and a source is such that "illuminant"

149

represents a theoretical definition and "source" represents a practical, or real, source of light.

II. PLANCKIAN RADIATOR AND CORRELATED COLOR TEMPERATURE

The physical law that defines the radiant exitance of nonselective objects that have been heated to specific temperatures is used as a fundamental reference for all types of light sources. The radiant exitance is given by Planck's law:

$$M_{e,\lambda}(\lambda, T) = c_1 \lambda^{-5}(e^{c_2/\lambda T} - 1)^{-1} \text{ W m}^{-3} \tag{1}$$

where $c_1 = 3.741 \times 10^{-16}$ W m^2, $c_2 = 1.4388 \times 10^{-2}$ m K, λ is the wavelength in meters, T the temperature in kelvins, W stands for watts, and m for meters.

Planck's law is used in three ways to characterize different light sources. Nonselective sources are defined by distribution temperature; that is, the absolute temperature T of a full radiator for which the ordinates of its spectral distribution curve of radiance are essentially proportional in the region of wavelengths from about 360 to 830 nm, to those of the distribution curve of the source under consideration.

Selective radiators are specified in color temperature units symbolized as K. The symbol K stands for temperature in kelvins and is an index of "color temperature." In the chapter on colorimetry we saw that color temperature is the absolute temperature of the full or Planckian radiator that emits radiation of the same chromaticity as that of a source under consideration.

Finally, when a selective radiator has chromaticity coordinates that are not the same as any full radiator but are not greatly different from those of a full radiator, the source is characterized by correlated color temperature (i.e., the color temperature of the full radiator that is visually most similar in appearance to the color of the source under consideration). Only distribution temperature may be considered to imply something about the spectral power distribution of a light source. Color temperature only implies a chromaticity match with some one of a series of full radiators. Correlated color temperature implies only a general similarity of chromaticities. Neither provide any information about spectral power. These facts are illustrated in Figs. 1–3.

In each figure, a spectral power distribution for a test source is compared with a Planckian distribution. Figure 1 indicates that the tungsten filament lamp provides a good match for the Planckian radiator. Figure 2 shows that the daylight illuminant has a different spectral power distribution from that of the Planckian radiator, even though both have the same color temperature. Figure 3 illustrates the fact that the fluorescent lamp's spectral power distribution is very different from that of the Planckian radiator even though they have the same correlated color temperature.

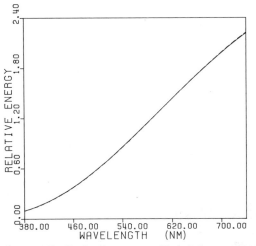

Fig. 1.   Spectral energy distribution for tungsten filament lamp at 3000 K. The Planckian radiator distribution is shown as a dashed line.

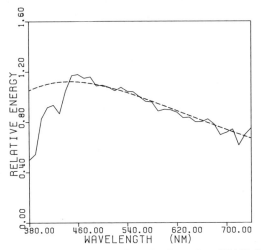

Fig. 2.   Spectral energy distribution for CIE Illuminant D_{65} 6504 K. The Planckian radiator distribution is shown as a dashed line.

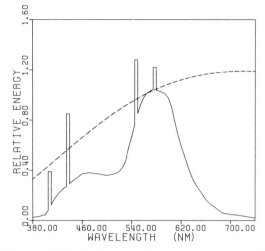

Fig. 3. Spectral energy distribution for cool white fluorescent lamp—4000 K. The Planckian radiator distribution is shown as a dashed line.

III. INCANDESCENT SOURCES

The tungsten incandescent lamp is the most widely used source of artificial light. It is represented in CIE colorimetry as both CIE Illuminant and Source A, which is defined as a full or Planckian radiator operating at an absolute temperature of 2857 K. The emissivity of a real tungsten incandescent lamp differs slightly from that of a black-body radiator of the same color temperature. This discrepancy in most instances, however, is small and is ignored, since it is less than 1% throughout the visible portion of the spectrum and approaches 2% only in the long-wave infrared portion of the spectrum. For critical work, carefully aged lamps are used as standards for the determination of the spectral energy distributions for other light sources.

Ordinary tungsten filament lamps are not operated too close to the melting point of the metal (3650 K), which evaporates from the filament and condenses on the inside of the glass envelope. This is the most frequent cause of failure in incandescent lamps; the life and stability of ordinary tungsten are inverse functions of the operating temperature. Zubler and Mosby (1959) introduced a tungsten filament lamp that was enclosed in a quartz envelope and contained a small amount of a halogen vapor. The advantage of this lamp is that it can be operated at somewhat higher temperatures with consequent higher light output, since the halogen forms a complex with the evaporated tungsten that is then broken down to tungsten metal when it comes into contact with the hot filament and the halogen is regenerated. Because of the quartz envelope and the halogen regeneration, the lamp can be used at higher temperatures than conventional tungsten filament lamps. The inside envelope temperature should be about 600°C for the cycle to operate effectively. Accordingly, too much cooling will prevent recycling and actually shorten lamp life. The spectral power distribution of such tungsten–halogen lamps is very similar to that of a Planckian radiator and gives more energy than ordinary tungsten lamps near ultraviolet, where regular lamps are deficient. This is largely the result of the greater ultraviolet transmission of the quartz envelope rather than glass. These lamps have found widespread use in colorimetric instruments, especially when measurements are made of fluorescent whitening agents.

IV. DAYLIGHT

Although daylight varies with time, place, weather condition, and direction, it is nevertheless the common reference and presumed model to simulate for lighting for color. The artist is insistent on lighting his studio with daylight; the purchaser of clothing wants to know how it looks in daylight; and most commercial colorimetric specifications are made with respect to some phase of daylight. These reasons and many others have fostered interest in determining just what daylight is, how it varies, and how it can be defined. There has been serious consideration of the character of daylight since the classical period of the Greeks, but precise determinations have only been done since about 1960, when automated radiometric equipment became available. Before the era of modern instrumentation the Smithsonian Astrophysical Observatory contributed a great amount of the information about daylight energy distribution in Washington and at the site of several astronomical observatories in the United States from about 1900 to shortly after World War II. The observatory was started in 1890 under the direction of Langley, who was succeeded by Abbot in 1907. Gibson (1940) reviewed the Smithsonian energy distribution data and prepared a typical set of data based on the Abbot *et al.* (1923) publication. This was published in the "Handbook of Colorimetry," by Hardy (1936) and came to be known as the Abbot–Gibson daylight energy distribution.

The need for a laboratory reference light source that simulated the Abbot–Gibson daylight distribution was met by the work of Davis and Gibson (1931), who developed a set of liquid filters that are used in conjunction with the CIE standard Illuminant (and Source) A. These filters are made up according to the formulae given in Tables V and VII of the chapter on colorimetry. In 1931 the CIE adopted the filter–light-source combinations and gave them the official CIE notations of B and C.

Extensive studies of daylight in the visual spectral region have been published by Henderson and Hodgkiss (1963, 1964), Condit and Grum (1964), Judd *et al.* (1964), Das and Sastri (1965), and Nayatani *et al.* (1967). Much of this work was directed to answering the problem that had been posed within the CIE to learn what was the most "suitable energy distribution · · · for natural daylight." What evolved was not a single set of values for daylight but several illuminants that were representative of various phases of daylight that were useful for different branches of commerce. At first the CIE wished to supplement the long-established CIE Illuminants A, B, and C. It was recommended that 5500, 6500, and 7500 K distributions be developed to represent phases of daylight that would provide more realistic illuminants than the CIE Illuminants B and C, which are deficient in violet and near-ultraviolet light, a particular problem when attempting to evaluate many fluorescent materials. All these distributions are listed in Tables III, IV, VI, and XI of the chapter on colorimetry.

As indicated earlier, enough data were available during the 1960s from several laboratories so that the task was to bring it together and make order out of the whole. Judd *et al.* (1964) evolved a method that gave the

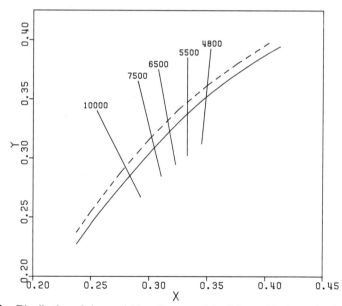

Fig. 4. Distribution of chromaticities of measured daylight used in determination of daylight. (After Judd *et al.*, 1964).

variance–covariance vectors that could represent all the radiometric data obtained by several workers calculated by the method of Simonds (1963). The data that were used in the computations were those of Henderson and Hodgkiss, Condit and Grum, and Budde (reported by Judd *et al.*). An important characteristic of natural daylight is that its chromaticity is generally on the "green" side of the Planckian locus. This tendency describes another locus that results from the variance–covariance vectors and is termed the "daylight locus." It is shown in Fig. 4 along with the chromaticity points of the data used in the Judd study. The locus of daylight can be described by

$$y_D = -3.000x_D^2 + 2.870x_D - 0.275. \tag{2}$$

Several calculations have been reported that presume to elucidate the variability of daylight in specific locations around the world. It is sufficient to say that not enough is known at this time and that the place-to-place variation in spectral energy distribution for any specific color temperature is not nearly so great as the variability with time and weather. Therefore, standardization is absolutely essential for the interpretation of colorimetric data.

V. DAYLIGHT SIMULATORS

When the sun goes down at night, man's activities do not cease; artificial sources of light are used. Primitive man used fire both to warm him and to allow him to continue seeing when the sun's light was no longer available. After the advent of the industrial revolution, it became commonplace to continue manufacturing operations all through the night, which meant that artificial lighting was widely used. In the textile industry in Great Britain, dyeing and color matching continued during the night shifts. This practice provoked some difficulties in characterizing daylight color matches correctly with lights such as the oil lamp or later some of the gas mantle lamps, which were of considerably lower color temperature than natural daylight. When electric power became more widely available even the incandescent lamp was not a good source to replace the commonly used "north light." Around the turn of the century (see Henderson, 1970) many gas discharge lamps were developed to substitute for the low color temperature of the incandescent lamp. These were not very successful. Ives (1912) did some of the earliest systematic work to find a suitable replacement for natural daylight, which was presumed to be much the same in spectral distribution as a black-body radiator at 5000 K. A light booth was constructed from mixed sources of incandescent and mercury discharge together with glass and gelatin filters. Another artificial source considered for standardization was a gas mantle as the primary source but with a special mixture of rare earth oxides. This was used by dyers as a substitute for lower color temperature light sources. In 1915 the Macbeth Daylighting Company was formed to manufacture and distribute a light fixture for simulating daylight. Later versions consisted of an incandescent lamp with a blue glass filter that had been specially designed by Corning Glass Works to raise the color temperature of the source from 3000 to 5000, 6500, or even 7500 K, depending on the thickness used. This was a much better duplication of "north light" and ultimately became the accepted artificial light used for judging color matches in the textile industry. This lamp, with many modern improvements, is still widely used for visual grading where the quality of illumination is critical to the visual judgment.

A study was done under the auspices of the Inter-Society Color Council to establish the type of artificial daylight that would "satisfy preferred conditions of daylight matching." The committee that collaborated on the problem was under the chairmanship of Nickerson (1948). They concluded that 7500 K on the average was preferred by several textile mills. This condition has become the industry practice in the United States, where it is commonplace to use a filtered incandescent lamp as an artificial daylight source. Figure 5 indicates the extent to which filtered incandescent lamps typically can approach the spectral power distribution of a CIE Illuminant D. Fluorescent lamps with mixed phosphors are now often used as a rela-

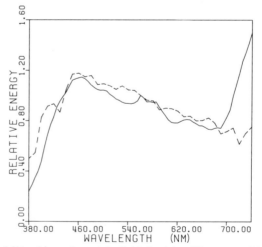

Fig. 5. Typical filtered incandescent daylight simulator. The comparable daylight distribution is shown as a dashed line.

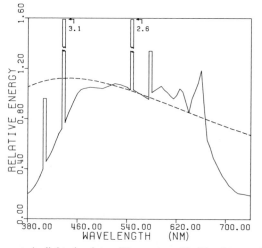

Fig. 6. Fluorescent daylight simulator. The comparable Planckian radiator distribution is shown as a dashed line.

tively inexpensive substitute for filtered incandescent. Figure 6 illustrates fluorescent lamp spectral power distribution.

In the United Kingdom, fluorescent lamps are more generally used than the filtered incandescent lamp for color examination, although a commercial filtered incandescent lamp was made there at one time. A British Standard, BS 950: Parts 1 and 2 describing a daylight fluorescent lamp was adopted for the textile and printing industries in 1967. In recent years high- and low-pressure xenon gas-discharge lamps have been used in conjunction with optical filters to simulate daylight. (See the following discussion.)

An important aspect of simulation of daylight with artificial sources is the desire for standardization. When the CIE adopted phases of daylight for colorimetric calculation, there was no source available that provided a generally satisfactory duplication of the standard spectral distributions of D_{55}, D_{65}, and D_{75}. Wyszecki (1970) has published a survey of a number of the available simulators based on the spectral data collected from various laboratories throughout the world. The simulators reported in the study were based on three types of sources: xenon-arc lamps, tungsten–halogen incandescent lamps, and fluorescent lamps of various types. The assembled data were analyzed in several ways in an attempt to find an expression that would indicate the most desirable or "best" simulation of daylight with an artificial source. There is no standard method of directly assessing the similarity of spectral power distributions. The CIE color-rendering method has sometimes been used as an indirect method of assessment, but as indicated in the discussion of color rendering in the chapter on colorimetry, the method evaluates color matching similarity and not similarities of spectral power distributions. Wyszecki (1970) has proposed several types of assessment that fall into three general categories: (1) statistical criteria; (2) metameric index criteria; (3) CIE color-rendering index determinations.

The spectral power distributions of incandescent and xenon light sources were modified with glass filters that had been selected to give a reasonable fit for one of the CIE Illuminants D. Only the filtered xenon and fluorescent lamps could be judged for conformance to the CIE Illuminants D in the ultraviolet region between 300 and 400 nm, since the incandescent lamps were all deficient there. Also, fluorescent lamps give some distortion in the spectra because of the mercury lines, which are not present in the D illuminant spectra. It was not possible to obtain a single index of performance from this study, but the numerical indices that were developed indicated that either several criteria must be taken together or a new method of assessment would have to be developed. Nayatani and Takahama (1972) used a metameric index obtained from comparing 12 metameric grays (Wyszecki and Stiles, 1967) illuminated by the test source and a reference distribution. The criterion used was simply the average metameric index taken for all samples. Berger and Strocka (1973) also investigated the principle of judging light sources with metameric samples and concluded that it was a reliable test for the assessment of artificial sources but is restricted to the visible range. They had several series of metameric samples that typified commercial samples and dye combinations. In order to calculate a metameric index it is best to have no color difference for the metameric pair when calculated for the reference light source. To eliminate small differences in the two samples under the reference source, ratios of the two tristimulus values were calculated and then multiplied by sections of the spectral reflectance curves to adjust the tristimulus values to equality. These authors also proposed a simple test with a fluorescent whitening agent to assess the duplication of a test distribution compared with a reference.

Another means of assessment of artificial daylight sources was proposed by Crawford (1959, 1963) and is contained in the British Standard, BS 950, Parts 1 and 2 (British Standards Institution, 1967). This so-called band method evaluates power in defined spectral regions. It describes limits within which the chromaticity of a given lamp must fall to meet the specification. There are also permissible deviations in the spectral power distribution in terms of specific bands in both the visible and ultraviolet regions. The chromaticity limits are given as a 12-sided figure that imitates an ellipse and is shown in Fig. 7.

At present there is still no recommended standard procedure for assessing the "goodness of fit" of a test spectral power distribution to that of a refer-

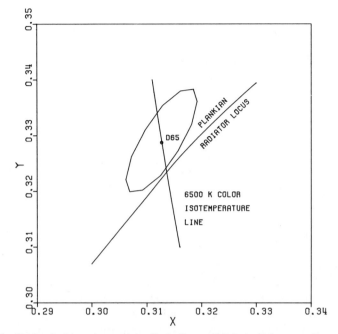

Fig. 7. Dodecahedron chromaticity limit for artificial daylight according to British Standard 950.

ence illuminant. A subcommittee of the CIE Committee on Colorimetry continues to investigate this problem. In the meantime it is probably best to assume that a high index of congruence from a battery of tests is sufficient evidence of a good fit between pairs of spectral power distributions.

VI.   FLUORESCENT LAMPS

There is a continuing yearly increase in the use of fluorescent lamps for general, industrial, institutional, and commercial lighting. The advantages of these lamps over incandescent sources are that they give more luminous flux for a given amount of power, illuminate large areas more evenly, and produce less heat. However, there are many types of fluorescent lamps available on the market and this variety makes for a lack of standardization both among users and manufacturers. In the last 10 years there has been a trend for general illumination purposes to adopt a type of fluorescent lamp called "cool white," which is a type made by many manufacturers.

The fluorescent lamp is a type of gas-discharge lamp that provides spectral power distributions that are all marked by the presence of some of the emission lines of mercury vapor, which is inside the lamp envelope and provides the excitation of the phosphors that are the major source of the light from the lamp. Many combinations of phosphors are available at present that can be used in mixture to produce a light that appears to the average observer as "white." Certain compounds produce specific bands in the spectrum, such as those shown in the accompanying tabulation. These

Phosphor	Central wavelength of emittance	Phosphor	Central wavelength of emittance
Calcium tungstate	440	Cadmium borate	615
Magnesium tungstate	480	Calcium-strontium	640
Zinc silicate	540	phosphate	
Calcium halophosphate	590	Magnesium arsenate	660
Calcium silicate	610		

are only a few of the many phosphors that are now available and more are being developed every year. Most phosphors tend to emit light in a broad band of wavelengths that are normally distributed in *frequency* about the centroid. In addition to centroids of emission differing among phosphors, there are also variations in the bandwidths of emission.

Special lamps that have been developed with mixtures of phosphors provide almost any desired chromaticity of light. Although the fluorescent lamp does not produce a simple continuous spectrum, it can resemble that of a thermal radiator and be made to approximate, say, "daylight" or "blue" or "incandescent" light, even though the spectral resemblance is only superficial. This is not to suggest that a useful light cannot be obtained, but the actual spectral power distribution is not very similar and a figure of correlated color temperature will not generally imply anything about the spectral power distribution of a fluorescent source.

Most common fluorescent lamps are particularly deficient in the long-wavelength end of the visible spectrum. When the long-wavelength emission is needed, the output of the fluorescent lamp is sometimes supplemented with an incandescent lamp. Similarly, most common fluorescent lamps lack radiation in the near-visible ultraviolet range of wavelengths. Special phosphors have been developed to alleviate these problems of long- and short- or ultraviolet-wavelength emission. Generally, these phosphors are both expensive and inefficient. For these reasons they usually will be found only in special lamps designed for color assessment.

A number of lamp manufacturers around the world offer lamps for use in color assessment. Choice of any one such lamp should not be based simply on correlated color temperature, nor even on what appears to be a suitably high CIE General Color Rendering Index. Instead, the CIE Special Color Rendering Indices should be evaluated, together with the lamps' spectral power distribution, in terms of the specific requirements of a particular application. In the final analysis, the question of whether or not a particular light source is satisfactory to a task will be determined by careful observation and experience about how well it seems to work. The point here is simply that the entire subject is one of great complexity, for which no simple solutions have yet been developed.

VII. ARC LAMPS

The carbon arc lamp was used in the late nineteenth and early twentieth centuries as a source of high-intensity, high-correlated-color-temperature radiation for scientific instruments, street, and theater lighting. Although it had desirable characteristics, it was unstable even for short periods of time. Its modern-day heirs are the large variety of commercial arc lamps, primarily enclosed arcs that are based on a plasma created between two electrodes in an atmosphere of either a rare gas element or a metal vapor such as mercury or sodium. Street lighting is still the province of the arc lamp, but it has been greatly improved with high-pressure sodium vapor lamps or mercury vapor lamps that have been enhanced with several other metals as well as phosphors that round out the spectral distribution. These lamps are important for general industrial lighting because of their luminous efficiency, but their color-rendering properties are compromised in order to achieve high luminous efficiency. Since most arc lamps require somewhat more elaborate starter ballasts and operate well at higher electrical frequencies than those found in the household or office, their use is generally restricted to special applications where much light is needed. None of these arc lamps are used in colorimetry, but the effect of such illumination on colored objects should be recognized.

The exceptional arc lamp that has found considerable use in colorimetry and in the laboratory is the high-pressure xenon lamp. Operating gas-discharge lamps at high pressure provides a more complete emission spectrum than is the case for low-pressure arc lamps. At high pressure the spectrum tends to "fill in" between the emission lines that are characteristic of a particular gas or vapor, with the result that the emission spectrum *tends* toward a continuum. The xenon high-pressure lamp is available in special high-silica glass envelopes that allow a large amount of ultraviolet light to be transmitted. One of the primary advantages of the xenon lamp is its high correlated color temperature of about 5000 K. When color temper-

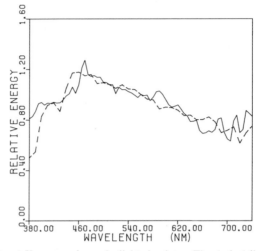

Fig. 8. Filtered Xenon arc-lamp daylight simulator. The dashed line represents the distribution for CIE Illuminant D_{65}.

ature conversion filters are used (see Grum, 1968; Grum *et al.*, 1970; Wyszecki, 1970), close simulation of CIE Daylight Illuminants, D_{55}, D_{65}, and D_{75} can be obtained without undue loss of light by filter absorption. Xenon lamps are used in several colorimetric instruments and offer a useful source of spectral radiation below about 340 nm, where the tungsten–quartz-halogen lamp is not effective. There are several designs of the xenon arc lamp that have virtually eliminated arc wander and greatly reduced generation of ozone by the emission of short ultraviolet radiation through the selection of the envelope materials. Like all other arc lamps, special electrical equipment is needed to start the xenon lamp as well as to operate it with a low-voltage, high-power, direct current supply. Much of the power spectrum of the high-pressure xenon arc lamp radiation is essentially a smooth continuum, as is shown in Fig. 8, but it is clear that there are emission lines that center around 475 nm that represent some distortion. The National Research Council of Canada (see Xe-6 in Wyszecki, 1970) has succeeded in partially absorbing these with a very thin didymium filter, Corning Glass Works 5120. Properly filtered xenon arc light offers one of the best simulations for the CIE Daylight Illuminants if the ultraviolet radiation is of considerable importance. The density of the filter system used with xenon is less than that required by a tungsten source, which is operated at a much lower color temperature.

REFERENCES

Abbot, C. G., Fowle, F. E., and Aldrich, L. B. (1923). *Smithson. Misc. Collect.* **74,** No. 7, 1. (Publ. No. 2714.)
Berger, A., and Strocka, D. (1973). *Appl. Opt.* **12,** 338.
British Standards Institution (1967). "Artificial Daylight for the Assessment of Colour," Br. Stand. BS 950, Part 1 and 2.
Condit, H. R., and Grum, F. (1964). *J. Opt. Soc. Am.* **54,** 937.
Crawford, B. H. (1959). *J. Opt. Soc. Am.* **49,** 147.
Crawford, B. H. (1963). *Trans. Illum. Eng. Soc.* **28,** 50.
Das, S. R., and Sastri, V. D. (1965). *J. Opt. Soc. Am.* **55,** 319.
Davis, R., and Gibson, K. S. (1931). Filters for the Reproduction of Light and Daylight and the Determination of Color Temperature. *Natl. Bur. Stand. (U.S.), Misc. Publ.* No. 114.
Gibson, K. S. (1940). *J. Opt. Soc. Am.* **30,** 88.
Grum, F. (1968). *Appl. Opt.* **7,** 183.
Grum, F., Saunders, S., and Wightman, T. (1970). *Tappi* **53,** 1264.
Hardy, A. C. (1936). "Handbook of Colorimetry." M.I.T. Technology Press, Cambridge, Massachusetts.
Henderson, S. T., and Hodgkiss, D. (1963). *Br. J. Appl. Phys.* **14,** 125.
Henderson, H. T., and Hodgkiss, D. (1964). *Br. J. Appl. Phys.* **15,** 947.
Henderson, S. T. (1970). "Daylight and its Spectrum." Am. Elsevier, New York.
Ives, H. E. (1912). *Trans. Illum. Eng. Soc.* **7,** 62.
Judd. D. B., MacAdam, D. L., and Wyszecki, G. (1964). *J. Opt. Soc. Am.* **54,** 1031.
Nayatani, Y., and Takahama, K. (1972). *J. Opt. Soc. Am.* **62,** 140.
Nayatani, Y., Hitani, M., and Minato, H. (1967). *Denki Shikenjo Iho* **31,** 1127.
Nickerson, D. (1948). *Illum. Eng.* **43,** 416.
Simonds, J. L. (1963). *J. Opt. Soc. Am.* **53,** 968.
Wyszecki, G. (1970). *Farbe* **19,** 43.
Wyszecki, G., and Stiles, W. S. (1967). "Color Science." Wiley, New York.
Zubler, E. G., and Mosby, F. A. (1959). *Illum. Eng.* **54,** 734.

5

Color Order

FREDERICK T. SIMON

Clemson University
Clemson, South Carolina

I. INTRODUCTION

Color order attempts to provide a system of reference whereby the relation of one color to another can be quickly perceived and the position of that color can be established with respect to the universe of all colors. Consequently, color order is not the special province of the scientist or the artist; it must satisfy the needs of both. It is also useful between the buyer and seller, the designer and producer, the dyer or paint tinter, and the customer, as well as the teacher and pupil.

Color perception is highly personal and is subject to individual interpretation, which presents difficulties when ideas or information on color are to be conveyed from one person to another. For this reason there has been a continuing effort since at least the time of Aristotle to interpret how one sees color and how any given color relates to another. Much of that which has been written attempts, on the one hand, to resolve theory and even the physiology of the visual process and, on the other hand, to take into consideration how the gamut of all colors, irrespective of origin, relate to each other. Unfortunately, there is no theory that completely explains how humans see color and how the colors of various objects and light interact with the sensation of sight.

This lack of complete understanding does not preclude developments that allow us to convey information about color in useful ways. In the past 300 years many approaches and solutions to particular or specialized practical problems have been offered as the means to describe color in a general sense. Color order, then, can be defined as a systematic means to communicate information about these personal visual perceptions from one individual to another in a convenient and rational manner. Color order is not restricted to sets of materials; it also includes consideration of how we see differences among samples that are both similar and dissimilar. The simplest type of order is to assign names to things. We do this regularly and will continue to use language to convey meaning or to differentiate among things. However, naming colors does not satisfy the many needs of color description because it is seldom systematic enough to convey information without ambiguity.

Along with the many methods used to attain a rationale for color, there have been efforts to bring together what is known about color systems and to compare them (Foss, 1949; Evans, 1948; Wyszecki, 1960; Billmeyer and Saltzman, 1966; Birren, 1969; Libby, 1974; Judd and Wyszecki, 1975). The discussion presented here will imply that color order comes from a knowledge of the means for producing color by additive, subtractive, or partitive color mixing. These methods will be used to classify various systems. In addition to describing the principles of each system, an attempt will be made to provide historical perspective for each of the methods, especially in the context of the time of its publication.

A. General Principles of Color Order

The principles of color mixture derive historically from the mixture or addition of lights. Isaac Newton (1704) was the first to show that the dispersed spectrum of sunlight could be recombined to form the white light of daylight. He further showed that isolated portions of the dispersed spectrum could be combined to form differently colored lights. In the extreme, Newton's findings show that oppositely hued spectral lights may be combined to form a white-appearing light, even without the presence of all forms of spectral light dispersed from the white light of the sun, although

Newton did not clearly recognize this implication of his work. Newton introduced the concept of color mixture by additive combination of colored lights according to the principle of center-of-gravity mixture. That is, if the spectral lights are arrayed in a circular fashion, a line drawn between any two spectral positions on such a diagram will describe the locus of all proportionate mixtures of those two spectral lights. Further, the position of a given light mixture along such a line will be determined according to the proportions of the two lights in the mixture; a mixture of equal parts will be represented at a point on the locus equidistant from the two spectral positions, and so on.

About a century later, Thomas Young (1802) reasoned that for color mixtures to occur, the visual mechanism need not contain many different receptors, each sensitive to a different spectral light. He proposed that, instead, the visual mechanism consisted of a limited number of receptors; he gave as a logical example the number as three receptors. James Clerk Maxwell (1856) demonstrated that colored photographs containing many hues could be made by using only three primary lights. Maxwell also adopted the hue (or spectral light) circle of Newton together with the center-of-gravity principle of representing light mixtures. By locating the three spectral primary lights on the circumference of the hue circle, Maxwell formed a triangle that contained an area representing all possible mixtures of the three primary lights. This was an early color mixture diagram, so called because it provides a kind of metric system, in diagrammatic form, for characterizing the additive mixture of lights.

Hermann Ludwig Ferdinand von Helmholtz (1852) systematically investigated the mixture of three colored lights, or "trichromatic mixture" as it has come to be called, and his contributions to our understanding of color mixture were so extensive and important that the entire subject is now known as the Young–Helmholtz trichromatic theory of color vision.

The basic principles of color mixtures of lights, based on Newton's observations and subsequent experimentation, were formally stated as a set of three rules by Hermann Günter Grassmann (1853). They have come to be known as Grassmann's laws of color mixture, although they refer only to the additive combination of colored lights. Those laws are now paraphrased to state that

(1)   Three suitably different lights are necessary and sufficient to match in *hue* the colors of all other lights,

(2)   If one of two components of a light mixture is continuously altered, the color of the mixed light is also continuously changed,

(3)   When two lights that match in color are added to two other lights that also match in color, the resultant light-mixture will match in color.

These three "laws" are usually combined with what is referred to as Abney's law (Abney and Festing, 1886) to form the complete basis for additive color mixture. Abney's law states that

(4)   The total intensity of any light mixture is equal to the sum of the intensities of the component lights.

Together these four rules provide all the corollaries that form the principles of additive color mixture and colorimetry. It is important to note that these rules apply only to the additive combination of lights. Combining colorants, materials that absorb light or materials that scatter, refract, or polarize light involves different sets of rules that are generally more complicated than those for mixing lights by addition. Although such other mixtures may be described in ways that relate to light addition, their relationships are so complex that these mixtures are most often described in terms of different sets of rules. Accordingly, we will call on several sets of rules to discuss the various forms of color mixture.

The rules that govern color mixture comprise one of the three aspects of color order systems: the manner in which the stimuli or samples are synthesized to provide examples of the color order. There are, in addition, two other aspects to any color order system. These are the plan of the order and the geometry by which the order is represented either in concept or in example.

Color orders may be structured according to intervals or magnitudes. That is. positional elements of the order may be arranged according to equality of color differences of some kind or according to equality of ratios of some aspect of color. There may, of course, be ordering according to some plan of systematic inequalities of either intervals or ratios. The differences among elements of the order may represent various kinds of equalities as well. For example, a perceptual plan may array elements so that differences among samples or elements all represent the same extent of perceived difference in color. The Munsell color notation attempts to provide such an ordering. Similarly, the Swedish Natural Color System attempts to order color elements such that ratios of color attributes are all perceptually equal. Some color orders strive to equate proportions of colorant concentrations, others normalize attributes of the colorant mixture process, such as half-tone dot area in photomechanical printing. The point is simply that regardless of the color mixture method used to construct a color order system, there may be differences that relate to the plan of ordering. These differences may not always be obvious but they can significantly affect the utility of a color order system in a particular application. The user should be aware, then, of both the color mixture principles on which an order is based and the plan of the order.

He should also know the geometrical plan by which the order is constructed, and this is fortunately somewhat more obvious than either of the other two aspects of a color order system. For example, some systems, such as the Munsell system, are arrayed in cylindrical polar coordinates. Others, such as the Hickethier and Pantone systems, use rectangular coordinates. Some systems follow the geometrical configuration of spheres and hemispheres; others are pyramidal or doubly pyramidal. The Optical Society of America color system is an example of a cubo-octahedral sampling of uniform visual intervals in color space. The point here is that the geometrical array for which the plan of the order (e.g., equality of intervals) is correct should be borne in mind by the user if he is to avoid confusion about the implications of the color order system.

In the following discussion of color order systems, we will first describe the color mixture plans used and provide some general information about the differences among results arising from the choice of color mixture method. Differences among order plans and geometries of representation will be pointed up in discussions of various particular color systems. In most instances the resultant color spacing implied or produced by a given color ordering can be described or specified according to the rules of colorimetry based on additive color mixture. We will draw on this descriptive utility to provide a common base for comparing the various results.

II. FORMS OF COLOR MIXTURE

Color stimuli may be produced according to three methods whereby light is modulated through mixing according to what are here called "additive," "subtractive," and "partitive" principles. It is these principles that govern any systematic mixture of color stimuli to achieve color order. The several forms of color mixing are related but operate differently insofar as the resulting effect is concerned. A practical understanding of how a particular color is achieved is as important to the artist and technologist as how a color is perceived by the human observer is to the psychologist. Only additive mixing includes the extreme limit of all color stimuli, the spectral lights. This factor alone recommends light mixing as the most generalized technique for evolving a systematic relationship among colors, and we have seen that the earliest rules of color mixture evolved from the addition of lights. Consequently, although light-based additive systems of colorimetry do not strictly apply to subtractive or partitive color mixing, the systems devised from the additive phenomena are used to represent the relations among colors produced by any means.

A. Additive Mixing

Additive color mixing is characteristic of combinations of light and is obtained when lights from different sources are merged into a single beam that cannot be resolved by the eye into the components. With the additive system, one can synthesize white light as a singular achromatic sensation if the correct primaries are mixed in proper proportions. Additive properties of the light in the visible spectrum were demonstrated by Newton in his experiments showing not only that the white light of the sun was refracted into the spectrum but that the spectrum could be recombined optically into white light. Beyond the discovery of the spectrum, Newton developed a center-of-gravity construction method for calculating mixtures of colored lights that is the conerstone of modern colorimetry. Newton used the term "intensity" in an ambiguous manner permitting him to make diagrams of a visible spectrum in the form that fitted his experimental observations. These diagrams are not correct in view of present-day knowledge of the physical properties of the spectrum. However, when a valid representation of the spectral locus is used, such as the CIE chromaticity diagram, the spatial relations for color mixing according to Newton's center-of-gravity construction give correct values for the relative amounts of mixtures of colored lights. The proof of the validity of additive mixing is based on Grassmann's laws.

In Fig. 1 Newton (1704) gave us the first additive color order system in the form of a circular diagram that arranged colored lights from the spectrum in a logical manner so that complements of any color could be shown. However, there was one great omission: the mixture of red and violet that

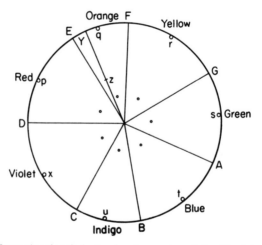

Fig. 1.   Newton's color circle showing the spectrum (after MacAdam, 1970).

gives purple. Newton's circle was divided according to his concept of seven colors of the spectrum: red, orange, yellow, green, blue, indigo, violet. The seven initials of these hue names form the "Roy G. Biv" mnemonic of many high school physics students. It was suggested that the seven spectral colors were analogous to the seven notes of the diatonic musical scales to justify the choice of that number. Since names are a form of color order, considerable interest has centered around the identification of indigo in the spectrum. That color name is now associated with a dull blue textile dye and is hardly regarded as saturated enough to be considered a spectral hue. R. A. Houstoun (1873) argued that Newton had very good color vision and thus saw indigo in the spectrum. O. N. Rood (1895) rejected the name "indigo" because of its modern connotation and substituted "ultramarine blue" as a suitable name for the part of the spectrum between blue and violet. It is this author's opinion that Newton divided the spectrum into seven parts according to the colors of the rainbow named by Maurolycus (1575), whose work surely was known to Newton. If Newton only followed traditional naming, the controversy over the word "indigo" should be put to rest.

Despite the fact that there was little immediate acceptance of Newton's theory of color, especially among his contemporaries in the Royal Society, the strength of his arguments and demonstrations prevailed and acknowledgment came within Newton's lifetime. One of his great champions outside of England was the French writer Voltaire, in *Elémens de la Philosophie de Newton* (Birren, 1963). Curiously, this was published in Holland without the writer's permission.

Goethe (Matthaei, 1971), the great German poet and writer, attempted to discredit Newton and his color theory and returned in part to the Aristotelian idea that color emanates from blends of light and darkness, black and white. Probably the greatest reason for the disagreement between Goethe and Newton was that Goethe's approach to natural phenomena was mainly subjective and used subtractive color mixing as a basis for argument. Newton's concepts were those of a scientist looking objectively at the physical world for phenomenological explanations. In discussing color Newton based his statements on additive mixtures of light. Therefore, when Goethe described mixing yellow and blue pigments to give a green by subtraction, he could never agree with Newton, who mixed yellow and blue light to give white. We look to Newton, not Goethe, as an early developer of color order, with his spectral circle, despite the fact that Newton missed the nonspectral color, purple. When his principles of center-of-gravity calculation are used with a properly spaced diagram, additivity of lights follows as further proof of the correctness of his thesis.

George Palmer (1777) was also a critic of Newton's theory but clearly distinguished between additive properties of light and subtractive pigment mixing. Even before the time of Thomas Young (1802), Palmer expounded a three-primary system for light as well as surface colors and then went on to relate the trichromatic principles to attempt to explain how human vision operates.

The discoveries of Newton had to be assimilated by the European world for about 50 years before the more explicit works of Grassmann and Maxwell were published. Grassmann's laws extended Newton's theory and gave precision and substance to additive mixing. James Clerk Maxwell (1872), on the other hand, was an experimentalist who was equipped with or invented several devices that expanded Newton's physical concepts of light mixture and coupled these with Young's theory of color vision. He firmly established the principle of trichromacy and quantitatively developed Maxwell's idea of representing mixtures of three primary colors in an equilateral triangle with primary colors at the apices.

It is generally accepted that Newton discovered the origin of the visible spectrum and demonstrated by numerous experiments the validity of his findings leading the way to a new understanding of color as it related to light. However, he could only postulate that when lights were mixed together a center-of-gravity principle could be followed to predict the effect of the mixture from the components. He was purposely vague in describing "intenseness" because he had neither the equipment nor technique to quantify the parameter. It remained for later scientists, such as Grassmann (1853), Helmholtz (1924), Maxwell (1860), and Schrödinger (1920), to give precise proof for the postulation and to describe laws that governed the method.

Grassmann, Maxwell, and Helmholtz allowed us to give "intenseness" a precise designation as radiant flux and to repeat those experiments with better data that verify the truth of Newton's hypothesis and later expansions regarding additivity. Simply, if we have a plane diagram that allows us to place the coordinates of any flux on the diagram, the relation of mixtures to the components can be calculated graphically by Newton's center-of-gravity method. Newton's method can be used for plotting the chromaticity of lights on a diagram or for plotting any additive system such as integrated absorbance data (Flaschka, 1960) or reflection function derivations (Ganz, 1965) on a plane diagram.

However, Grassmann, who gave the explicit extension to the Newton method, did not recognize the origin of purple, and as a matter of fact duplicated Newton's circle, but did note in his writing that purple is the complement of green. He evidently did not understand the concept of trichromacy, and even when Helmholtz began to publish his views on the

Young color vision hypothesis this did not cause Grassmann to alter his published views.

Helmholtz set forth two diagrams to explain color order that are more in line with the way that we conceive color relations at this time. Figure 2 is a triangle showing a hypothetical green primary that is necessary to contain the entire visible spectrum within the bounds of the area shown. The other figure (Fig. 3) that Helmholtz showed is a truncated circle derived from Newton's idea but more realistic with respect to the nonspectral color, purple; however, it is not spaced well as far as the spectrum itself is concerned and thus cannot be used with any confidence with center-of-gravity calculations. It does show that if a triad of violet, red, and green (V, R, G)

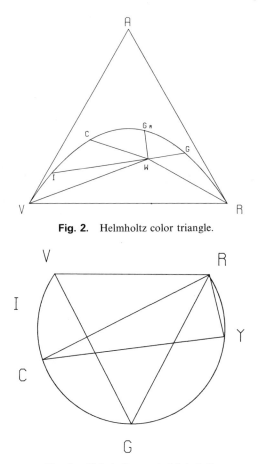

Fig. 2.   Helmholtz color triangle.

Fig. 3.   Helmholtz spectral "circle."

lights is mixed, indigo (I), cyan (C), and yellow (Y) are excluded from that mixture. Similar results are obtained with a mixture of red, cyan, and yellow, which cannot form violet, indigo, or green. Helmholtz gave an accurate explanation of the origin of purple and indicated that any color could be described in terms of luminosity, hue, and saturation. He further stated that a color chart could be constructed as Lambert (1774) did, with an inverted pyramidal form having black at the bottom. Helmholtz retained the seven colors for the spectrum as Newton and Grassmann did before him probably because he had no reason to change the concept. Helmholtz also clearly indicated that mixtures of pigments do not give results that are comparable to those resulting from mixing lights. Most of Helmholtz' attention was drawn to study of the mechanism of color vision, but he did use center-of-gravity calculations and the spinning disks that are mainly associated with Maxwell (1857).

Helmholtz also was the first scientist to use the line element to describe the sensibility of the human observer to differences in perception in various parts of the spectrum. Unfortunately, all that was available to him for basic spectral response data was the work of König and Dieterici (1886) and he assumed a Weber–Fechner scaling as the least perceptible response for each of the three visual mechanisms, red, green, and blue. Since both the basic color-matching functions of König and Dieterici and the perceptibility concept were in error, Helmholtz' conclusions are doubtful. Schrödinger, like Helmholtz, realized that to match all real colors with three positive color mixture values, one, two, or even three must lie outside the spectral color limits and therefore must be regarded as "unreal" or imaginary. This concept presents no difficulty for the mathematician, but to the uninitiated it may be a problem.

In Fig. 3 it is shown that the spectrum locus J is drawn inside a color mixture triangle. This was based on König's experiments matching the spectrum with three primary lights: 400, 505, and 700 nm. These observations required that in order to achieve a match for any spectral light, it had to be mixed with one desaturating primary and then the combination could be matched with the other two remaining primary lights. This experimental method was followed by both W. D. Wright (1928–1929) and J. Guild (1931) and is the underlying precept for obtaining trichromatic mixture curves that are now the basic 1931 CIE color mixture data for the standard observer (see Chapter 3).

Helmholtz' color mixture triangle is formed by two real primaries, red and violet, and an imaginary green primary. The spectrum locus is line J.

Practical examples of mixing of colored lights are not plentiful. Color television, stage lighting, and the mixed light from several phosphors in fluorescent lamps are the most common present-day instances of additive mixing of colored lights. Nonetheless, it remains a powerful and important scientific tool, since the addition of lights of different spectral compositions can be calculated unequivocally with the use of Grassmann's laws.

1.   GENERALIZED MATHEMATICS FOR ADDITIVE COLOR MIXING

If two colored lights, represented as X_a and X_b, are added together by superposition, the mixture is X_m. The relative amounts (radiant flux) of each, a and b, can be calculated from the following equation:

$$X_m = aX_a + bX_b. \tag{1}$$

It follows from this equation that the amount of light in the mixture is the sum of the two and is greater than that in either of the original lights.

B.   Subtractive Color Mixing

Examples of subtractive color mixing in practice abound, since the specific color of objects results from "subtraction" through the selective absorption of portions of the light that are incident on them. Plastics, printing ink, dyed textiles, vegetation, minerals, glass, and photographs are colored because of subtraction of some part of the light incident on them. The characteristic of the subtractive system is that a black can be obtained with the proper proportions of a mixture of colorants. With different total amounts of colorants, but in the right proportion, a series of grays, rather than black, can be obtained but there is no way that white can be obtained with any positive mixture of (subtractive) colorants.

Subtractive color mixing has been known from prehistoric times as humans saw life around them and tried to imitate it by drawing scenes and objects to depict their observations. Knowledge of coloring matters most likely evolved by trial and error; in an effort to get special color effects, mixtures were made of the available colorants. The art of dyeing textiles similarly evolved from very early civilization, with the origin of the use of mixtures of dye colorants lost in the dim unrecorded past. Although the formality of subtractive color order systems is believed to have originated with Moses Harris (1766) in the eighteenth century, the knowledge and recording of recipes describing effects obtained by mixing colorants preceded that time by at least 1000 years.

Aristotle, the Greek philosopher, had much to say about the origin and arrangement of color. His influence extended well into the eighteenth century, when Goethe expounded his principles. It was said that all colors were derived from blends of black and white, depending on the relative strength or ratio of each. Because black objects when heated in a fire change to a crimson (red) color, it was reasoned that crimson came from black.

Many confusing, and sometimes conflicting, statements are attributed to Aristotle based on translations of "De Coloribus" (Aristotle, 1913).

The puzzle of the way in which Aristotle related all colors to black may not be so mysterious, considering the principles of subtractive color mixing. Black is obtained when the proper proportions of primary colorants red, yellow, and blue are mixed. This must have been known by Aristotle; thus, the reverse of the mixture could be regarded as the source of (subtractive) colors. We readily concede that all spectral light is contained in the white light of the sun, so black could be regarded in an analogous fashion. The statement in *Meteorologica* (Aristotle, 1923) that, "The colors of the rainbow are those which almost alone painters cannot make. For they compound some colors; but scarlet and green and violet are not produced by the mixture, and these are the colors of the rainbow" seems to support the view that Aristotle understood to some degree the principles of both subtractive and additive color mixing.

In his writings, Pliny referred to Aristotle's theory but suggested that there were three principal colors, red, purple, and heliotrope. But yellow was excluded from this system because "it is reserved exclusively for the mystical veils of females" (Birren, 1963).

Many of the surviving fine examples of ancient art in the form of cave drawings, ceramics, painted surfaces, textile fabrics, and colored glass are testimony to a fundamental understanding of the mechanism of subtractive coloration. Early references to mixing colorants come to us mainly from the classical period in Greece, a little from the Middle Ages, and the remainder from Renaissance painters (Halbertsma, 1949). It was recognized from very early times that the widest selection of colors could be obtained by using the colorants that were the most saturated (the least grayish), and these were considered to be primary or primitive colors. These could be mixed in different proportions to suit the needs of the artist; however, no palette was complete without the achromatic colorants, black and white, to give the range of tints, shades, and tones used. The textile dyer, on the other hand, generally had transparent colorants that automatically included the near-white substrate in the subtractive coloration system.

Robert Boyle, a scientist who preceded Newton by 50 years, gave some of the foundation for the discovery of the spectrum. Boyle's (1664) publication repudiated Aristotle through experimentation and demonstration. He made mixtures of white and black that did not produce a "blue, a yellow, or a red." He also went on to distinguish additive colors as light or atmospheric colors, which he named "apparent," and he described surface colors in the following way: "Genuine Colours seem to be produced in opacous bodies by reflection." A substantial amount of Boyle's treatise is speculation but credit must be given to his insight into the nature of color as we know it today. Later writings often could have profited from the clear statements of Robert Boyle.

Newton had used a triangle to represent the orderly relation of colors and the principle was taken up by Maxwell to represent surface colors. However, the triangular arrangement as a color order system did not originate with Maxwell, because Tobias Mayer in 1745 and Johann H. Lambert in 1772 each had proposed a triangular color solid (Birren, 1963) based on mixtures of colorants. Lambert even described a pyramid of planes (see Fig. 4) of mixtures of colorants having increasing amounts of white, with pure white at the top of the pyramidal stack. Maxwell's triangle was unique in that it was geometrically exact, being based on Newton's additive concepts of light. In succeeding years, following Maxwell's lead, many colorant-ordered systems have been developed; these are consistent only within the concentration range of the three colorant primaries that have been used (vide infra).

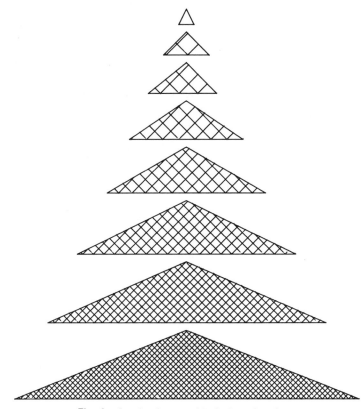

Fig. 4. Lambert's pyramid of triangular planes.

Helmholtz was the younger contemporary of Maxwell and Grassmann who seemed to be able to bring together Young's theory of color vision and pair it with the findings of Newton on additive color and subtractive mixing of pigments. He clearly demonstrated by experimentation that the two types of color mixing gave different results (Helmholtz, 1924). He further discounted Goethe's misconceptions and showed that the nonspectral color, purple, was the result of mixing the violet and red ends of the spectrum. In one of his most important works, Helmholtz collaborated with Arthur König, who performed critical experiments to which Helmholtz referred in his "Physiological Optics." König also collaborated with Dieterici (König and Dieterici, 1886) to provide the first set of color-matching functions, which ultimately were incorporated into the OSA colorimetry system, used until the adoption of the 1931 CIE system.

During the Helmholtz era W. von Bezold (1873) published a scientific paper on what is now termed the Bezold–Brücke color effect, as well as a popularized book (Bezold, 1876) on color theory and its application to design. Von Bezold has been criticized (Birren, 1963) for producing a pyramid that tops out with black and is opposite from Lambert's. However, he did mark the generation that clearly distinguished between color as light and color as pigment in order systems (Herbert, 1974). It was Höfler in 1883 (Gerritsen, 1975) who proposed a double cone with white on one end and black on the other with various (primary) hues arranged around a bulge between them. Rood (1895) also wrote about the same double-cone principle. This idea is the basis of many other color systems that have evolved over the years. Outstanding among the systems is the Ostwald system of 1917, but Titchener in 1887, Ebbinghaus in 1902, Pope in 1929, Hesselgren in 1955, Hård in 1968, Hickethier in 1940, Küppers in 1972, and Gerritsen in 1975 (Gerritsen, 1975) have all suggested double pyramids of some shape or other. In order of precedence, however, William Bensen (1868) conceived the idea of a cube set on a corner as representative of a color solid. White was placed in the uppermost corner and black on the base; red, yellow, green, sea green, blue, and pink were located on the remaining corners. Hickethier's (1952) color solid is organized similarly to Benson's except that all colors are derived from red, yellow, and blue inks.

A color solid based on a sphere (see Fig. 5) having white at the top and black at the bottom with spectral hues distributed around the equator was first proposed by Runge in 1810 (Birren, 1969). A hemispherical arrangement was adopted by M. Chevreaul (1890) wherein a hue circle with three primaries, red, yellow, and blue, set at the ends of equidistant radii was subdivided into 72 parts. The purest colors were placed on the first chromatic circle and nine more concentric circles with increasing amounts of black were part of the system. The charts conceived by Chevreaul were not exe-

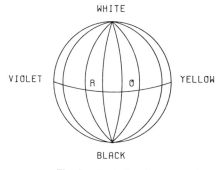

Fig. 5. Runge's sphere.

cuted as such, but he succeeded in organizing color harmony so that it could be taught to students of art. Chromatic colors were arranged so that complements, triads, and other chromatic effects could be demonstrated. These principles are still useful in arranging color harmoniously, even though they are not as scientifically logical as other concepts.

There are no general relationships that apply to subtractive coloration such as Grassmann's laws for mixing lights. In the special instance where light is passed successively through more than one transparent colored filter glass or clear plastics, the resulting amount of light is diminished by the value obtained by multiplying together the transmission values of the individual components of the stack, wavelength by wavelength. Furthermore, the concentration of colorants in many solutions of dyes and other colored materials follows the precise relation of Beer's (1852) law so that the effect of mixtures of colorants in solution can be calculated. The most common cases of surface colors, such as in dyed textiles, paints, or pigmented plastics, do not lend themselves to such rigorous solutions. In general, the analysis of Kubelka and Munk (1931; see also Judd and Wyszecki, 1975) is followed but is more complicated and less exact than Beer's law. Several examples are given below of the application of subtractively produced color in order systems that relate one color to another in a quantitative geometric fashion. In general, it is more advantageous to make subtractive mixture systems on some basis, such as amount of colorant, visual perceptibility, or instrumental responses, that is not necessarily directly translatable to colorimetric coordinates.

1. GENERALIZED MATHEMATICS OF SUBTRACTIVE MIXING

Three cases are given to illustrate the mathematical relations of several forms of subtractive mixing. Not all types of subtractive color mixing are covered in these examples, but they serve to show that the mixture is always less reflective or transmissive than the least component. Note that all quantities given are in terms of decimal fractions.

(1) Successive filters. If light of a given wavelength is passed through two stacked filters that have transmissions τ_a and τ_b, the resulting transmission is τ_m

$$\tau_m = \tau_a \cdot \tau_b \tag{2}$$

or

$$\log \tau_m = \log \tau_a + \log \tau_b. \tag{3}$$

(2) Mixtures of colorants in solution. If light of a given wavelength is passed through a solution containing two colorants that have separate internal transmittances τ_a and τ_b at unit concentration but in fractional amounts a and b, the transmittance of the mixture τ_m follows:

$$\log 1/\tau_m = a \log 1/\tau_a + b \log 1/\tau_b. \tag{4}$$

(3) Mixtures of colorants in surface colorations. For the simplest case take a mixture of two colorants that have separate reflectances β_a and β_b at unit concentrations but that are in fractional amounts a and b, respectively; the reflectance of the mixture β_m is as follows:

$$(1 - \beta_m)^2/2\beta_m = a(1 - \beta_a)^2/2\beta_a + b(1 - \beta_b)^2/2\beta_b, \tag{5}$$

where all reflectance values β are obtained on a sample layer that is opaque.

C. Partitive Color Mixing

Gerritsen (1975) introduced the term "partitive" to describe the phenomenon of mixing color by (1) spinning sectored disks; or (2) combining small areas of colors that can be visually separated at close range (50 cm) but blend when viewed from some distance (2 m or more). Aside from Maxwell's spinning disks, examples of partitive mixing are fairly commonplace—for example, three- or four-color halftone printing, textiles with differently colored yarns in close proximity, ceramic and glass mosaics, and pointillistic paintings. The fusion of the effect of the different colors in the eye is proportional to the area of each color on the surface of the material. The light reflected from the surface is added after it leaves the object. Because light additivity is involved, this category of color mixing is sometimes included in the mode of addition (Judd and Wyszecki, 1975) or called by other names, such as medial (Birren, 1969) or juxtapositional (Luckiesch, 1915). We must note the progress of color science with the experimental method of spinning disks used by Maxwell to establish mathematical relations between surface colors based on partitive principles. The

Maxwell disk method was mentioned by Ptolemy in the second century and revived by Musschenbroek in 1762 (Rood, 1895). It was used much later by Nickerson and co-workers (1978 personal communication) to check the Munsell system chips and describe the colorimetry of agricultural products. The importance of partitive color mixing should not be underestimated, since it was the best method for comparison of surface colors before the advent of photoelectric instrumentation.

It is interesting to speculate on the origin of what we now call the Munsell disk method of color mixing, which was so effectively used by Maxwell. Perhaps it was only a toy, but there could have been some experimentation on partitive color mixing done in very early times that has not been recorded in writings of that period. This is not to ignore the fact that the artists who constructed the great mosaics of the ancient Egyptian, pre-Hellenic, and Byzantine cultures must have understood the principles of partitive color mixing.

A form of partitive color mixing that is somewhat out of the ordinary is the mixture of differently colored fibers in a textile yarn. This is fairly common in fabrics made from blends of fibers that dye differently or when the same fiber is used in several colors to achieve special effects. Partitive mixing achieved in this way does not follow the same rules as other cases, presumably because the individual fibers are not completely opaque.

1. GENERALIZED MATHEMATICS FOR PARTITIVE MIXING

If light is incident on two areas of color with reflections R_a and R_b and they are small and contiguous or are sectors of a disk spun rapidly, the resultant reflection for their apparent mixture R_m can be calculated knowing the relative area a and b of each color if we use the following equation:

$$R_m = [a/(a + b)]R_a + [b/(a + b)]R_b. \tag{6}$$

III. THE STRUCTURE OF COLOR ORDER

A. Additive-based Color Order

The CIE system of colorimetry described in the chapter on colorimetry is a method for specification of color either as light or as subtractive surface color. Although it is not a color order system in the strictest sense, the chromaticity diagram is the most convenient vehicle for representation of the relationships among colors. In fact, the CIE system comes closest to being used in place of an order system for additive color mixtures; the predictability of the additivity of mixtures of light flux based on Grassmann's and Abney's laws has probably discouraged the development of a color order system for light mixtures.

B.   Subtractive-based Color Systems

Most of our experience since the beginning of time is with color that comes from subtractive or absorptively produced mixtures that are the basis of surface color. As an illustration of continued interest in subtractive color, we note that several ancient civilizations have left examples of artistry in which the color was achieved by mixtures of colorants rather than by single colorants. Thus, there has long been empirical knowledge about the effects of mixtures in altering appearance and increasing possibilities over what a single colorant could achieve. It is this body of knowledge coming originally from art that has been given order or organization to transmit information.

1.   COLOR MAPPING

Although the chromaticity diagram was specifically developed with lights allowing for additive mixtures to be correctly spaced according to their proportionate radiances when plotted on the diagram, the location of any coordinate point on the diagram indicates the chromaticity of the color whether, it is a light, a surface color, or the result of partitive mixing. This unique property makes it extremely useful for mapping, that is, showing the relation of one color to another and at the same time relative to the visible (light) spectrum and to the illuminant that was used in calculating its chromaticity coordinates. This is analogous to a geographic map showing the borders of a "country," color, defined by the spectral locus plus the purple limit line.

Many maps of the color domain of surface colors have been shown using the chromaticity diagram for display. The most commonly reproduced map of surface colors comes from Kelly and Judd (1955) (Fig. 6), which assigns names to various regions of the chromaticity plane. Although such a map is somewhat idealized for surface colors, as is discussed below, it serves to show the relation of all colors, irrespective of luminance, as they are projected onto a common plane and named according to a regular naming classification, discussed in Section III.E.4.

Many maps have been drawn for the display of chromaticity changes with change in concentration of colorants. This has meant that the luminance for all samples is not necessarily given and the chromaticity of the various samples has been projected to a common plane. It is to be expected that colorants exhibit individual behavior insofar as change in hue and saturation with concentration is concerned. A typical set of data on individual pigments is from Ruth M. Johnston (1973) and is shown in Fig. 7. Ralph M. Evans (1948) gives a three-dye grid for transparent mixtures of cyan, magenta, and yellow dyes that have been used in photographic emulsions (see Fig. 8). These diagrams are composite pictures of the range of chroma-

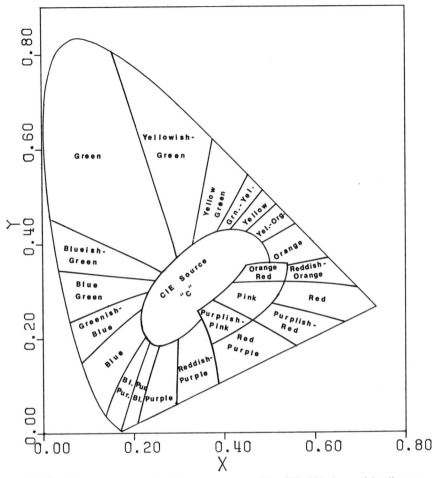

Fig. 6. Common names assigned to various areas of the CIE 1931 chromaticity diagram (after Kelly and Judd, 1955).

ticities that are possible with colorants but must be considered with the idea that only certain colors can be achieved at a given luminance. This is discussed in the next section.

2. OPTIMAL COLORS

Several principles limit the gamut of absorptive colors and make them fall short of the spectrum locus that is light. Once the third parameter, luminance factor, is considered, the chromaticity is limited for real colors at every level of luminance factor. Conceptually the limits for reflecting or

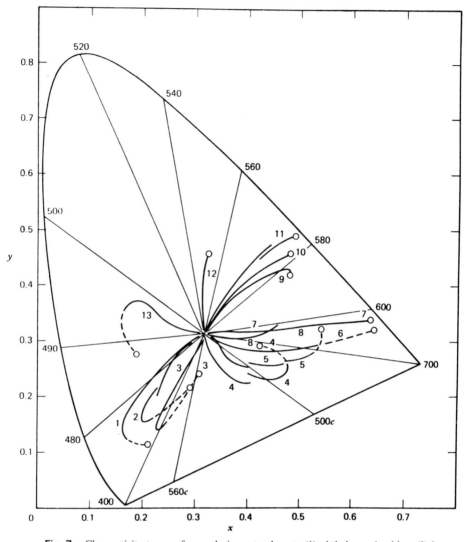

Fig. 7. Chromaticity traces of several pigment colorants (1) phthalocyanine blue, (2) indanthrone blue, (3) carbazole dioxazine violet, (4) quinacridone magenta, (5) lithol rubine (transparent), (6) BON red, (7) molybdate orange, (8) vat orange, (9) flavanthrone yellow, (10) light chrome yellow, (11) yellow FGL (monoazo), (12) chromium oxide green, (13) phthalocyanine green (after Johnston, 1973).

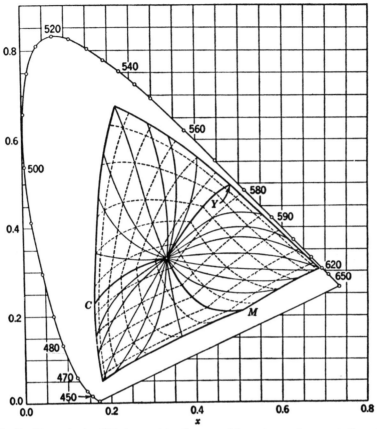

Fig. 8. Traces in the CIE chromaticity diagram of dyes at several concentrations (after Evans, 1948).

transmitting colors are based on what are termed ideal or optimal colors that have straight-line spectrophotometric curves. These curves have reflectance or transmittance values of either 0 or 1.0, going from one to the other as straight lines perpendicular to the wavelength axis. Four types of ideal curves can be constructed, as shown in Fig. 9, where the wavelength of the line perpendicular to the abscissa moved along the wavelength axis to achieve different optimal colors.

Optimal color limits in chromaticity were computed by Schrödinger (1920) and Rösch (1928). MacAdam (1935) reported on the maximum possible chromaticities that could be achieved for materials with given luminances under CIE illuminants A and C. Limiting contours are shown in Fig. 10 for CIE illuminant D_{65}. Atherton and Peters (1955) refined the optimal limit contours for pigmented surfaces by using ideal color curves

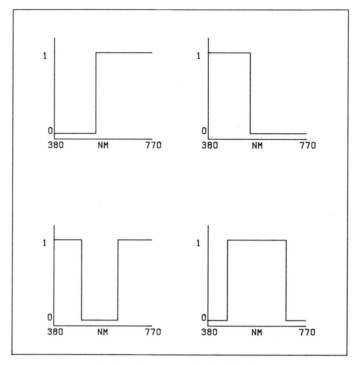

Fig. 9.   Four types of optimal colors.

that bottomed out at 4%, rather than 0, to account for surface reflection. Depending on the material, the minimum reflection that can be practically achieved is from about 1 to 4% regardless of how dark a surface color is produced. On the other hand, transparent objects such as photographic films, glass filters, transparent plastics, and the like, can achieve nearly total absorption (or zero transmission) in certain instances.

When the contours for the optimal colors at each luminance factor are joined, a color solid is obtained that describes the bounds of reflecting and transmitting colors that fall considerably short of the spectrum at higher luminance levels. Luther (1927) and Nyberg (1928) developed color solids for optimal colors before the 1931 CIE color-matching functions were adopted. The idea of an optimal color solid is incorporated into the DIN-6164 (1964) color cards with modifications developed by M. Richter (1953, 1955).

Ideal colors are also the fundamental assumption of Ostwald (see Foss *et al.*, 1944), which implies that the Ostwald system should in fact be based on optimal colors; however, all the collections of Ostwald systems ter-

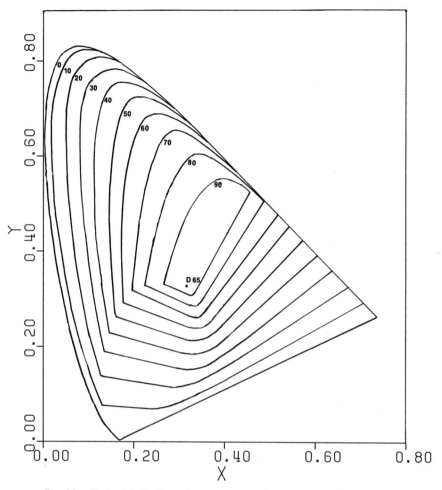

Fig. 10.   Optimal limits for surface colors at various luminance factor (D_{65}).

minate in the most saturated colorants availabe, considerably lower in purity than the optimal color limits at any given luminance factor.

C.   Colorant Mixture Ordered Systems

1.   PRIMARY COLORS

If we think only about the effect of color as we see it, the most dominant characteristic of any colored object is whether it has hue (i.e., is red, yellow, orange, blue, etc.) or whether it is without hue (i.e., is gray, black, or white).

Once the distinction is made that there are chromatic and achromatic, or hueless, colors, a deeper insight into the phenomenon of color is possible; it is the chromatic colors that can be arranged in some sort of order, as Newton did with his seven colors of the spectrum. Hue is therefore primary to the systematic description of color. Whether Newton selected seven primary hues because of the influence of Maurolycus, whether he was making an analogy to music, or for any other unknown reason, the number of hues is not important except that it defines the limits of colors that are obtainable by mixing adjacent ones taking them two at a time.

Any number of hues greater than three could be chosen to describe a color gamut bounded by mixtures of the pairs of them. Many three-primary systems have been suggested, but the most usual combination is red-yellow-blue or red-green-blue (athough not necessarily the same red and blue hues in both instances). Using either set of three primaries, a set of three secondary colors gave six hues—red, orange, yellow, green, blue, and violet—that were the bases of many methods of teaching color in schools in the late nineteenth and early twentieth centuries (see Prang *et al.*, 1893; Rood, 1895; Jacobs, 1927). In these instructions, reference is frequently made to the visible spectrum and the matter of differences between light mixing and surface colors is generally explained. But the neutral point is frequently called "white" for subtractive mixtures when the result is actually gray or black.

Munsell (1905) selected five hues for his primaries and constructed a circle of hues divided into 100 intervals. Although the choice of five was unusual, a decimal-based division was obviously convenient for creating subdivisions. The spacing of the hues in the Munsell system was carefully done, with many observations and adjustments. Indow (1974) used multidimensional scaling techniques to show that the spacing could be improved. Some of the other hue circles had resulted mainly from the selection of colorants and the intermediate spacing was in even proportions of the adjacent pairs.

The Munsell system is used to teach color principles to students, particularly in training for the fine arts, as described by Luke (1976). Another group of art teachers could be classified as adherents to the Ostwald system (Judson, 1935; Birren, 1944; Ostwald, 1916; Pope, 1968), which is based on the fundamentals of opponent color theory given by Hering (vide infra); this system usually involves a 24 hue circular arrangement of primary and secondary hues. Color theory based on subtractive color mixing can be learned through any reasonable system of primaries. The question that should be answered is, "What is the maximum color gamut that can be produced by mixing"? The artist and technologist are always faced with the problem of minimizing the number of materials with which they work so as to have an understandable and economical group.

Leonardo da Vinci described the minimum working pallette of pigments as red, yellow, green, blue, white, and black. But he mentioned orange and special turquoise pigments, since he was not above using special effects. This indicates that no practitioner of color mixing will restrict himself to primary colors to satisfy all the effects that can be reasonably conceived.

2.   UNITARY HUES

For a long time there have been attempts to find the best primary hues. Moses Harris (1766), who is credited with the first color order system, described the primaries red, yellow, and blue as "primitive" colors, which are mixed to give the color gamut. Ewald Hering (1964) described four primary colors with the specification that primary yellow (*Urgelb*) is a yellow that shows no trace of red or green and primary blue (*Urblau*) is similarly described. There is also a primary red (*Urrot*) and a primary green (*Urgrün*), which are neither bluish or yellowish. These are defined as unique hues. Sometimes they are called "unitary hues." Their uniqueness lies in the fact that no other hues are singular nor unitary in appearance; all other hues may be seen, with varying degrees of ease, to be combinations of two such hues. These other, mixture hues are often called "binary hues," in contrast to the unitary hues. Hering referred to the unitary hues as the basis of the Natural Color System. Burnham *et al.* (1963), Judd (1951a), and Newhall (1953) summarized the spectral wavelengths found for unitary hues under neutral or dark adaptation: unitary yellow = 577 nm; unitary blue = 473 nm; unitary green = 513 nm; and unitary red = 521C nm, (C is the complementary wavelength of a purple color). These values are only means and vary from observer to observer and experiment to experiment for the neutral adaptation condition. The wavelengths that elicit unitary hue responses change with variations in chromatic adaptation. Each of these wavelengths is only indicative of a range of dominant wavelengths that are equally acceptable as uniquely different from others. The concept can even be extended to regions of the chromaticity diagram to accommodate the range of wavelengths necessary to describe hue differences in words (see Simon, 1972). The unitary hue names of Hering have found widespread use as general names for hues.

Although the choice of specific primaries is highly subjective, this principle has been adopted by Ostwald and the Swedish Natural Color System. The success of the application of the concept is limited by the choice of available colorants and their stability when viewed under various common light sources. On the other hand, the concept is extremely useful even though the actual unitary hues are not exactly achieved in practice. There are many color systems that relate to the red-green and blue-yellow unitary hues for opponent stimuli since this principle is both conceptually simple and easily reduced to instrumental practice.

3. COMPLEMENTARY COLORS AND CONSTANT HUE LINES

Complementary colors are of special interest to the artist for producing harmonious combinations and to the scientist for consistency with the concept that complementary pairs of colors are "opposites." By definition, complementary colors produce a neutral (hueless) color when mixed in proper proportions. This is straightforward in terms of additive systems because the line connecting the chromaticities of two components in a mixture is straight and the mixture is predictably the color of the neutral reference light to which the observer is assumed to be adapted. A similar linear construction for subtractive mixtures does not exist, since the hue of a surface color purity series does not plot as a straight line on a chromaticity diagram. The dominant wavelength line that intersects the illuminant point, the chromaticity coordinates of a sample color, and the spectrum locus is useful for specification of additive mixtures but does not describe complements for subtractively produced colors. When colorants are mixed, their chromaticity loci are curved, and the amount and kind of curvature are related to physical properties of the particular colorants used in a given subtractive mixture. Different pigments or dyes yield loci of different curvatures.

There may be other reasons for curvature of loci in a chromaticity diagram. For example, the Munsell color order arrays samples along three dimensions or attributes, one of which is hue. When constant Munsell hue is plotted in a chromaticity diagram (see Fig. 11), the resultant loci are generally quite curved. That curvature does not relate to the physics of subtractive color mixture but to the variation in chromaticity required to maintain constant hue; the curvature would be the same for additively mixed colored lights viewed under the same conditions (Fig. 11). Theoretical support for the curvature comes from Schrödinger (1920) and further supporting data are based on Stiles (1946) line element transformation to the CIE x, y chromaticity diagram. Hue lines are not straight in the x, y diagram because they are projections of uniform geodesics (lines of the shortest path for a surface with curvature) and the x, y diagram is not visually uniform. Since the CIE 1931 x, y, Y space is not one of equal visual intervals, it represents a curved surface in visual cross section on which the shortest *visual* distance between two points is a curve. Using color discrimination data, MacAdam (1971) developed a metric, xi-eta, by transformation from CIE so that constant hue gave straight lines in this diagram.

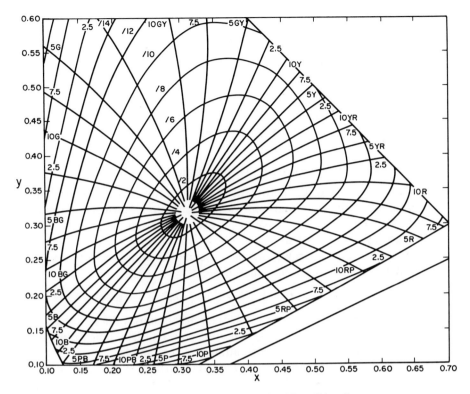

Fig. 11.   Chromaticity diagram showing Munsell hue lines.

Much attention was given to complementary colors in Maxwell's time (Maxwell, 1857) using spinning disks. Central disks were of black and white matched to chromatic ones on the periphery. Maxwell called them color tops, as he used them. This is an accurate method for obtaining complementary colors, but it is dependent on the choice of available disks describing the colors and neutral that can be obtained.

A small historical note is taken from Rood (1895), who devoted much attention to obtaining complementary colors for "strongest possible contrasts." Reference is made to a schistoscope developed by Brücke to see complementary colors. This is a polariscope combined with a low-power microscope. If a thin piece of mica is placed over a square opening in a mask between the crossed polarizer and analyzer of the schistoscope, two images are seen that give a precise complement of each other. These are the complementary colors obtained by interference. The range of color obtained is a function of the thickness of the piece of mica. Although the schistoscope is useful for observing complementary interference colors, it does not satisfy the problem of complementary hues for surface colors. Rood describes a method using the double imaging obtained when observations are made through calcite (iceland spar). If two presumably complementary colored samples are imaged coincidentally through a calcite prism, they should blend to make a neutral gray. This is a quick test for checking complementary colors, but it requires an assortment of samples to make matches for gray. The Maxwell disk method is preferred because of its flexibility.

4. Complementary Tristimulus Colorimetry

The nonadditivity of the chromaticity diagram for mixtures of colorants has prompted some work that improves this situation. Normal colorimetry is done with the calculations based on equations given in the chapter on colorimetry. An abbreviated form is followed here. However, if the quantity $\log 1/\tau$ is substituted for τ, we then have a complementary set of tristimulus values based on the following equation:

$$T' = \int \bar{t}_\lambda S_\lambda \log 1/\tau_\lambda \, \Delta\lambda, \tag{7}$$

where $T' = X', Y', Z'$, the complementary tristimulus values; $\bar{t} = \bar{x}, \bar{y}, \bar{z}$, the color-matching functions; $S = $ spectral power distribution for an illuminant; $\tau = $ spectral internal transmittance of the sample.

Flaschka (1960) showed that colorant data in this system are additive and that points on a chromaticity diagram in terms of the chromaticity coordinates x' and y' calculated from

$$x' = X'/(X' + Y' + Z') \quad \text{and} \quad y' = Y'/(X' + Y' + Z') \tag{8}$$

place any colorant at a unique point on the x', y' diagram. Then mixtures of colorants at unit concentration can be calculated using Newton's center-of-gravity method. The specific concentration of the dye must then be calculated from the relative contribution to the Y' value. Rounds (1969) reported on the same method used for soluble dyes.

These coordinates are similar to conventional x, y coordinates with a gray point in place of the illuminant white point. They derive the name *complementary* from the fact that if the spectral locus is traced onto the x', y' diagram, the dominant wavelength for a given colorant is approximately the complement of the same colorant point calculated for the x, y diagram. However, the term "dominant wavelength" is meaningless in the complementary chromaticity diagram defined by Eq. (9).

Ganz (1965) developed what he called the Ciba-Q-method, which uses complementary tristimulus values based on the substitution of the Kubelka–Munk (K/S) function for log $1/\tau$ in Eq. (9):

$$XQ = \sum \bar{x}_\lambda S_\lambda (\text{K/S})_\lambda \, \Delta\lambda,$$
$$YQ = \sum \bar{y}_\lambda S_\lambda (\text{K/S})_\lambda \, \Delta\lambda, \qquad (9)$$
$$ZQ = \sum \bar{z}_\lambda S_\lambda (\text{K/S})_\lambda \, \Delta\lambda,$$

where XQ, YQ, ZQ are the complementary tristimulus values; $\bar{x}, \bar{y}, \bar{z}$ are the color-matching functions; S is the spectral power distribution for an illuminant; $(\text{K/S}) = (1 - \beta_\infty)^2 / 2\beta_\infty$; β_∞ = reflectivity, that is, the reflectance of a colorant layer so thick that further increase in thickness does not change its reflectance.

A chromaticity diagram xQ, yQ results from the following:

$$xQ = XQ/(XQ + YQ + ZQ) \quad \text{and} \quad yQ = YQ/(XQ + YQ + ZQ). \quad (10)$$

Individual colorant chromaticity values are plotted on the diagram, which permits graphical calculation by the Newton center-of-gravity method. Although it can be hypothesized that any colorant will plot at the same place relative to the gray point, surface colors do not behave as predictably as transparent colorants, and shifts in hue are quite common. Nevertheless, reasonable approximations can be achieved by the graphical method. A similar method was published by Hoffmann (1971) under the name HOE-Q.

Although graphical techniques are feasible, they have given way to numerical methods based on least-squares or iterative numerical solutions to the problem of color formulation, which are used to calculate the concentrations of combinations of colorants in mixtures.

5.   THE OSTWALD SYSTEM: WHITE, BLACK, AND FULL COLOR

Wilhelm Ostwald began his intense interest in color at the age of 60, and through his writings he (Ostwald, 1916, 1917, 1918–1940) gave the world a system that is partly based on scientific principles but to a larger degree represents the creative thinking of a dedicated mind. His major effort in developing a color system occurred during the height of Germany's involvement in World War I, when Ostwald was more or less isolated from contact with other workers. In part, the Ostwald arrangement of color space uses the concept of a double pyramid with white on the top and black on the bottom and a circle of full colors, called semichromes, around the equator of the solid. Like many workers before him, Ostwald based his system on Hering's principles of color, using the four unique primaries—red, yellow, green, and blue—but he divided the hue circle into 24 parts, including secondary hues. A scale of grays was made the central pole and these were spaced according to Weber–Fechner principles so that the steps were the geometric mean between upper and lower neighbors. This means that the white content of the gray scale differs according to the antilogarithm of the numbers 1.0, 0.9, 0.8, 0.7, 0.6, . . .; or 0.95, 0.85, 0.75, 0.65, . . .; the difference between the amount of white and 1 is the black content of the gray scale step. Each of these steps is designated with a lower case letter starting with *a, c, e, g, i, l, n, p*; these form a typical series of grays that use every other possible step. Then if the full color, one of the 24 hues, is placed at the point of a triangle formed with the gray scale, the outline obtained with white, black, and full color encompasses the color gamut obtainable with these three elements. The spacing used in the *Color Harmony Manual* (Container Corporation of America, 1942), which is a collection of Ostwald-arranged chips, includes eight steps between white and black, eight between full color and white, and eight between full color and black. The white to full-color series is called a tint series and the black to full color is called shade; the mixtures of black and white are grays, which are the terms used to describe the types of compound colors given by the system. All mixtures of full color, black, and white are simple arithmetic combinations of all three components so that the sum of the three equals one:

$$W + B + C = 1, \tag{11}$$

where W is the fraction of white in the mixture, B the fraction of black in the mixture, and C the fraction of full color in the mixture.

Despite the fact that the gray scale is spaced according to Weber–Fechner fractions, the system is ordered by simple colorant mixture and does not attempt to achieve visually equal suprathreshold increments between adjacent chips.

Ostwald principles are based on ideal colors that can be represented graphically as spectrophotometric curves made from straight lines, as in Fig. 9. The optimal colors that would be the hue limits also follow the optimal color limits for colors that contain neither white nor black. Unfortunately, optimal colors are not achievable in real life and this distorts any color solid developed from Ostwald principles. This is a basic premise that has never been satisfactorily resolved by the adherents of the Ostwald principles, although it is conceivable that a system might be based on imaginary end points. This has not been done in any execution of the Ostwald system probably because it destroys the symmetrical triangular arrangement of any page of the collection.

Some artists are particularly fond of the Ostwald system because of its superior portrayal of the harmonious relationships among colors (Birren, 1944). The fact that full hue colors (semichromes) are not scaled according to an equal luminous parameter such as Munsell value is a feature that is particularly useful to show harmony by geometric analogy. For instance, the arrangement of Munsell requires that for a harmonious combination of yellow and blue a diagonal slice must be taken through the color solid, whereas Ostwald isovalent colors—that is, those with equal white and black contents—are given the same designation for all hues. The other advantage to an artist is that modification of any hue is achieved simply by mixing a hue with white to obtain tints and mixing with black to yield shades; this simplifies the task of creating and visualizing color effects rather than organizing all colors into a perfect geometric arrangement.

From a scientific point of view, Ostwald satisfied the meticulous definitions of a gray scale limited by an ideal white equal to barium sulfate and an ideal black conforming to Kirchhoff's absolute black of a box with a hole in it. Even though the white is only about 98.5% as reflective as absolute white, there is little scale distortion because of this difference. On the other hand, using actual colorants to bound the color solid rather than using ideal colors does not permit the inclusion of some real colors, as was first pointed out by Nyberg (1928). Many scientists are not satisfied with the use of alphabetic rather than numerical designations for color scales, which precludes easy interpolation to achieve in-between and precise description of the entire color gamut. It would seem, therefore, that the Ostwald system has missed its mark in satisfying both the scientific and art worlds at one stroke.

6. HALFTONE REPRODUCTIONS

The most inexpensive way to reproduce graduations of color mixtures is with the halftone method of color printing. This commercial method employs patterned blocks of engraved dots that vary in density (dot area) from one side of the plate to the other and from the top to the bottom. The dots are small enough to be unresolved by eye. First, one pure color is printed on the paper; then the second color is printed on top with a screen that is opposite to the first in dot density progression. Since the dot pattern is not precisely registered between the first and the second color patterns, there are both subtractive color mixing when they coincide as well as partitive color mixing when the dots do not superimpose. In most instances, small patches of the base white paper are also mixed into the composite color seen by the eye.

The most famous halftone system is that of Maerz and Paul (1950), "Dictionary of Color," which consists of 7056 small samples, 12 × 15 mm, arranged in a 12 × 12 array on each page. There are eight base colored inks and seven base gray inks fundamental to the system. The gray inks are used to achieve a series of colors that are the binary chromatic ink mixtures except that seven additional sets of charts are obtained by first printing one of the grays and then the other two chromatic inks in various proportions on top of the gray. With a set having only white paper as the background, eight steps of grayness are obtained. The dictionary is cross-referenced with several thousand color names and is a very useful compendium of color names.

Hickethier (1952) published a manual of 1000 printing ink samples obtained from mixtures of yellow, red, and blue inks. The proportion of each colored ink in the mixture is represented by a positional digit in a three-digit number. Thus, the amount of yellow is indicated by the first digit, 000–900; the amount of red by the second digit, 000–090; and. the amount of blue by the third digit, 000–009.

The Villalobos and Villalobos (1947) "Color Atlas" is a halftone system that covers the color solid in much greater detail. Unlike the Maerz and Paul dictionary, in which an entire page is covered with a single gray printing, the Villalobos appears to have 20 rows of grays on every page graded from the white of the paper to black. The density of dots for each of the hue inks is decreased from the top of the page to the bottom, whereas the black dots run counter to this. Thus, the scale from saturated color to gray or black is obtained through 12 steps. There are 38 hues, one on each page, which are broken down into 12 distinct color names. The Atlas has 7279 color chips, which are 1-cm square with a 4-mm-diameter hole in each sample that permits close comparison with an unknown sample. The system is quite well executed but does not have the benefit of colorimetric analysis, which could relate it to the CIE system. It is no longer available in the United States.

The Wilson-Color System (British Colour Council, 1942), developed for designating the color of flowers, had the basic concept of designating hues according to 64 divisions of the most saturated hue circuit, given by numbers. Derivations from the 64 hues were described in two ways: from light to dark with a prefix number, 4–10, going from white to black and with additional prefix zeros for increasing desaturation. The individual samples are 4.4 × 3.2 cm and the collection contains about 800 of the possible 2000 colors designated by the system. The Royal Horticultural Society has reissued the Wilson-Color System in a revised form as a looseleaf booklet of narrow pages, providing easy comparison of the colored samples with flowers.

7. PIGMENT MIXTURES

The simplest way to cover a region of color space with surface colors is to assemble several bright colorants and mix them, systematically altering the concentration of each colorant by small increments. White and black pigments are used, together with the chromatic colorants, to achieve the maximum range of color. Even though black can be achieved with proper mixtures of chromatic colorants, carbon black or iron oxide black pigments are inexpensive and generally stable so they are used whenever it is possible to achieve the color. A white pigment such as titanium dioxide is both an efficient opacifier and a lightening component for colors that are less than completely saturated and can be used in simple mixtures with black to produce a gray scale. Therefore, most of the colors in a system of pigment mixtures are combinations of one or two chromatic colorants with amounts of black and white, depending on the lightness and saturation of the color. The increments of concentration of the colorants among adjacent samples are purely arbitrary but are usually in the range 10–20% change from sample to sample. Naturally, large steps among chips give fewer samples in the collection; experience with any sampling of color space will dictate the reasonableness of the steps among the samples.

At the present time, paint manufacturers are the most enthusiastic users of pigment mixture systems. Many collections are made to show the possibilities that can be obtained using 7–11 chromatic pigments, plus black, that are available through commercial dispensing systems. The formula for obtaining a color similar to the sample is given so that so much of each pigment can be added to a quantity of "tint base," which consists of the paint chemicals and white pigment. A great variety of colors can be obtained in this way. The color gamut is usually not sampled in any uniform way, but areas that have sold well are represented by more samples and those of small interest are stepped over at relatively infrequent intervals. Collections of samples for display usually number between 500 and 1200 individual chips. Such a procedure allows the retail outlet for the paint manufacturer the luxury of having many colors available to the customer, yet requires an inventory of only primary materials. Color cards for paint mixing stations are generally in some looseleaf form so that new colors, together with their formulae, can be added from time to time as the market preferences change.

Custom color formula collections are also popular in printing ink sales. Two of the earliest examples of this type of color system were the "I.P.I. Color Finder" (International Printing Ink, c. 1950) of the company now known as Inmont Corporation and the "Glenn–Killian Color System" of the Glenn–Killian Ink Company (1950). These were collections of colored chips that showed the effect of mixing various proportions of six or eight primary ink pigments. The formula associated with each chip gives the proportions of each of the primary colorants. If the requisite colorants are in hand, the recipe can be followed to give the intended color. Alternatively, a mixture could be ordered from the manufacturers.

A system of printing ink colors has been developed by Pantone, Inc. (1978), that consists of a base collection of 563 colors printed on several paper stocks. This collection, called the "Pantone Formula Book," gives the basic, numbered standards that are the key to the system. The Pantone set of color chips is inexpensive and has been licensed by many ink manufacturing companies, who provide recipes for each Pantone color in terms of their own primary inks. In addition to ink and printing applications, special pens are available in Pantone colors as well as overlays and colored papers that allow the industrial designer great flexibility in making layouts using the system colors. When particular colors are needed that are not in the formula book, a computer-assisted formulation program is available as well. This is probably one of the most comprehensive color order systems that has been devised. However, because of the nature of printing ink colorants, the gamut of colors that can be achieved is not as broad as that normally encountered in textile and pigmented plastic materials. Many manufacturers use the Pantone system to designate colors that are to be formulated for pigmented colors other than for printing. Several of the Pantone collections of samples consist of pages of representative colorant mixtures which provide a means of color specification that promotes understanding between buyer and seller.

In the 1920s a system of paint mixtures was developed in Germany under the English name of the Baumann Color Guide (Baumann, 1925). It consisted of 1309 color chips with formulas. Each chip was given a three-part designation: The first letter signified the "excess of color" of one of 24 hues; the succeeding number, the luminosity; and the last number, the "bluntness or dullness." Lower numbers for a given hue indicated brighter and more intense colors and higher numbers, darker and duller colors. There were several editions of the Baumann system and an English-language edition was distributed in the United States by G. Plochere. During the World War II period, when the Baumann system was no longer available, Plochere and Plochere (1948; see also Middleton, 1949) brought out a counterpart to the Baumann Color Guide, but patterned more on Ostwald principles and physical arrangement. Eleven pigments were mixed, two at a time, to give 26 hues. These were systematically mixed with white and black to give 48 samples for each hue, or 1248 chips in the entire collection. Each sample was designated by three characters: a capital letter for hue; a number for black content; and a lower case letter for white content. Six levels of black content and eight levels of white content were used. The four principal hues of red, yellow, blue, and green used by Ostwald are augmented by Plochere with purple and orange as well as compound hue designations, such as Gb and Gbb to indicate greenish hues with increasing displacement toward blue.

Color guides such as Baumann and Plochere were among many systems developed to aid artists, designers, decorators, and other people who used color in applications of paint and decorative products. Unfortunately, the elaborate and somewhat expensive collections gave way to the color cards of paint manufacturers, who made them available at no cost for commercial promotion of their products.

8.   DYE MIXTURES

Many small collections of dyed fabrics and yarns have been made by dye manufacturers over the past years to display the possibilities of their products in producing what are termed "compound shades," or mixtures of two and three colorants. One notable publication that attempted to cover a considerable gamut of colors for a number of fiber substrates was the Colorthek (BASF, 1964). This was published in two forms; one is a collection of 1600 small samples dyed on wool fabric and the other a collection of about 1500 samples dyed on a filament acetate fabric. Formulas were given for all the colors dyed on other substrates such as acrylic fabric for the wool colors and nylon fabric for the acetate samples. Although the effect of metamerism is always present for matches other than the particular colorants used in the book sample, a starting formula could be useful in certain circumstances.

Collections of dyed patterns that attempt to cover the range of colors encountered in textile fabrics and yarns are so varied that dye manufacturers have abandoned issuing them in favor of instrumental-computer computation methods, which are in widespread use today.

Another method of designating color collections for textile materials comes from the world of styling and design. The Textile Color Card Association, now succeeded by the Color Association of the United States (1941), has been publishing seasonal and general color cards for more than 50 years. The ninth edition of the "Standard Color Card of America" contains 216 color swatches with cable numbers used for reference and cataloguing purposes. Although the collection has a somewhat random arrangement, it, together with the seasonal color cards, is of value in communicating information, particularly with respect to style trends. Swatches from the ninth edition and the United States Army Color Card (Textile Color Card Association of the United States, 1943) were measured spectrophotometrically and catalogued in cooperation with the National Bureau of Standards (Riemann et al., 1946).

D. Systems Ordered According to Uniform Perception

1. THE GRAY SCALE

Following the idea that all colors belong to two general classes, chromatic and achromatic, a simplified explanation of perceptual color scaling can be made if the gray scale is considered first. The concept of spacing a series of grays that are arranged according to how they appear to an average observer was first described by Plateau (1872), who constructed a five-step visual gray scale, which was shown to have different proportions of black and white, from simple mixtures that also produce a series of grays. This was the first of many attempts to perfect the psychological scale, lightness, which gives a visual correlate with the physical scale, luminance factor. The purpose of many of the investigations has been to develop a mathematical description of visual response.

a. Logarithmic Scales The general approach to scaling gray samples was the subject of earlier investigations by psychologists concerned with visual response as well as all other natural sensations. E. H. Weber (1834, 1846) had been credited with the principle "that equal relative stimulus increments correspond to equal threshold increments in sensation." When applied to visual phenomena, this principle is called the Weber–Fechner law because of Fechner's (1889) derivation of a measurement equation based on Weber's principle (see the chapter on colorimetry). According to the law in mathematical terms, a series of gray samples with corresponding CIE Y values, will appear to be separated visually when taken two at a time according to the following relationship:

$$L = (Y_1 - Y_2)/Y_{avg}, \tag{12}$$

where L is the lightness scale and Y the CIE luminance.

If the steps of $(Y_1 - Y_2)$ are small enough to be considered dY, then from calculus the integration of the increment gives a scale

$$L = \int dY/Y = k \log Y. \tag{13}$$

Delboeuf (1872) modified the simple Weber–Fechner law to take care of the negative and infinite values of the logarithm of Y that occur when Y approaches zero:

$$L = c \log [(Y + k)/k]. \tag{14}$$

Several workers of this period confirmed Delboeuf's findings and ultimately Plateau retracted his power function lightness scale in favor of that of Delboeuf.

Godlove (1933) did a careful comparison of the actual Munsell sample data with several of the logarithmic and power function forms of lightness scales then in existence. Since the Munsell gray scale was established by visual scaling under carefully controlled observational conditions, the mathematical fit of the measured reflection data was important to relate physical data to psychological observation. This would also give the means for making precise interpolations for intermediate values. Godlove also attempted to settle whether the Munsell value scale was "better" than logarithmic scales for describing gray scales.

The Ostwald system uses the logarithmic scale of lightness. The gray scale of the Color Harmony Manual, according to Foss et al. (1944), is not based on mixtures of ideal white and black but on a practically achievable white, $Y = 89.1\%$, and a good commercial black, $Y = 0.89\%$. Then the relation is patterned after Delboeuf:

$$L = K[(20 \log Y) - 1]/4. \tag{15}$$

Richter (1953) also followed the Delboeuf form of relationship in the DIN–Dunkelstufe scales except that the scale numbers are reversed from the usual convention of following the same order as the luminance values

$$D = 10 - 6.1723 \log(40.7h + 1), \tag{16}$$

where D is the Dunkelstufe value and h the $Y/Y_{\text{optimal color}}$.

b. *The Munsell-Value Function* A. H. Munsell in the early 1900s began experimentation with his system, which was based on the three-dimensional concept of hue, value, and chroma, discussed in full below. One of his main concerns was the relation of all colors to an achromatic gray scale that ranged from black to white and had the fundamental principle of equal visual steps between adjacent samples. Probably the idea developed from his association with members of the scientific community in Boston, who certainly must have influenced his thinking. The original gray scale developed from many attempts to produce a set of samples that "looked right," although little was known at that time about standard illumination and the effect of background.

Over the next 20 or 30 years after Munsell's original work, many refinements were made to the first set of grays that Munsell developed. Yet at all times the principle of equal visual spacing was maintained and certainly improved as its usefulness was demonstrated. It was Dr. I. G. Priest of the National Bureau of Standards who first related the Munsell gray scale to the physical measurement of reflection, reporting this in 1920 (Priest *et al.*, 1920). This report gave the earliest quantitative description of a perceptual visual effect. Priest went on to define the Munsell gray scale as an equation:

$$V = 10R^{1/2}, \tag{17}$$

where V is the Munsell value and R the reflectance (0–1.0).

Although this is not a very precise definition of the Munsell value, it has been a sufficient scale to be used by many workers for years. Hunter (1940, 1942) adopted the relation in the early tristimulus photoelectric colorimeters and continues to use the square-root derivation of CIE luminance Y as a psychological lightness scale. This use is common in other colorimeters produced at present and the scale is used in many industries that have to make small color difference measurements. The simple conversion of CIE responses of the colorimeter is accomplished electrically with simple parallel resistance circuitry, which provides the output:

$$L = 10Y^{1/2}, \tag{18}$$

where L is the lightness scale that is analogous to Munsell value and Y the CIE luminance for a particular illuminant, usually CIE illuminant C.

Subsequent studies (Munsell *et al.*, 1933; Godlove, 1933) indicated that the square-root formula applied best to gray samples viewed against a white-appearing surround. It was suggested that when samples are viewed, instead, against a gray surround of about Munsell value 5 (actually, 0.191 luminance factor), a better representation would be

$$V = (1.474Y - 0.00474Y^2)^{1/2}. \tag{19}$$

This equation was probably the best functional simulation of the dependence of Munsell value on luminous reflectance in terms of measurements then available. As more accurate and precise reflectance data on the Munsell samples became available, it was decided that changes should be made to the value equation and these were carried out by a subcommittee of the Optical Society of America (OSA).

When the OSA subcommittee that worked on the spacing of the Munsell system published its report (Newhall *et al.*, 1943), a mathematical definition was given to the value scale, as follows:

$$Y/Y_{\mathrm{MgO}} = 1.2219V - 0.23111V^2 + 0.23951V^3 - 0.021009V^4 + 0.0008404V^5. \tag{20}$$

This was established by fitting a curve to the plot of the luminance values determined from the spectrophotometric measurement of a standard set of Munsell gray samples. Once this relationship was defined, the psychophysical scale of lightness enabled the scientist to assign perceptual scaling values to data obtained from measurements made on physical instruments.

The form of Eq. (21) as a quintic parabola (fifth-order polynomial) has two drawbacks: (a) it is not easily solved without a calculating aid and (b) it is not algebraically invertible with V given as a function of Y. To aid in the use of the value function, tables were prepared by Nickerson (1950) and Billmeyer (1963) that gave the user a simple means of looking up V or the corresponding Y, depending on which was given. Since the early 1960s, digital computers have become widely used in colorimetric computations, and it is now commonplace to solve the quintic parabola on a computer to obtain luminances from Munsell value. To overcome the problem of obtaining value V from luminance Y, the Newton–Raphson iterative technique (Rheinboldt and Menard, 1960; Ganz, 1962; Alexander, 1971; Stenius, 1977) is used on a computer, which quickly gives precise value data from luminance measurements.

   c.   Cube-Root Lightness Scales   Ladd and Pinney (1955) and Glasser *et al.* (1958) found that a much simpler relation could be used to relate luminance and value based on a weighted cube-root scale. The Glasser *et al.* equation is

$$L = 25.29(Y/Y_n)^{1/3} - 18.38, \qquad (21)$$

which is easily inverted to give

$$Y/Y_n = [(L + 18.38)/25.29]^3, \qquad (22)$$

where L is the perceptually uniform interval lightness scale and Y/Y_n the CIE relative lightness for an illuminant n (0–100).

These authors showed that there was a very small difference in lightness between the cube-root calculation and the quintic parabola equations. Accordingly, the equation found widespread acceptance among industrial users of colorimetric methods. By 1965, however, complicated equations such as the quintic parabola value function presented no problem because of use of the digital computer. Nonetheless, the CIE adopted the cube-root scale for lightness as part of the 1964 color difference formula (W^*) and subsequently the standard 1976 CIE color difference formulas (L^*).

It is now possible with modern electronic circuitry to build a cube-root scale into a colorimeter instrument. This allows CIE L^* scale values to be obtained directly as a colorimeter output without having to resort to digital computation. However, with the rapid development of inexpensive microprocessors, which can be made a part of instruments, the need for simple equations to express relationships has become much less important.

2. THREE-DIMENSIONAL COLOR ORDER

a. Collections of Samples   (1)   Munsell System.   The Munsell system is the most widely used and the most thoroughly investigated color order system. It is a collection of about 1450 chips that are contained in the "Munsell Book of Color" (Munsell Color Co., 1943). This is a looseleaf set of charts arranged according to three dimensions of perceptual spacing conceived by A. H. Munsell, hue, value, and chroma. The hue scale involves five primaries: red, yellow, green, blue, and purple. The scale is further divided into secondary hues: yellow-red, green-yellow, blue-green, purple-blue, and red-purple. All of these primary and secondary hues are arranged in what is supposed to be an even visual spacing around a hue circuit that can be related to 100 divisions. The value dimension is the gray scale described before and ranges from perfect white, denoted as 10, to black, denoted as 0. Any decimal from 0 to 10 describes a gray that can be related to the CIE luminance scale with Eq. (20). The remaining dimension, chroma, is a perceptual scale that indicates the relative colorfulness of a color with respect to the brightness of the white similarly illuminated and, thus, its distance from the comparable gray. This means that the chroma is "open ended." It could be used to designate any color and is not bound by the limits of available colorants. The description of a color in Munsell terms is given as hue value/chroma, which is abbreviated H V/C. Since the hue circuit can be subdivided in a decimal fashion, all hues are denoted by their one- or two-letter abbreviations such as R for red or YR for yellow-red preceded by a number from 0 to 10. Only the ten principal and secondary hues are indicated by the letters; the subdivisions of hue are indicated by the decimal number. The value scale and the chroma scale are given as simple decimal numbers. There is no actual limit to the possible interpolation between chips, but the practical limits are about 0.1 hue step, 0.01 value step, and 0.1 chroma step.

Munsell showed remarkable foresight in the instructions that he gave for constructing a psychophysical color solid. The principles (Tyler and Hardy, 1940) were followed in developing the system and are worth noting here:

Hue—When a chromatic color is mixed additively with a neutral (white, gray, or black), the hue of the mixture is the same as that of the chromatic color.

Value—When two colors whose values are V_a and V_b occupy relative areas a and b on a Maxwell disk, the value of the mixture is given by the equation

$$V^2 = aV_a^2 + bV_b^2. \tag{23}$$

Chroma—When two complementary colors occupy areas that are inversely proportional to the product of value by chroma, a neutral gray results.

In 1928 the Bureau of Standards examined 70 samples for the Munsell Color Company, using visual spectrophotometric measurements to assess the consistency of the system with the principles laid down by Munsell and to check the gray scale. Although the results were not published until 1940 (Nickerson, 1940), reports were made to Munsell that led to improvements in the "Munsell Book of Color" (see Nickerson, 1940, 1969, 1976). The first complete colorimetric measurements and colorimetric calculations were completed by Glenn and Killian (1940), who completed their data in 1934 as a bachelor's thesis on about 400 Munsell samples. This work led the OSA subcommittee chaired by Newhall (Newhall *et al.*, 1943) to analyze the data with respect to spacing of the Munsell system, hoping that it would provide a basis for further improvement of the Munsell system as a whole. Newhall presented these findings as a preliminary report of the OSA subcommittee on spacing. During 1943 Kelly (Kelly *et al.*, 1943) published the Bureau of Standards tristimulus specifications for 421 standard Munsell samples; these were significant for their care and accuracy. Granville *et al.* (1943) supplemented the Bureau's work with another 561 samples, measured in the Interchemical Laboratories in New York City. It was on these data that the final report of the OSA Subcommittee on Munsell Spacing was based. This work described an ideal Munsell system as completely as it could be described using visual interpolation of graphs drawn for the colorimetric data obtained for the 421 and 561 samples. A series of charts and tables are the body of the report and give the recommended Munsell notations, called "renotations," which were obtained from the committee's work. Only data for values 1/–9/ are given in the report, with recommendations that in-between points be linearly interpolated. Judd and Wyszecki (1956) extended the 1943 final report data for Munsell renotation for very dark colors, down to a 0.1/ value. The basic definition for Munsell value in the form of the quintic parabola, Eq. (21), is also given in the OSA final report (Newhall *et al.*, 1943).

The Munsell Color Company began to issue the "Book of Color," Gloss Edition, based on the renotation, some time after the 1943 OSA report and this has become the basis of the present-day system. Several editions of the "Book of Color" are available, including a pocket edition with 13×16 mm chips of painted paper arranged as pages of a looseleaf book with each of the 40 hues on one page of white paperboard. A larger publication, called the library edition, comes in two volumes having 17×22 mm matte chips inserted in slotted pages with the notation printed on each chip. This is also available as a gloss edition. The specifications of the high-gloss Munsell collection were reported by Davidson *et al.* (1957); they conform to the renotations given by the OSA subcommittee.

There are many applications of the Munsell system in which subsets of the entire collection have been gathered together for special purposes. For example, there is a set of chips used to match the red of the tomato with its several variations. A series of browns, tans, and dark reds are brought together to describe the color of soils. And a special collection of very light near-whites and yellows is offered for use in describing near-white papers and other products. Much of the success of the Munsell color system, aside from its perceptual spacing, comes from its adaptability to interpolation and open endedness of the chroma scales. It will never lack for a position for any surface color either presently known or yet to be conceived. Furthermore, it is well founded in the scientific correlation of colorimetric specification and is easily adapted to Maxwell disk colorimetry. It is noteworthy that the Japan Color Research Institute also published a Munsell system along the same lines as the U.S. version.

Some of the utility of the Munsell system is lost if conversions of colorimetric data to Munsell notation are so laborious that one must resort to the charts of the 1943 OSA report and graphical and arithmetic interpolation to obtain sufficiently precise notations. Rheinboldt and Menard (1960) have devised a digital computer program to obtain Munsell notations from tristimulus values that is based on mathematical interpolation between tables of values of the published Munsell renotation data. Alexander (1971), on the other hand, developed a simple algorithm for an approximate Munsell conversion from CIE tristimulus data; this has been useful for identifying collections of samples for cataloguing purposes. This method has been refined by Simon and Connelly (1979) to give better precision for the hue attribute in the Alexander transform.

After a Munsell notation is obtained, the resultant dimensions are in perceptual terms necessary for color difference calculations. Nickerson (1936), recognizing this, proposed an *ad hoc* equation to satisfy the results obtained on a set of samples that had changed from their original color by being exposed to light. This empirical relation is given by

$$I = (C/5)(2\,\Delta H) + 6\,\Delta V + 3\,\Delta C. \tag{24}$$

Balinkin (1941) modified the Nickerson index of fading to make the formula conform to Euclidean spatial relations:

$$I = \{[(2/5)C\,\Delta H]^2 + (6\,\Delta V)^2 + [(20/\pi)\Delta C]^2\}^{/2} \tag{25}$$

where I is the Nickerson index of fading, H the Munsell hue, V the Munsell value, and C the Munsell hue.

If the Munsell hue is given its pure decimal notation, from 0 to 100, where 0 is 0.0R and 100 is 10.0RP, this can readily be converted from the normal polar coordinates to a set of rectangular coordinates that are designated a'' and b'' for convenience:

$$H° = 0.36H/(\pi/4),$$

$$a'' = C \cos H°, \qquad (26)$$

$$b'' = C \sin H°$$

where $H°$ is Munsell hue angle in radians, H the Munsell hue (scaled 0–100), and C the Munsell chroma.

Then it follows that a Euclidean color difference equation can be written in these coordinate terms:

$$\Delta E = [(\Delta a'')^2 + (\Delta b'')^2 + (\Delta V)^2]^{1/2}, \qquad (27)$$

using the additional Munsell value term for the lightness parameter.

When the Munsell renotation was made, it was the best that could be done with the existing mathematical techniques, but without the aid of the digital computer. Since that time several studies have been made suggesting that a better spacing could be achieved based on the experience of the last 25 years. Nickerson and Judd (Nickerson, 1969) were considering a Munsell "re-renotation" at the time of Dr. Judd's death. There is some evidence of anomalies in the hue spacing when it is examined by multidimensional scaling as reported by Indow (1974). Certainly, experimental work could be done to improve the Munsell system and provide a better perceptually spaced set of surface colors for the benefit of the many users of the system.

An outstanding use of the Munsell system is for communication of information about color from one person to another. The textile industry has made considerable use of the system for relaying information about placement and selection of colors in patterns, both woven and printed. Moon and Spencer (1944a,b) have made an analysis of color harmony with an adaptation of the Munsell system, and Luke (1976) and Libby (1974) use Munsell to teach color to artists and art students.

(2) Munsell Renotation and Transformations. The final results (Newhall et al., 1943) of the OSA subcommittee establishing the renotation values of the present Munsell system gave numerical specifications for each of the ideal Munsell renotation samples and, most importantly, a set of chromaticity charts that have been the basis for obtaining Munsell (re)notations from CIE chromaticity data for any sample. Sets of these charts have been available from the Munsell Color Company since the publication of the report and are used to find the Munsell hue and chroma of samples from CIE x, y coordinates.

The first step in obtaining a Munsell notation from colorimetric data is to convert the CIE tristimulus Y to Munsell V through a table taken from Nickerson (1950), which gives the conversion of CIE Y to Munsell V. The hue and chroma are then obtained graphically from the charts. Since charts are available only for integral Munsell values, the hue and chroma must be obtained from the chromaticity graphs for values on both sides of the sample value. The exact hue and chroma are obtained by interpolation.

Unfortunately these charts are all given in terms of CIE Illuminant C, which provides slightly different values for the notation than those that would be obtained with the present CIE-recommended D_{65} illuminant. If very precise conversions are needed, a new set of charts must be drawn, which is an extremely arduous task. On the other hand, the first-order correction could be made to tristimulus values of samples obtained with integrations made using D_{65} illuminant rather than Illuminant C, by first multiplying the tristimulus values by a set of values correcting the data to C illuminant values, derived by the method given in Judd and Wyszecki (1975). The set of equations used to obtain the C illuminant adaptation is as follows:

$$X_C = 0.995 X_{D_{65}} + 0.015 Y_{D_{65}} + 0.02 Z_{D_{65}},$$

$$Y_C = Y_{D_{65}}, \tag{28}$$

$$Z_C = 1.088 Z_{D_{65}}.$$

It was recognized fairly soon after the adoption of the CIE colorimetry system that a simple test for the validity of any transformation of the system to another set of coordinates to give better perceptibility spacing should give circular contours for constant chroma data of the Munsell system. It was also understood that the Munsell hue spacing should give equal angular displacement for comparable hue differences. Burnham (1949) examined several color order systems with these criteria in mind. Glasser *et al.* (1958) gave direct information on the use of the cube-root transformations of CIE tristimulus values to a new set of coordinates that yielded better Munsell spacing and that in fact could be used to give an approximation very much like the Munsell system itself. Nickerson (1978b) was first attracted to the work of Adams because of the better display of Munsell data on Adams earliest $(X-Y)$, $(Z-Y)$ transformation, which ultimately led to the publication by Adams (1942) of the chromatic value space, which was a direct result of Nickerson's encouragement. Some improvement in perceptibility spacing is achieved over the nonlinear transformation of Adams by making a further linear projective transformation according to Wyszecki (1954):

$$\xi = 1.438 V_X - 1.282 V_Y - 0.157 V_Z - 0.107$$

$$\eta = 0.342 V_X - 0.787 V_Y + 0.483 V_Z + 0.138. \tag{29}$$

Alexander (1971) recognized the need for filing colored samples system-atically so that the formulas that produced them could be retrieved quickly by a computer search. This effort led to an investigation of the available transformation systems from CIE to Munsell in order to find a filing method adaptable to computer searches as well as a visual file of the samples them-selves that would place them in a logical order. Three algorithms were needed, one for each of the Munsell dimensions, hue, value, and chroma. The hue and chroma algorithm stemmed from the Glasser *et al.* (1958) concept of dividing the transformed chromaticity coordinates into four quadrants and then calculating approximate hue and chroma from the new coordinates. But rather than use the Glasser cube-root derivation of CIE tristimulus X, Y, Z, a Newton–Raphson iterative method was used to obtain the set V_X, V_Y, V_Z from X, Y, Z. Several simple calculations must be made with the derived chromatic values, as follows:

$$a_1 = V_X - V_Y, \qquad b_1 = V_Y - V_Z; \tag{30}$$

$$\theta = \arctan(b_1/a_1) \tag{31}$$

for first and third quadrants:

$$a = 105a_1 \qquad b = 53.6b_1, \tag{32}$$

for second and fourth quadrants:

$$a = 125a_1 \qquad b = 30.5b_1, \tag{33}$$

$$\text{chroma} = 0.00509 + 0.08005(a_1^2 + b_1^2)^{1/2}, \tag{34}$$

$$\text{hue}^\circ = \arctan(b/a). \tag{35}$$

The chroma and hue angle calculated by the preceding equations are only approximations but suffice for many applications. Value, of course, is precise, being calculated from the defining equation given in Eq. (20). The Alexander approximation for chroma is reasonably accurate and it can be used to calculate chroma for many applications. The hue transform is shown in Fig. 12. This transform was improved by Simon and Connelly (1979), first by making the transition from quadrant to quadrant into smooth curves with revised coefficients for a_1 and b_1 based on transcen-dental coefficients:

$$f_1 = 90 + 32 \sin\theta, \qquad f_2 = 37 + 4\cos(\theta + \pi) \tag{36}$$

θ is defined in Eq. (30)

$$\text{hue}^\circ = \arctan(f_1 a_1 / f_2 b_1). \tag{37}$$

After the hue° is determined by Eq. (37), a line that approximates the actual hue curve is calculated with a series of straight-line segments based on

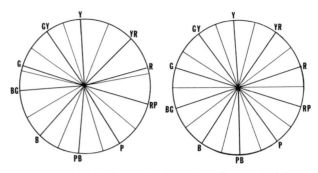

Fig. 12. The hue spacing with the simple Munsell transform on the left and the hue spacing with the revised coefficients on the right.

coefficients stored in the computer program. Another set of coefficients is used to obtain more precise chroma data than that calculated with the Alexander transform. This transform results in hue calculations that do not differ by more than 0.4 in Munsell hue from the actual and by more than 0.3 in Munsell chroma. The Simon and Connelly transform is now in use as a filing and identification scheme in several industrial applications.

Rheinboldt and Menard (1960) developed a complicated digital computer conversion method using vector interpretation to convert CIE X, Y, Z data to precise Munsell hue, value, and chroma. They were the first ones to report the Newton–Raphson method for calculating Munsell value.

2. THE DIN-COLOR SYSTEM

M. Richter (1953, 1955) and co-workers developed over a period of about 10 years a color system that is the official German standard color system; it is still in use and is available as a collection of samples (DIN-6164, 1964) based on a colorimetrically spaced color solid. Conceptually, the color solid is founded on the work of Luther (1927) and Nyberg (1928). The three dimensions of the system are DIN–Farbton, which is somewhat analogous to Munsell hue except that lines of DIN–Farbton have constant dominant or complementary wavelength, unlike Munsell hues, which do not follow straight lines in the CIE 1931 x, y chromaticity diagram; DIN–Sättigung, which is analogous to Munsell chroma; and DIN–Dunkelstufe, which is analogous to Munsell value. The system is open ended and the DIN-6164 charts give correlate data for the samples in CIE chromaticity coordinates as well as translations to both Munsell and Ostwald color systems. The Dunkelstufe scale is a modified Delboeuf formula and is based on a derivation from Weber–Fechner principles. It utilizes the logarithm of relative lightness where the ratio of the luminance of the sample is related to that of the optimal color for a given hue. Therefore, Dunkel-stufe is not directly translatable to Munsell value but depends on the equation

$$D = 10 - 6.1723(\log 40.7h + 1), \qquad (38)$$

where $h = Y/Y_0$; Y is the luminance of the sample, and Y_0 is the luminance of the optimal color of the same hue as the sample.

DIN–Farbton is a 24-step hue scale that begins at yellow and proceeds through the visible spectrum and purple boundary. Any given Farbton has a constant dominant wavelength rather than constant hue. This denies the concept that constant hue lines for surface colors are curved in the CIE x, y chromaticity diagram. The Farbton primary hue selections are similar to the Ostwald hues but are spaced a little differently, attempting to give equal perception steps between hues of spectral stimuli and still maintain the Hering opposite-hue concept. The Sättigung scale is derived from the Luther–Nyberg work, which results in a series of concentric ovals around the CIE illuminant C origin where S, Sättigung, is equal to zero. Figure 13 shows the various Farbtons as straight lines emanating from the illuminant point with several Sättigung ovals surrounding the illuminant point. Neither CIE excitation purity nor Munsell chroma, which are satu-ration analogs, give simple correlated values to Sättigung for a given dominant wavelength or hue (Kundt and Wyszecki, 1955). Budde *et al.* (1955) gave the physical transformations of DIN-6164 to Munsell notation through instrumental measurement but did not conduct observational ex-periments that would lead to a comparative evaluation of the two systems.

The DIN system occupies a singular position in the field of color order systems, having been officially adopted by a governmental agency. No other system enjoys this position in the realm of color specification although limited sets of Munsell are used to define the quality of certain agricultural products in the United States.

3. OPTICAL SOCIETY OF AMERICA UNIFORM COLOR SPACE

After about 30 years of work by a Committee on Uniform Color Scales (UCS) (see Judd, 1955; MacAdam, 1974; Nickerson, 1975), an announce-ment was made in 1977 (Optical Society of America, 1977) that sets of 558 colored samples were offered for sale to demonstrate the color order sys-tem developed by the committee. The substance of this work was the pro-duction of a set of perceptually uniformly spaced samples based on a regular rhombohedral lattice spacing rather than the polar arrangement found in Munsell, Ostwald, DIN-6164, and other less widely known color order systems. The rhombohedral arrangement, published by Wyszecki (1954,

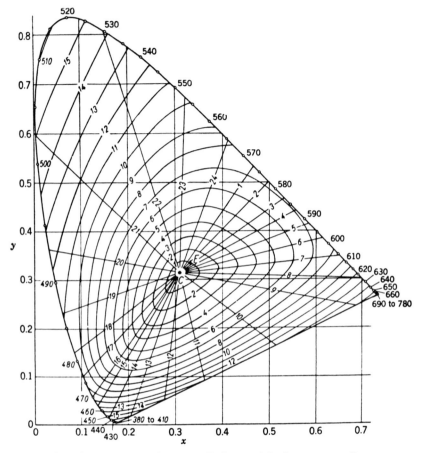

Fig. 13. DIN-6164 color system Farbton and Sättigung contour lines.

1960), has the advantage that any color is surrounded in three-dimensional space by 12 equidistant neighboring colors. This array of points in (color) space is the cubo-octohedron shown in Fig. 14.

The lattice is made of close-packed layers; therefore, when a color solid is derived from a series of observations, the individual colors have to be fairly close to each other. The system is developed from paired comparison of color differences, which means that if three colors are placed in an equal perceptual arrangement on a plane, an equilateral triangle is formed; a fourth color added to this scheme forms a congruent triangle with two of the other colors if they lie on the same lightness plane or an equilateral tetrahedron if the fourth color is on the adjacent lightness plane. The entire

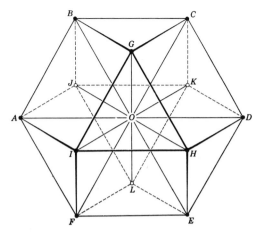

Fig. 14. Cubo-octohedral lattice of points.

array of colored samples, which are spaced equidistantly by observers, can be arranged this way. If the scales are uniform, in the same sense as the Munsell system, the Munsell space can be embedded into it, using Munsell to understand the theoretical arrangement of the lattice. The positions of the samples are indicated by three decimal numbers designated as $\mathscr{L}$, j, and g. Only lightness, $\mathscr{L}$, is directly analogous to Munsell value in a simple way.

$$\mathscr{L} = 2(V - 6), \tag{39}$$

where $\mathscr{L}$ is the OSA–UCS lightness scale(plane) and V the Munsell value.

All scales originate from Munsell N 6/, neutral gray, where $\mathscr{L} = 0, j = 0$, $g = 0$, and they radiate in all directions with the integrally numbered samples spaced on lines that form equilateral triangles. All positions in the lattice are indicated by the Euclidean axis numbers, which are patterned after the Wyszecki (1954) treatment of the Munsell system shown in Fig. 15. The figure only illustrates an arrangement of the UCS system; the conversion of the actual samples published by the committee is given by Nickerson (1978a) (see the chapter on colorimetry). MacAdam (1978) published the final report of the Committee on Uniform Spacing, giving CIE colorimetric specifications for the 558 samples as well as the precise OSA–UCS designations for all samples included in the final edition.

The OSA–UCS system offers many fascinating opportunities to the scientist to study color difference scaling for color order systems since many sections can be made through the rhombohedral lattice to obtain uniform spacing. It also offers the designer and artist many varieties of color harmony obtained when plane slices are taken through the color solid, which, when

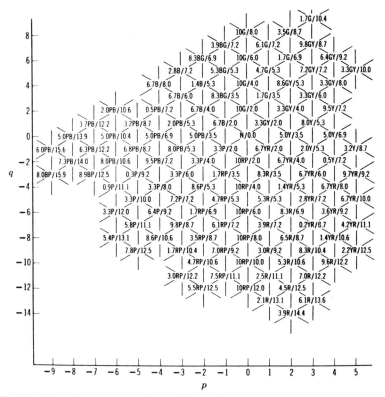

Fig. 15. Triangular lattice sampling of Munsell value 5 plane given in terms of OSA–UCS coordinates (after Wyszecki, 1954).

spaced in a regular manner, can achieve many pleasing combinations. The OSA–UCS is open ended, which means that new colors that are produced with more saturated colorants than those used in the 558 samples would fit into the system of notation. However, since the $\mathscr{L}$ scale of the OSA–UCS is tied to the physical quantity of luminous reflectance, it does not overcome the objection of one artist, Birren (1944), who finds the arrangement of Ostwald more understandable. Much work remains to be done in developing commercially available collections of samples based on the OSA–UCS concept before it has practical significance to the color-using world.

E. Descriptive Color Order Systems

Words, numbers, and symbols are all used to convey color information among individuals. The type of color names used can be grouped in several ways. The most fundamental set of names consists of those that have been termed by various authors as "primary," "primitive," or even "prismatic" colors (evidently referring to spectral colors). These have been discussed in detail in another section. Sets of primaries are limited to three, four, or five names. The next level of naming includes a few common names such as pink, brown, orange, violet, which usually result from mixtures of primary color colorants, and their frequent occurrence has led to a fairly well understood but inexact description. The third level of naming attempts to be more specific, using either generic adjectives to modify the common names or very specific words that attempt to convey a specific description for a precise definition. The name may impart meaning to some persons but be completely lost on others. Color order attempts to sort all of this out and explain in objective terms what particular color names that have been used by others mean in general, and therefore reference is made to the Munsell system, since it is widely used and understood.

1. COLOR NAMES

Many words used to describe colors are derived from names of the colorants that are used in painting and design. Pliny (1912) describes many pigments by their source in the ancient world. For example, indigo came at one time from India, armenium (blue) from Armenia, melenium (a white marl) from Melos, and, later, sienna and umber from places in Italy. These terms are still used, for example, to name artists' pigments even though the material may now be manufactured synthetically in any part of the world. Another group of color names is related to the color of natural vegetation and flowers. Although the actual colorant from these sources is rarely if ever used, the name is generally understood even if it is somewhat exotic or rare.

One of the simplest systems of naming colors uses only the four unitary hue names plus light (white) and dark (black) of the Hering opponent-color concept. If this restricted set of names is used as nouns and descriptive adjectives, the communication of color information is made quite simple. Any group of common color names can be used that first utilizes the words "red," "yellow," "green," and "blue," supplemented with "white" and "black," but also can include the commonly understood words "gray," "orange," "violet," "brown," "pink," and so on. The list may be longer or shorter than that cited, but the important thing is that the unitary hue names are the reference descriptors. Thus, if a color is to be named it is always given first, if possible, in terms of the unitary primary color names, which can be modified by any of the primary words used as adjectives— for example, reddish-yellow, bluish-green, greenish-yellow; or, in the extended name set, yellowish-brown, bluish-gray, or reddish-violet. A major difficulty in any color naming is when the less well-known words are used to name a color and the concept invoked by a less commonplace term is not shared by the giver and receiver of particular information. The word "rufous" may have meaning to an ornithologist but its meaning is lost on most people; "navaho" is better known as an American indian tribe than a bright orange; and "mustard" comes in many flavors as well as colors. The more exotic names have a place in promotion and styling but can add to confusion in communication.

An elaboration of the principles just described has been used in industrial color difference description. However, two other parameters are needed to complete the two other color dimensions, chroma and lightness. If the simple terms "lighter" and "darker" are applied to lightness differences and "brighter" and "duller" to chroma differences, a complete vocabulary is at hand. Simon (1972) used this set of descriptors (i.e., "lighter" and "darker," "brighter" and "duller," "redder," "yellower," "bluer," "greener" to describe in words the direction in addition to the numerical sense of color differences. Since it is useful to know more precisely when to use which hue descriptor word, the CIE chromaticity diagram was divided up as in Fig. 16, which delineates those areas where the unitary hue name is to be used and those areas where the unitary hue is used only as an adjective. These limits are only an estimate but serve to define the provinces of any color name in such a way that can be programmed into a computer in terms of hue angle as given earlier.

2. Color Dictionaries

Several authors have attempted to give systematic definitions to groups of names that either are found in literature or are in common usage in a particular field such as ornithology or botany. The most general dictionary involving a collection of color samples is the Maerz and Paul (1950) "Dictionary of Color," which was described before in connection with collections of samples produced by halftone reproduction. Munsell notations have been given for the dictionary by Nickerson (1947). Taylor *et al.* (1950) published a cross-reference supplement to the "Color Harmony Manual," which gave color names in common usage for the chips contained in the manual. The names were identified with spectrophotometric measurements and the comparable Munsell notations were also given. The Ridgway (1912) and the British Colour Council (1942) systems are collections of samples that also have a large number of names associated with the samples in the collection. Although both of them are aimed at the colors of natural objects, they have been of interest to scientists because many of the names are also used by others for color description. Hamly (1949) has correlated

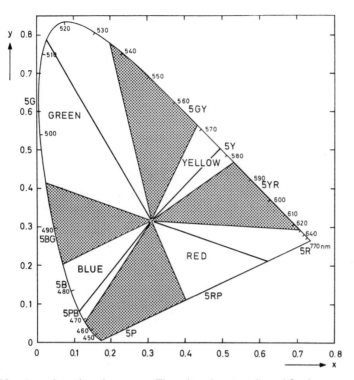

Fig. 16.   Areas for unitary hue names. The unitary hue name is used for those areas which are shaded and is used as an adjective descriptor in the unshaded areas.

the Ridgway system with Munsell, and Middleton (1949) has given Munsell notations for the Plochere collection.

Kelly and Judd (1955) have brought together many of the diverse sources of color names into one dictionary that groups all the major information on color names into one systematic classification: the ISCC–NBS Dictionary of Color Names. This generic naming system is described next.

3. ISCC–NBS Color Names

The National Bureau of Standards (NBS) developed a method of assigning generic names to blocks of the Munsell color solid based on a system devised by Kelly and Judd (1955). The idea was to apply a small number of common color names to various blocks of the color space and then subdivide those blocks into smaller sections by means of descriptive adjectives that would modify and refine the simple color name. Thirteen color names, including black and white, were chosen as the nucleus of the system.

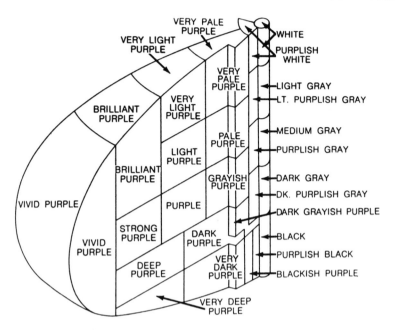

Fig. 17.   Example of ISCC–NBS color names.

These names are hue descriptions and are modified by 10 adjectives that can be used alone or in pairs. An example of a section of the Munsell solid is shown in Fig. 17, which is a cutaway part of the Munsell color solid.

Although the generic color naming idea came from the NBS, a subcommittee of the Inter-Society Color Council worked cooperatively with Kelly and Judd for many years and the final publication is a result of the effort of several committee members. It was published by the NBS as Circular 553 (Kelly and Judd, 1955).

One of the outgrowths of color naming was the concept that once a generic name was agreed on, it was useful to see what it looked like. Again Kelly of the NBS led the effort, calculating a color coordinate point that was about at the center of gravity of the volume of color space described by the ISCC–NBS name and then having a Munsell chip made to represent that centroid point. This resulted in a set of 267 chips that were displayed on 31 charts, published by the NBS, which showed the centroid colors of the system. In publishing the centroid colors, the chips were mounted on paper that was graded through halftone engraving to give the effect of an increasingly dark background from the top of a page to the bottom. This was an attempt to produce a graded scale of value for the viewing background for centroid chips of different Munsell values. Although the idea was reasonable, the resulting charts were not very well executed. The Munsell Color Co. presently sells sets of the charts that cover the gamut of centroid colors feasible with commercial colorants.

The concept of the ISCC–NBS color names is not too easily understood and the proposed generic names have not been widely used. Aside from the somewhat obscure adjectives used, perhaps the major reason for a lack

	Color Name Designations			Numeral and/or Letter Color Designations		
Level of Fineness of Color Designation	Level 1 (least precise)	Level 2	Level 3	Level 4	Level 5	Level 6 (most precise)
Number of Divisions of Color Solid	13	29	267*	943-7056*	≃ 100,000	≃5,000,000
Type of Color Designation	Generic hue names and neutrals (See circled designations in diagram below)	All hue names and neutrals (See diagram below)	ISCC-NBS All hue names and neutrals with modifiers (NBS-C553)	Color-order Systems (Collections of color standards sampling the color solid systematically)	Visually inter-polated Munsell notation (From Munsell Book of Color)	CIE (x,y,Y) or Instrumentally Interpolated Munsell Nota-tion
Example of Color Designation	brown	yellowish brown	light yellow-ish brown (centroid #76)	Munsell 1548* 10YR 6/4**	9½ YR 6 4/4¼ **	x = 0.395 y = 0.382 Y = 35.6% or 9.6YR 6.4₁/4.3**
Alternate Color-Order Systems Usable at Given Levels	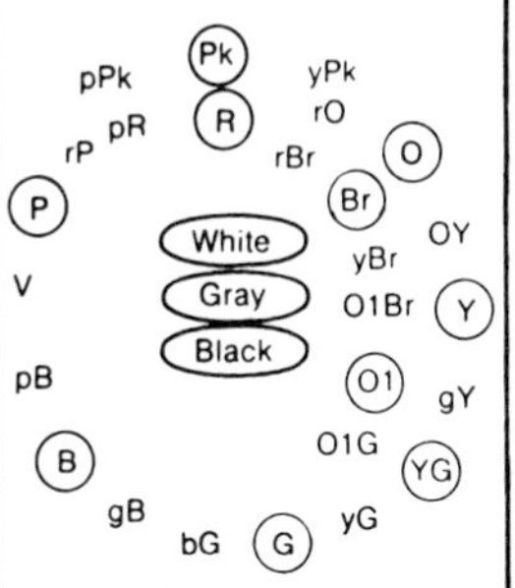		SCCA 216* (9th Std.) 70128 HCC 800* H407	M&P 7056* (1st Ed.) 12H6 Plochere 1248* 180 0 5-d Ridgway 1115* XXIX 13 b CHM 943* (3rd Ed.) 3 gc		
General Applicability		→ → → Increased Fineness of Color Designation → → →			← ← ← Statistical Expression of Color Trends (roll-up method) ← ← ←	

*Figures indicate the number of color samples in each collection.
**The smallest unit used in the Hue Value and Chroma parts of the Munsell notation in Levels 4 (1 Hue step, 1 Value step and 2 Chroma steps), 5 (½ Hue step, 0 1 Value step and ¼ Chroma step) and 6 (0.1 Hue step, 0.05 Value step and 0 1 Chroma step) indicates the accuracy to which the parts of the Munsell notation are specified in that Level

Fig. 18. Schematic diagram illustrating the six levels of the Universal Color Language.

of interest in generic names is that style-oriented people like to name colors with words that imply "excitement" or interest rather than use dispassionate words that have scientific accuracy but little else.

4. UNIVERSAL COLOR LANGUAGE

Although no color language can be completely universal, Kelly and Judd (1976) have developed a methodical approach to color naming that encompasses the entire possible gamut of surface colors through an approach shown in Fig. 18. The six levels of increasing complexity give the user a choice of level to work with, requiring no more complication than needed. Naturally, the complexity of the Munsell system is needed in the upper four levels to give a sufficiently fine distinction to avoid a contradiction in designation given a color. The ISCC–NBS color names are built into the Universal Color Language at level 3, which is the smallest collection of samples given in the system. Level 4 embraces all the common collections of color chips, irrespective of how they have been put together, and allows a freedom of choice that is up to the user. No implication is made that one system translates into the other even though the UCL makes no distinction among them. A major proponent of the Universal Color Language is the Color Marketing Group. Although all types of people who work in color probably use some level of the Universal Color Language, its recognition is not widespread.

F. Acceptability Limits

A numerical description of color immediately suggests its specification in precise, objective terms. It is only necessary to give some set of three descriptive terms that will define a given color exactly; this gives rise to the concept that color can now be handled as any engineering quantity using the particular dimensions unique to the system used. For example, a spectrophotometric curve will define the reflectance or transmittance of a material in an exact and unambiguous manner for a given geometry of viewing and illumination, aside from the errors that could be introduced by the measurement itself. However, this is not defining "color" per se, since in order to have color, a light source and an observer must be included as part of the definition. Therefore, although the spectral curve may be physically descriptive, it is only a partial definition of the factors affecting color. It is therefore usual in the general case to specify color in terms that are analogous to the CIE tristimulus values, XYZ. These will give a complete definition in the simple form or in any transformed set of values such as a uniform color scale. The principle of trichromacy must be observed in order for the description to be complete. Thus, most of the color order systems given earlier in this chapter could be used. In addition to specifying the exact location of a color in three-space with the trichromatic specification, any other similar color can be related to it using simple plus and minus tolerances or differences that surround the "standard" color in the given three dimensions. This area of the application of color order is of great interest to industrial, governmental, and commercial users of color measurement.

1. INDUSTRIAL TOLERANCES

One of the earliest examples of specification of colors with the CIE system was that given for the American flag colors—red, white, and blue—defining the colors purchased by the U.S. federal government. Although the numerical values were given, the purchasing agency had to determine the acceptability tolerances that surrounded the physical description of the flag colors. In fact, without the tolerances or the measurement method being given, only a referee institution such as the National Bureau of Standards could give adequate assessment of whether the specification was being met.

Since that time, many more realistic specifications have been promulgated such as many of those given by the American Society for Testing Materials (1966), which provide both methods of test and specific tolerances for many colored materials and colorants themselves. In the paper industry the matter of the color of white paper is so important to commerce that a colorimetric specification, called TAPPI brightness, has been in general use for many years, having been developed by the Technical Association of the Paper and Pulp Industries. This is a number equivalent to the reflectance of a sample of paper at 489 nm. The method calls for the use of a specific instrument, which removes some of the uncertainty of the measurement. Any tolerance or deviation from the given TAPPI brightness value is generally agreed on between buyer and seller.

The chemical industry has many specifications that are used to describe the colors of nearly colorless liquids. These are usually not given simple tristimulus specifications; instead, they are given in arbitrary one-dimensional scales that are graduated sets of mixtures of chemicals such as complexes of platinum and cobalt or arbitrary yellow-colored glasses of increasing density. There have been recent efforts to develop test methods based on tristimulus colorimetry but the replacement of well-established methods is not easily accomplished.

One of the more successful indirect methods of describing color tolerances is that used by the textile industry to quantify color fastness test results. The American Association of Textile Chemists and Colorists (1977) and the Society of Dyers and Colourists (Great Britain) have worked for many years to establish a set of references in the form of graded physical scales that can be used for visual comparison of the results obtained when fastness tests are performed. These are usually sets of gray chips scaled according to a psychological scale such as Munsell or Weber–Fechner (so-called geometric) and they serve as the visual arbiter. Other sets of chips have been developed in several hues at fairly high lightnesses to simulate the staining of white fabric as in a wash-fastness test. All of these reference scales are used visually but are checked with instruments to assure conformity to designated standards.

Many standards and tolerances have been in use in other industries and the instrumental mainstay of these specifications is the tristimulus colorimeter. For the most part, color specifications must be negotiated between the buyer and seller of articles of commerce, and the acceptability criteria are not well defined in most industries. Since instrumental methods are largely continued to select applications, visual methods are still used by large consumers such as the automotive manufacturers or the military procurement agencies. These organizations supply visual acceptability tolerance samples for guidance. Although these indicate the range of acceptability, there is no reasonable way in which to provide enough samples to envelope a standard in all directions from the center. Furthermore, any set of tolerance samples must be obtained from the assortment of samples available at the moment and this can be highly prejudicial in certain color directions that are not immediately available. It is becoming increasingly evident that numerical tolerance limits developed from instrumental measurements are the only feasible answer to this problem.

Nutting (1935) published the first comprehensive discussion of all the methods available to describe and compare samples that were presumably matches for a standard. He gave both visual and instrumental techniques that we.e used at the time, and although the instrumental methods have greatly improved with modern technology, the visual procedures are still appropriate. Nickerson (1936) reviewed the methods that were in use and recommended at that time that the Munsell system be the framework for expression of color specifications and tolerances. These ideas were later refined in a paper by Nickerson and Stultz (1944) that treated colorimetric data rather than visual comparisons. After the discrimination data of MacAdam (1942) were published and instrumental methods became more popular, the use of numerical methods to express color tolerances was adopted by a number of segments of industry. The procedures were first used for internal process control and inspection and gradually formed the body of information and procedures that could be used between buyer and seller. Since the objective numerical methods could lend themselves to inspection methods, the role of colorimetry became that of another procedure that could move out of the laboratory and into the production environment. Simon (1961) described a statistical quality control method that was used for checking continuously produced spun-colored fibers.

One of the major objectives of recent scientific work in trying to find better color difference expressions is to get a metric that will be truly uniform in all regions of the color solid. Although this has yet to be achieved, several modern tools give reason for hope that the objective can be achieved in the future. In the first place, past experience has indicated that the problem is somewhat complicated, and since it must simulate human experience, the more advanced techniques of psychometric testing are now used to advantage (see (Morley *et al.*, 1975; Rich *et al.*, 1975). Furthermore, the computer, with its power to handle large statistical problems, allows the interpretation of masses of observational data that could not be handled in former days.

The CIE subcommittee on color differences has formulated a program that has been generally followed (Wyszecki, 1968) that gives some specifications for the procedure of performing visual tests that will be useful as the basis of establishing a new color metric. Robertson (1978), as the present CIE subcommittee chairman, has amplified the procedural outline for performing these tests so that observational data of various workers can be compared. The reader is referred to the complete published directions, but the essentials are as follows:

Size of difference—1–10 CIE $L^*a^*b^*$ units.
Field size—more than 4° subtense.
Surround—uniform and equivalent to D_{55}, D_{65}, or D_{75}.
Luminance of sample—5–500% of the surround.
Observers—only normal color vision.

An important question that must be answered in connection with any color metric, no matter how uniform it may be, is whether color acceptance is visually symmetric about the aim. Experience has shown that there are specific asymmetric tolerances that are widespread in industry. For example, it is well known that lightness differences among samples are often more acceptable than hue differences. This probably stems from the fact that the lighting on any colored surface gives the surface a different appearance, depending on the angle. Therefore, experience has taught us that lightness variability is to be expected. On the other hand, hue differences either to the red or yellow side are not as acceptable, since they may connote a defect or undesirable deterioration of the colored material. Such differences are not parallel to the limits of our visual perceptibility under ideal conditions, but it is important to reckon with them because they affect the world of commerce. McLaren (1969) made a quantitative study of the acceptability judgments made by textile dyers and concluded that there was a great amount of randomness to it, although a pattern could be discerned. A year later (McLaren, 1970) he turned the question around and postulated that an instrument such as a colorimeter could give consistent judgmental data and might be preferable in some instances for shade passing. Schultze (1971) studied several color difference formulas, with the objective of testing them for use in establishing tolerance scales for acceptability.

The ideal color difference equation is not available as yet. Many industrial users have need for an interim method for specification and means for subjective evaluation of color differences. A decision could then be made more easily about whether that number is tolerable in any particular situation. In general, the 1976 CIELAB formula seems to fulfill the current needs and will probably continue to be used until a better formula is adopted.

2. SHADE SORTING

An interesting, relatively new application of industrial color tolerance has come into use in the fabric and apparel manufacturing industries. Shade sorting is based on the idea that the volume around a color standard can be divided into a number of small subvolumes that are sized according to acceptability limits. These subvolumes are each identified according to a three-digit number that correlates with the three dimensions of color space. This idea was first developed by Simon (1961) in connection with spun-dyed fiber that could not be produced within commercial limits but could be classified and later blended to make an acceptable product. The same problem prevails in the production of most textile fabrics. It is an unfortunate fact that the process variability does not allow "one color" to be manufactured. This makes it necessary to sort fabric pieces into groups, so that only one group is cut up into garment parts and sewed together.

In this system the standard color is identified with the number 555. Any sample that differs from the standard only in lightness will be designated by changing the first of the three digits, higher if lighter, such as 655 or 755, or lower if darker. The other two digits, respectively, indicate differences in hue and saturation. The size of the subvolumes or shade blocks is determined by experience. It has been found that a typical relationship among tolerances for variation in hue, saturation, and lightness is in the ratio of $1:2:4$. This means that hue differences are one-fourth as tolerable as lightness differences and that saturation differences are intermediate to them.

In order to have a simple means to calculate color differences and to express color tolerances, the CIE 1976 $L^*a^*b^*$ metric is generally used. Since the psychological-parallel expressions of metric hue and metric light-

ness are defined by the CIE, the only special parameter needed is a saturation dimension that is perpendicular to the metric hue line in the CIE
$L*a*b*$ space. A typical set of tolerances are hue = 0.2, saturation = 0.4,
lightness = 0.8. Figure 19 shows a set of tolerance blocks used in shade
sorting where the tolerances for hue, saturation, and lightness are equal.

Shade sorting has become a fairly well accepted practice in the plain-
dyed shade textile industry. The original CIE tristimulus values can be
obtained on a colorimeter and the shade sort numbers are usually calculated

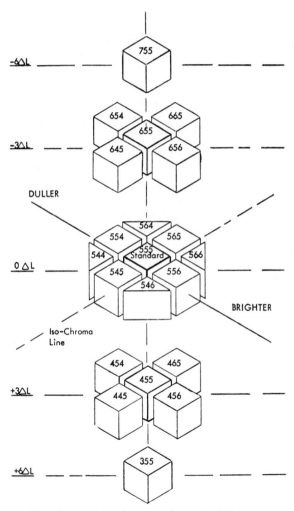

Fig. 19.   Shade sorting according to the 555 system.

on a programmable calculator or microprocessor. Some colorimeters have been designed to give the shade sort number directly from the measurement with a simple "push-button" operation.

G. Partitive Color Order

Partitive color mixing has only received limited attention from the scientific community in recent years. This is not to say that modern color technology does not use the effects produced partitively, but the most important application of partitive mixing in halftone paper printing usually produces color effects not by simple partitive mixing but by a combination of partitive and subtractive mixing. About the only time that almost purely partitive effects are obtained is when very coarse halftone screens are used to produce black and white images such as those found in newspaper publishing. The size and density of the black dots combine partitively with the white of the paper to produce a scale of grays. In three- and four-color process printing the size of the dots and the fact that one color is partially printed over the other do not make a clear case for either partitive or subtractive mixing; therefore, this area has not received adequate study.

In weaving of colored textile yarns into patterns where different colors appear side by side, there seems to be little appreciation of the partitive effects that are obtained. For the most part designers are likely to try out different combinations, usually not knowing that the combined color effect is predictable by simply knowing the relative area apparent for each colored yarn to determine the final color of the woven fabric. It would be expeditious for the designer to select Munsell papers that match the textile yarns and spin the combination as Maxwell disks. This technique could be taught to students learning to design colored fabrics.

Ceramic mosaics are still being produced as a form of art. However, since this is not very popular at this time, skilled artisan can learn only by trial and error what composite effects occur when the spots of complementary and adjacent hues are placed together. Since conventional painting provides no clue to the results that can be expected, this must be a reason for the lack of interest in this attractive art form.

1. THE MAXWELL DISK

Maxwell (1857) was an innovative scientist who developed the spinning disk method of partitive mixing to the level that was possible in the nineteenth century with the equipment that was available to him. He used it to obtain a systematic understanding of how colors mixed, following the principles of Newton, and how human vision acted, according to the ideas of Thomas Young. The equipment that he used is shown in Fig. 20, which

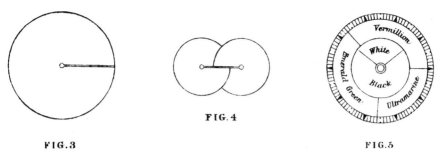

FIG. 4

FIG. 3   FIG. 5

Fig. 20.   Maxwell disks for partitive color mixing.

is taken from his publication. Certain papers by D. R. Hay and T. Purdie were used in Maxwell's experiments. These included highly saturated, pure artist's pigments such as vermilion, chrome yellow, ultramarine blue, and emerald green. The general procedure was to match combinations of outer chromatic disks to some binary combination of black and white, which were smaller disks located concentrically with the chromatic disks. This is shown in the figure. The results of a match between the adjusted proportions of the three chromatic disks to the mixture of black and white could then be written as an equation. For example:

$$0.37 \text{ V} + 0.27 \text{ U} + 0.36 \text{ EG} = 0.28 \text{ SW} + 0.72 \text{ Bk}, \tag{40}$$

where V is a vermilion paper, U an ultramarine paper, EG an emerald green paper, SW a snow white paper, and Bk an ivory black paper.

Since the disk background was divided into 100 parts, it was simple to read off the relative proportions exposed for each disk.

Maxwell was able to substitute other colored papers for his set of primaries, vermilion, ultramarine, and emerald green, together with black and white, and deduce the position of any color with respect to the primaries. An arbitrary color space was laid out with the three primaries forming an equilateral triangle with the proportions given by Eq. (40). This method is basically Newton's but has the equilateral triangle principle that is often coupled with Maxwell's name.

The Maxwell spinning disk method was taken seriously by Helmholtz (1924) and Rood (1895), who performed experiments similar to those of Maxwell to confirm that neutral colors could indeed be produced in this manner. Since Rood was a physics teacher, his students probably became familiar with the Maxwell disk method and the equilateral triangle for describing a color gamut.

Nickerson (1935) described use of the Maxwell disk method to obtain tristimulus values for agricultural products generally with four disks that were fairly close to the product that was to be evaluated. Since the tristim-

ulus value was known for each of the disks, it was a simple matter to calculate the intermediate tristimulus value from the proportion of each disk
that was finally determined to match the unknown. Nickerson had the
advantage of an improved apparatus built by the Keuffel and Esser Company that viewed the sample through a rotating prism that in effect mixed
the light reflected from the disks while the disks themselves remained
stationary. This speeded up the colorimetric determinations since adjustments could be made to the proportions of each disk without having to
start and stop as in the traditional Maxwell apparatus.

Although the partitive method seems to be of only academic interest,
equipment is still available from the Munsell Color Company that is used
for the Maxwell method. An adjustable-speed electric motor drives a spindle
on which various Munsell papers can be mounted. Since the papers are
furnished with Munsell notations, the combination of the notations according to the apparent proportion in a mixture will give a fairly reliable notation for an unknown sample. This is more accurate than visual interpolation between adjacent Munsell chips. The major disadvantage is that unless
the unknown is a sheet material that can be mounted concentrically, the
error due to positional differences can be fairly large.

2. NOECHEL–STEARNS WOOL-MIXING FUNCTION

Noechel and Stearns (1944) developed an empirical relation to satisfy
the need for a method to compute the amount of various colored fibers
that when blended together would give a specific color. This is a function
much like that used for colorants in subtractive mixtures except that blends
of fibers behave in a special way in that they still retain an amount of integrity but are also partly transparent. In order to treat this somewhat
complicated optical situation, a specific function was developed that is
neither partitive nor subtractive:

$$F_w = (100 - 100R)/(15R + 0.85). \tag{41}$$

This function has been used by a few textile manufacturers who produce
cloth from blends of wool or other fibers. The factor, 0.85, given in the equation is adjustable for different fiber blends and must be determined by
experimentation.

Connelly (1970) developed a special derivation of the Noechel–Stearns
function that imbeds the function into the determination of a type of tristimulus value. In place of the usual tristimulus integration with β for reflecting samples, the Noechel–Stearns function is used yielding a type of
complementary tristimulus value that would be analogous to the Ciba-Q
method of Ganz (see the preceding discussion). Connelly also adapted the
calculation to the digital computer.

REFERENCES

Abney, W. deW., and Festing, E. R. (1886). *Philos. Trans. R. Soc. London* **177**, 423–456.

Adams, E. Q. (1942). *J. Opt. Soc. Am.* **32**, 168.

Alexander, J. F. (1971). "A Scientific Method for Cataloguing Colored Samples," M.S. Thesis. Text. Dep., Clemson Univ., Clemson, South Carolina.

American Association of Textile Chemists and Colorists (1977). *Technical Manual* (published annually).

American Society for Testing Materials (1966). "Recommended Practise for Spectrophotometry and Description of Color," ASTM E308-66. (Also Am. Natl. Stand. Z138.2-1969.)

Aristotle (1913). "De Coloribus." *In* "The Works of Aristotle" *Opuscula*, (W. D. Ross, T. Loweday, and E. S. Forster, eds.), Vol. VI. pp. 791–792. Clarendon, Oxford.

Aristotle (1923). (Translated by E. W. Webster.) "The Works of Aristotle." *Meteorologica* **III**, p. 372a, Clarendon, Oxford.

Atherton, E., and Peters, R. H. (1955). *Congr. FATIPEC III* p. 147.

Balinkin, I. A. (1941). *J. Opt. Soc. Am.* **31**, 461.

BASF (1964). "Colorthek" (K. Thurner). Ludwigshafen, West Germany.

Baumann Color Guide (1925). Prase, Hamburg.

Beer, A. (1852). *Ann. Phys. Chem.* **86**(2), 78.

Benson, W. (1868). "Principles of the Science of Color." Chapman & Hall, London.

Bezold, W. von (1873). *Poggendorfs Ann.* **150**, 221.

Bezold, W. von (1876). "The Theory of Color." Louis Prang, Boston, Massachusetts.

Billmeyer, F. W., Jr. (1963). *J. Opt. Soc. Am.* **53**, 1317.

Billmeyer, F. W. Jr., and Saltzman, M. (1966). "Principles of Color Technology." Wiley (Interscience), New York.

Birren, F. (1944). *J. Opt. Soc. Am.* **34**, 396.

Birren, F. (1963). "Color. A Survey of Color" University Books, New Hyde Park, New York.

Birren, F. (1969). "Principles of Color." Van Nostrand-Reinhold, New York.

Bouma, P. J. (1971). "Physical Aspects of Colour." Macmillan, New York.

Boyle, R. (1664). 'Experiments and Considerations Touching Colours." (Johnson Reprints, New York, 1968.)

British Colour Council (1942). "The Wilson Color Chart I, II" British Colour Council, London.

Budde, W., Kundt, H. E., and Wyszecki, G. (1955). *Farbe* **4**, 83.

Burnham, R. W. (1949). *J. Opt. Soc. Am.* **39**, 387.

Burnham, R. W., Hanes, R. M., and Bartleson, C. J. (1963). "Color: A Guide to Basic Facts and Concepts." Wiley, New York.

Chevreaul, M. E. (1890). "The Principles of Harmony and Contrast of Colors" (transl. by C. Martel). George Ball & Sons, London.

Color Association of the United States, Inc. (1941). "Standard Color Card of America," 9th Ed. New York.

Connelly, R. L. (1970). "A Computer Method for Determination of the Content of Fiber Blends," M.S. Thesis. Text. Dep., Clemson Univ., Clemson, South Carolina.

Container Corporation of America (1942). "Color Harmony Manual" Chicago, Illinois. (Other editions in 1946 and 1949.)

Davidson, H. R., Godlove, N. N., and Hemmendinger, H., (1957). *J. Opt. Soc. Am.* **47**, 336.

Delboeuf, J. (1872). *Bull. Acad. R. Belg.* **34**, 250.

DIN-6164 (1964). "DIN-Farbenkarte Deutschen Normenausschuss." Beuth-Vertrieb Gmbh., Berlin.

Evans, R. M. (1948). "An Introduction to Color." Wiley, New York.

Fechner, G. T. (1889). "Elemente Psychophysik," 2nd Ed. Breitkopf & Haertel, Leipzig.

Flaschka, H. (1960). *Talanta* **7**, 90.

Foss, C. E. (1949). *J. Soc. Mot. Tel. Eng.* **52**, 30.

Foss, C. E., Nickerson, D., and Granville, W. C. (1944). *J. Opt. Soc. Am.* **34**, 361.

Ganz, E. (1962). Personal communication. (Newton–Raphson method for calculating Munsell value functions.)

Ganz, E. (1965). *Text. Rundsch.* **20**(9), 255.

Gerritsen, F. (1975). "Theory and Practice of Color." Van Nostrand-Reinhold, New York.

Glasser, L. G., McKinney, A. H., Reilley, C. D., and Schnelle, P. D. (1958). *J. Opt. Soc. Am.* **48**, 736.

Glenn, J. J., and Killian, J. T. (1940). *J. Opt. Soc. Am.* **30**, 609.

Glenn-Killian Ink Co. (1950). "Glenn-Killian Color System." Philadelphia.

Godlove, I. H. (1933). *J. Opt. Soc. Am.* **23**, 419.

Granville, W. C., Nickerson, D., and Foss, C. E. (1943). *J. Opt. Soc. Am.* **33**, 376.

Grassmann, H. G. (1853). *Poggendorfs Ann.* **89**, 69. [Also *Philos. Mag.* 7(4), 254 (1854).]

Guild, J. (1931). *Philos. Trans. R. Soc. London, Ser. A* **230**, 149.

Halbertsma, K. T. (1949). "A History of the Theory of Color." Swets & Zirtlinger, Amsterdam.

Hamly, D. H. (1949). *J. Opt. Soc. Am.* **39**, 592.

Harris, M. (1766). "The Natural System of Colors." (Privately reprinted by Faber Birren, New York, 1963.)

Helmholtz, H. von (1852). *Poggendorffs Ann.* **87**, 45.

Helmholtz, H. von (1924). "Physiological Optics," Optical Society of America, Vols. 1 & 2. Dover, New York. (Engl. transl.; "Physiologische Optik," 3rd Ed., 1905.)

Herbert, R. L. (1974). Yale University. *Univ. Libr. Gaz.* **47**, 1.

Hering, E. (1964). "Outline of a Theory of the Light Sense" (transl. by L. M. Hurvich and D. Jameson). Harvard Univ. Press, Cambridge, Massachusetts.

Hickethier, A. (1952). "Farbordnung Hickethier." Verlag H. Osterwald, Hannover.

Hoffmann, K. (1971). *Melliand Textilber.* **52**, 819.

Houstoun, R. A. (1873). "Light and Colour." Longmans, Green, London.

Hunter, R. S. (1940). "A Multipurpose Photoelectric Reflectometer." *Natl. Bur. Stand. (U.S.), Res. Pap.* No. RP1345.

Hunter, R. S. (1942). "Photoelectric Tristimulus Colorimetry with Three Filters." *Natl. Bur. Stand. (U.S.), Circ.* No. C429.

Indow, T. (1974). *In* "Handbook of Perception," Carterette, E. C., and Friedman, M. P. (eds.), Vol. 2, pp. 493–531. Academic Press, New York.

International Printing Ink Co. (1950). "I.P.I. Color Finder." New York.

Jacobs, M. (1927). "The Study of Colour." Doubleday, Page & Co., New York.

Johnston, R. M. (1973). *In* "Pigment Handbook (1973)" (T. C. Patton, ed.), Vol. 3, pp. 229–288. Wiley, New York.

Judd, D. B. (1951). *In* "Handbook of Experimental Psychology" (S. S. Stevens, ed.), Chap. 22. Wiley, New York.

Judd, D. B. (1955). *J. Opt. Soc. Am.* **45**, 673.

Judd, D. B., and Wyszecki, G. (1956). *J. Opt. Soc. Am.* **46**, 281.

Judd, D. B., and Wyszecki, G. (1975). "Color in Business, Science, and Industry," 3rd Ed. Wiley, New York.

Judson, J. A. V. (1935). "A Handbook of Colour." Manual Arts Press, Peoria, Illinois.

Kelly, K. L., and Judd, D. B. (1955). "The ISCC–NBS Method of Designating Colors and a Dictionary of Color Names." *Natl. Bur. Stand. (U.S.)*, No. 553.

Kelly, K. L., and Judd, D. B. (1976). "Color—Universal Color Language and Dictionary of Names." *Natl. Bur. Stand. (U.S.), Spec. Publ.* No. 440.

Kelly, K. L., Gibson, K. S., and Nickerson, D. (1943). *J. Res. Natl. Bur. Stand.* **31**, 55. (Also *J. Opt. Soc. Am.* **33**, 355.)

König, A., and Dieterici, C. S. (1886). *Sitzungsber. Preuss. Akad. Wiss., Phys.-Math. Kl.* **29**, 805.

Kubelka, P., and Munk, F. (1931). *Z. Tech. Phys.* **12**, 593.

Kundt, H. E., and Wyszecki, G. (1955). *Farbe* **4**, 289.

Ladd, J. H., and Pinney, J. E. (1955). *Proc. IRE* **43**, 1137.

Lambert, J. H. (1774). "Beschreibungen über Farben pyramide." Berlin.

Libby, W. C. (1974). "Color and the Structural Sense." Prentice-Hall, Englewood Cliffs, New Jersey.

Luckiesch, M. (1915). "Color and its Application." Van Nostrand, New York.

Luke, J. T. (1976). *Color Res. Appl.* **1**, 23.

Luther, R. (1927). *Z. Tech. Phys.* **8**, 540.

MacAdam, D. L. (1935). *J. Opt. Soc. Am.* **25**, 361.

MacAdam, D. L. (1942). *J. Opt. Soc. Am.* **32**, 247.

MacAdam, D. L. (1970). "Sources of Color Science." MIT Press, Cambridge, Massachusetts.

MacAdam, D. L. (1971). *Appl. Opt.* **10**, 1.

MacAdam, D. L. (1974). *J. Opt. Soc. Am.* **64**, 1961.

MacAdam, D. L. (1978). *J. Opt. Soc. Am.* **68**, 121. (Also addenda AIP Document No. PAPS JOSA-68-121-55.)

McLaren, K. (1969). *Farbe* **18**, 171.

McLaren, K. (1970). *J. Soc. Dyers Colour.* **86**, 189.

Maerz, A., and Paul, M. R. (1950). "Dictionary of Color." McGraw-Hill, New York.

Matthaei, R. (1971). "Goethe's Color Theory" (transl. by H. Aach). Van Nostrand-Reinhold, New York.

Maurolycus, F. (1575). "Photismi de Lumine et Umbra Venitii." Cited in Halbertsma (1949).

Maxwell, J. C. (1856). *Trans. R. Scot. Soc. Arts* **4**, 394.

Maxwell, J. C. (1857). *Trans. R. Soc. Edinburgh* **21**, 275.

Maxwell, J. C. (1860). *Proc. R. Soc. London* **10**, 404.

Maxwell, J. C. (1872). *Proc. R. Inst. G.B.* **6**, 260.

Middleton, W. E. K. (1949). *Can. J. Res.* **27**, 1.

Moon, P., and Spencer, D. E. (1944a). *J. Opt. Soc. Am.* **34**, 46.

Moon, P., and Spencer, D. E. (1944b). *J. Opt. Soc. Am.* **34**, 73.

Morley, D. I., Munn, R., and Billmeyer, F. W., Jr. (1975). *J. Soc. Dyers Colour.* **91**, 229.

Munsell, A. E. O., Sloan, L. L., and Godlove, I. H. (1933). *J. Opt. Soc. Am.* **23**, 394.

Munsell, A. H. (1905). "A Color Notation." Munsell Color Co., Inc., Baltimore, Maryland. (5th printing, 1946.)

Munsell Color Company (1943). "Munsell Book of Color." Munsell Color Co., Inc., Baltimore, Maryland. (Editions continue to be updated up to the present time in both matte and glossy form.)

Newhall, S. M. (1953). *In* "The Science of Color," OSA Committee on Colorimetry, Chaps. 4 & 5. Crowell, New York.

Newhall, S. M., Nickerson, D., and Judd, D. B. (1943). *J. Opt. Soc. Am.* **33**, 385.

Newton, I. (1704). "Opticks." Smith & Walford, London.

Nickerson, D. (1935). *J. Opt. Soc. Am.* **25**, 253.

Nickerson, D. (1936). *Text. Res.* **12**, 505.

Nickerson, D. (1940). *J. Opt. Soc. Am.* **30**, 575.

Nickerson, D. (1947). *Pap. Trade J.* **125**, 188.

Nickerson, D. (1950). *Am. Dyest. Rep.* **39**, 541.

Nickerson, D. (1969). *Color Eng.* **7**, 42.

Nickerson, D. (1975). *Opt. News* **3**, 8.

Nickerson, D. (1976). *Color Res. Appl.* **1**, 121.

Nickerson, D. (1978a). *J. Opt. Soc. Am.* **68**, 1943.

Nickerson, D. (1978b). Personal communication. (Letters to E. Q. Adams urging the publication of the X-Z planes work, 1942.)

Nickerson, D., and Stultz, K. F. (1944). *J. Opt. Soc. Am.* **34**, 550.

Noechel, F., and Stearns, E. I. (1944). *Amer. Dyestuff Reporter* **33**, 177.

Nutting, R. D. (1935). *Text. Res.* **6**, 107.

Nyberg, N. D. (1928). *Z. Phys.* **52**, 406.

Optical Society of America (1977). *J. Opt. Soc. Amer.* **67**, 515.

Ostwald, W. (1916). "Die Farbenfibel." Unesma, Leipzig.

Ostwald, W. (1917). "Der Farbenatlas." Unesma, Grossbothen.

Ostwald, W. (1918–1940). "I. Mathetische Farblehre; II. Physikalische Farblehre; III. Chemische Farblehre; IV. Physiologische Farblehre by Podesta; V. Physiologische Farblehre." Energie, Grossbothen.

Palmer, G. (1777). "Theory of Color and Vision." Leacroft, London. From MacAdam (1970).

Pantone, Inc. (1978). "Pantone Color System." Moonatchie, New Jersey.

Plateau, J. (1872). *Bull. Acad. R. Belg.* **33**, 397.

Pliny (1912). "Natural History" (transl. by H. Rackham), Book XXXV. Harvard Univ. Press, Cambridge, Massachusetts.

Plochere, G., and Plochere, G. (1948). "Plochere Color System." G. and G. Plochere, Los Angeles, California.

Pope, A. (1968). "The Language of Drawing and Painting." Russell & Russell, New York.

Prang, L., Hicks, M. D., and Clark, J. S. (1893). "Instruction in Color for Public Schools." Prang Education Co., Boston, Massachusetts.

Priest, I. G., Gibson, K. S., and McNicholas, H. J. (1920). "An Examination of the Munsell Color System," Technologic Papers of the Bureau of Standards, No. 167. Washington, D.C.

Reimann, G., Judd, D. B., and Keegan, H. J. (1946). *J. Opt. Soc. Am.* **36**, 128.

Rheinboldt, W. C., and Menard, J. P. (1960). *J. Opt. Soc. Am.* **50**, 802.

Rich, R. M., Billmeyer, F. W., Jr., and Howe, W. G. (1975). *J. Opt. Soc. Am.* **65**, 956.

Richter, M. (1953). *Farbe* **1**, 85.

Richter, M. (1955). *J. Opt. Soc. Am.* **45**, 223.

Ridgway, R. (1912). "Color Standards and Color Nomenclature." Hoen Co., Baltimore, Maryland.

Robertson, A. R. (1978). *Color Res. Appl.* **3**, 149.

Rösch, S. (1928). *Phys. Z.* **29**, 83.

Rood, O. N. (1895). "Textbook of Color." 3rd Ed. Appleton, New York.

Rounds, R. L. (1969). *Text. Chem. Color.* **1**(14), 297.

Schrödinger, E. (1920). *Ann. Phys. (Leipzig)* **63**, 397, 481.

Schultze, W. (1971). *Farbe* **20**, 13.

Simon, F. T. (1961). *Farbe* **10**, 225.

Simon, F. T. (1972). *In* "Color Metrics" (J. J. Vos, L. F. C. Friele, and P. L. Walraven, eds.), p. 308. AIC/Holland, Soesterberg.

Simon, F. T., and Connelly, R. L. (1979). To be published.

Stenius, A. S. (1977). *Color Res. Appl.*, **2**, 189.

Stiles, W. S. (1946). *Proc. Phys. Soc. (London)*, **58**, 41.

Taylor, H. D., Knoche, L., and Granville, W. C. (1950). "Descriptive Color Names Dictionary," Container Corporation of America, Chicago.

Textile Color Card Association of the U.S., Inc. (1943). "United States Army Color Card." New York.

Tyler, J. E., and Hardy, A. C. (1940). *J. Opt. Soc. Am.* **30,** 587.

Villalobos, C., and Villalobos, Y. (1947). "Atlas de los Colores." Liberia El Ataneo Editorial, Buenos Aires.

Vos, J. J., Friele, L. F. C., and Walraven, P. L. (1972). "Color Metrics," Proceedings of the Helmholtz Memorial Symposium, Driebergen, Netherlands, 1971. AIC/Holland, Soesterberg.

Weber, E. H. (1834). "De pulsu, resorptione, auditu et tactu annotationes anatomicae et physiologicae." Koehler, Leipzig.

Weber, E. H. (1846). *In* "Handworterbuch der Physologie" (R. Wagner, ed.), Vol. 3, Part 2, pp. 481–588. Viewig, Braunschweig.

Wright, W. D. (1928–1929). *Trans. Opt. Soc. (London)* **30,** 141.

Wyszecki, G. (1954). *J. Opt. Soc. Am.* **44,** 725.

Wyszecki, G. (1960). "Farbsysteme." Musterschmidt Verlag, Göttingen.

Wyszecki, G. (1968). *J. Opt. Soc. Amer.*, **58,** 290.

Young, T. (1802). *Philos. Trans. R. Soc. London* **92,** 12.

6

Colorimetry of Fluorescent Materials

FRANC GRUM

Research Laboratories
Eastman Kodak Company
Rochester, New York

I. INTRODUCTION: THE CONTENT OF THIS CHAPTER

First we shall define the problem: Why is it important to deal separately with color measurements of fluorescent material? After giving appropriate background information, terminology, and a discussion on fluorescent materials, we point out the problems associated with color measurement of such materials and suggest and describe the remedies for these problems.

It will be shown that color measurements of fluorescent materials must be made by a spectrophotometer in which the sample is irradiated with undispersed white light of controlled spectral-power distribution; the radiation reflected (transmitted) and emitted is then detected monochromatically. Such measurements are called total radiance factor measurements. (See Section V for definition.) Since the fluorescent (emitted) component in these types of measurements depends greatly on the spectral irradiance of the light source, the source must be well standardized.

A review of the instrumental requirements for color measurements of fluorescent samples will be given along with a discussion on sources of error associated with these measurements.

Furthermore, we will discuss various methods for separating the components from the combined measurement and how the separated component can be treated colorimetrically.

Also described will be methods that can be used for predicting total radiance factor for any desired illuminant from a source that is only close in spectral-power distribution to that illuminant.

Although the treatment described in this chapter is valid for both transparent and opaque materials, we will henceforth deal primarily with opaque materials, since these are more commonly encountered in practice.

II. BACKGROUND

In the past few years the use of fluorescent dyes in various articles on the market has increased markedly. Examples include paints, crayons and toys, hunting jackets, and safety warning devices. These dyes have also been used on billboards and in packaging via block printing. Fluorescent inks are used widely these days in the graphic arts industry as well as in other applications. In fact, nearly any place we turn we are confronted with fluorescent materials.

Fluorescent dyes exhibit spectral selectivity, and one can expect from such dyes more brilliant colors than from nonfluorescent dyes. In photography they provide reflection prints that resemble illuminated transparencies. Grum and Clapper (1969) reported that the fluorescence of both inks (yellow and magenta) and stock affect the color of halftone images. When colors on whitened stock are compared with the same colors on nonwhitened stock, an increase in saturation is observed if the samples are evaluated (measured) properly. Similar observations were made by Preucil (1959) and Wilson (1967). Fluorescent dyes, pigments, and inks enable us to use more efficiently the energy that must be absorbed by dyes to make images. A similar argument applies for the replacement of a dye-protecting UV absorber with a fluorescent whitening agent.

Fluorescent whitening agents are often used to enhance the overall whiteness appearance of papers, fabrics, and plastics. Chromatic fluorescent dyes, pigments, and inks have also found wide application as a means of increasing brightness and saturation. Therefore, new techniques and instrumentation are needed for physical assessment of such materials.

Visual appraisal of the color of fluorescent materials is made in heterochromatic light; hence, the combined subjective effects of reflectance and radiation from fluorescent additives are perceived by human visual observations. Thus, physical measurements on fluorescent materials must be made in a similar manner for the results to agree with visual assessment. These measurements yield the "total radiance factor" of the test samples.

III. FLUORESCENT WHITENING AGENTS

Fluorescent whitening agents nowadays are often used with paper, textiles, plastics, and so on, to give an apparent enhancement of their reflectance in the blue region of the visible spectrum. Undyed papers, plastics, fabrics, and white pigments have nearly flat spectrophotometric reflectance curves except for a sharp decrease of reflectance toward short wavelengths. The greater the decrease of reflectance in the blue spectral region, the yellower the sample and the less satisfying it is as a "white." The fluorescent whitening agents (FWA's) are rarely white themselves, and they cause tints that are almost as noticeable as those caused by blue dyes. This is shown in Fig. 1, which reproduces the absorption spectrum of a whitening agent at three different concentrations. Also given in the figure is the emission distribution of the medium-concentration sample. Thus, the amounts of the whitening agents that can be used are limited in the same fashion as the amount of blue dye that can be used to produce white. These problems are described by Harkavy (1958), Carr (1955), Morton (1963), and Stensby (1967a,b). Allen (1957) reported that the whitening effects are directly proportional to the effective fluorescence, defined as the sum of the CIE tristimulus values of the fluorescent light from the dye. In this case the

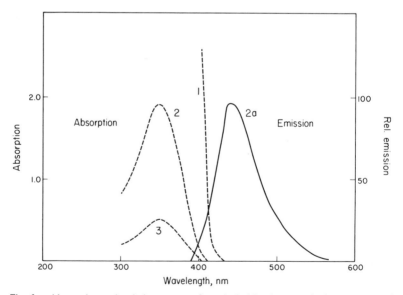

Fig. 1.   Absorption and emission spectra of a typical whitening agent in three concentrations $(1 - 1.4\ M, 2 - 4.0 \times 10^{-4}\ M, 3 - 1.8 \times 10^{-4}\ M$, all in 1 mm cell). Curve 2a in the emission spectrum of the $4.0 \times 10^{-4}\ M$ sample.

effective fluorescence (tristimulus sum) is claimed to be a reliable indication of the whitening power. Further discussion on FWAs will be given in the section on whiteness.

IV. FLUORESCENT DYES

Fluorescent dyes, because they are dyes, absorb visible light. Moreover, they must emit visible light, which can result from the energy absorbed either in the visible or in the near ultraviolet. It is necessary only that they both absorb and emit visible light. Classic examples of fluorescent dyes are fluorescein and rhodamine, whose absorption and emission spectra are given in Figs. 2 and 3. Chemical descriptions of typical fluorescent dyes are documented in the literature (Switzer and Switzer, 1950, 1958; Kazenas, 1957, 1960; Gaunt, 1958; Pringsheim, 1949). Some applications of fluorescent dyes, besides their use as fluorescent whitening agents, are given by Day-Glow Color Corporation (1971).

Discussion of an interesting possible application of fluorescent dyes in photography and printing follows. In color photography or in three-color printing, three dyes—cyan, magenta, and yellow—are normally required, the dyes used to monitor red, green, and blue light, respectively. However, such dyes also absorb various amounts of the other colors in addition to

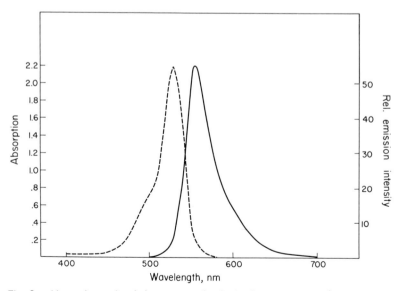

Fig. 2.   Absorption and emission spectra of a rhodamine solution (10^{-4} M in H_2O).

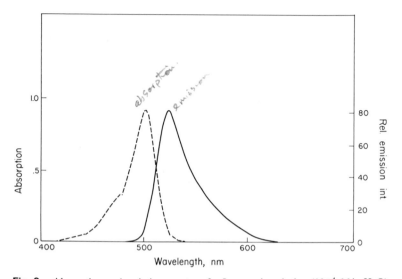

Fig. 3. Absorption and emission spectra of a fluorescein solution (10^{-4} M in H_2O).

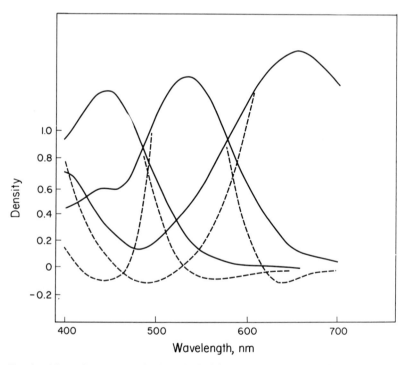

Fig. 4. Absorption spectra of a hypothetical fluorescent dye set (--- is the contribution of fluorescence).

their desired absorption in a particular spectral region. For example, the yellow dye absorbs not only blue light but some green as well. The magenta dye absorbs green light but has unwanted absorption in both the blue and the red regions; and the cyan dye absorbs some blue and green as well as the desired red. By using fluorescent dyes with emissions in the appropriate spectral regions, one would expect to be able to compensate for the unwanted absorptions. Therefore, fluorescent emission can be looked on as being "negative density", and this is the way the eye perceives fluorescence. To observe this, one need only look at a red fluorescent dye through a red filter; the sample appears to glow.

For the yellow dye we would want emission primarily in the green region but possibly extending into the red. A good rule of thumb is to have the dye fluoresce at the dominant wavelength, that is, the principal region of transmission of the dye. In this way one obtains fluorescent reinforcement of the reflected light. For the magenta dye one would require both blue and red emission, a requirement that may necessitate using two compounds, one of which absorbs green and emits red and one of which absorbs ultraviolet and emits blue (i.e., a fluorescent whitening agent). For the cyan dye a broad emission around 490 nm would be required. The absorption spectra of such a hypothetical fluorescent dye set are shown in Fig. 4. An example of using such a dye set for printing inks was published by Maycumber (1964).

V. TERMINOLOGY

The definitions of many terms associated with fluorescence and luminescence in general are not readily available; hence, it is advisable to give some definitions pertaining to colorimetry of fluorescent materials and fluorescence in general. The terms and symbols given in this section are those published by the CIE (1970) and those suggested by the CIE (1978) TC-2.3 Committee in Materials.

VI. GENERAL LUMINESCENCE TERMS

A. Luminescence

Luminescence is the phenomenon of the emission by matter of electromagnetic radiation that for certain wavelengths or restricted regions of the spectrum is in excess of that due to the thermal radiation from the material at the same temperature.

Note 1: The radiation depends on the particular host and activator materials, exciting wavelength, and temperature.

Note 2: By applying the radiometric and/or photometric procedures to the luminescence, the quantities of luminescence radiation and luminescence light are gained.

Note 3: In lighting, this term is generally restricted to the emission of radiation in the visible or near-visible spectrum.

B. Photoluminescence

Photoluminescence is luminescence caused by ultraviolet, visible, or infrared radiation.

Note: A special form of photoluminescence is the anti-Stokes luminescence, where an emission occurs at wavelengths shorter than the shortest wavelength of the exciting radiation.

C. Fluorescence

Fluorescence is photoluminescence where the decay is determined by an intrinsic time constant.

Note: This time constant is generally less than about 10^{-8}s.

D. Phosphorescence

Phosphorescence is luminescence delayed by storage of energy in an intermediate level with a time constant exceeding the intrinsic one.

Note: This time constant is generally longer than about 10^{-8}s. For organic substances phosphorescence means generally a radiative relaxation from an excited triplet level to ground level.

E. Electroluminescence

Electroluminescence is luminescence of certain substances, generally solid, under the action of an electric field.

F. Cathodoluminescence

Cathodoluminescence is luminescence excited by electrons striking the surface of the luminophore from vacuum or a gaseous atmosphere.

G. Radioluminescence

Radioluminescence is luminescence excited by ionizing radiation.

H. Chemiluminescence

Chemiluminescence is luminescence due to the energy liberated in a chemical reaction.

Note: *Bioluminescence* is a special form of chemiluminescence occurring in animal and plant life.

I. Thermally Stimulated (Released) Luminescence: Thermoluminescence

Thermoluminescence is luminescence produced in a previously excited luminophor owing to heating.

Note: The total emitted radiation is called *light sum.*

J. Efficiency of Luminescence

1. LUMINESCENCE (RADIANT) EFFICIENCY

Luminescence (radiant) efficiency is the ratio of the energy of radiation emitted to the energy of the radiation absorbed.

2. LUMINESCENCE QUANTUM EFFICIENCY

Luminescence quantum efficiency is the ratio of the number of photons emitted to the number of photons absorbed.

Note: The ratio of the number of photons emitted to the number of photons impinging on the luminescent material is called *external quantum efficiency.*

3. CATHODOLUMINESCENCE RADIANT EFFICIENCY

Cathodoluminescence radiant efficiency is the ratio of the emitted power to the power of the electron bean falling on the phosphor.

Note: The cathodoluminescence radiant efficiency, CRE, refers to the total power incident on the phosphor; thus, elastically scattered primary electrons and slow secondary electrons are neglected. Usually it is necessary to state separately the electron-beam current and the accelerating voltage.

VII. SPECTRAL QUANTITIES OF LUMINESCENCE

A. Excitation Spectrum

The excitation spectrum is the photon flux emitted, for a given emission wavelength, as a function of excitation wavelength. Symbol: $\overline{X}(\lambda')$

Note: If the emitted luminescence radiation is not linearly proportional to the excitation intensity, the excitation spectra should be measured keeping the emission flux constant.

B. Effective Radiant Excitation

Effective (radiant) excitation is irradiance supplied by the excitation unit at the sample position.

C. Luminescence Emission Spectrum

The luminescence emission spectrum is the spectral-power distribution of the radiation emitted by a luminescent material owing to monochromatic or heterochromatic excitation.

D. Stimulation Spectrum

The stimulation spectrum is the quantum (radiant) efficiency of a luminescent material as a function of the stimulating wavelength.

VIII. LUMINOUS MATERIALS

A. Radioluminescent Paint

Radioluminescent paint is a special type of luminescent paint, consisting of luminophors mixed with a binder.

B. Fluorescent Whitening Agent (FWA)

A fluorescent whitening agent is a compound that, by its presence in or on a near-white substrate, creates a *visual* whitening effect by virtue of fluorescence in the blue region of the visual spectrum.

C. Fluorescent Dye (Colorant)

Fluorescent dye is a material showing excitation in the ultraviolet (and sometimes in the visible) spectrum and emission, usually in a narrow band, in the visible spectrum.

D. Fluorescence Emission Intensity of a Fluorescent Whitening Agent

The fluorescence emission intensity of a fluorescent whitening agent is the difference between the Z (CIE standard observer) tristimulus value of a sample treated with an FWA and that of the untreated sample under standardized illumination conditions (D_{65}) and viewing conditions (CIE-approved geometry) for any specified substrate and specimen-presentation techniques.

IX. REFLECTION TERMS

A. Reflected Radiance Factor

The reflected radiance factor is the ratio of the radiance due to reflection of the medium to that of a perfect reflecting diffuser identically irradiated. Symbol: $\beta_S(\lambda)$.

B. Fluorescent Radiance Factor

The fluorescent radiance factor is the ratio of the radiance due to fluorescence of the medium to that of a perfect reflected diffuser identically irradiated. Symbol: $\beta_L(\lambda)$.

Note: The spectral fluorescent radiance factor is a characteristic of the material that depends also on the spectral distribution of the irradiation. The term "spectral" here refers to monochromatic assessment of the emission. This term should not be used in monochromatic irradiation.

C. Total Radiance Factor

The (total) radiance factor is the sum of the reflected and fluorescent radiance factor. Symbol: $\beta_T(\lambda)$.

D.   Conventional Reflectometer Value

The conventional reflectometer value is the apparent reflectometer value when a luminescent sample is measured relative to a nonluminescent white reference, using a reflectometer with monochromatic irradiation and heterochromatic detection. Symbol: $\rho_C(\lambda)$.

X.   INSTRUMENTATION IN FLUORESCENCE MEASUREMENTS

A.   Definition of the Problem

The measurement of spectral reflectance, which can be complex, has taken on a new measure of complexity owing to the common use of fluorescent additives.

The main objective of this chapter is to discuss the means and methods for proper measurement of fluorescent materials and to describe the instrumental methods available for such measurement. The physical evaluation of

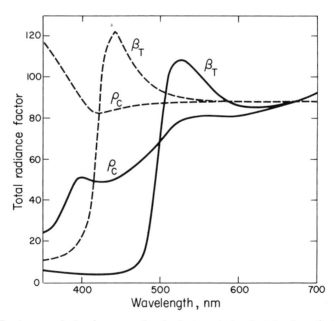

Fig. 5.   An example showing conventional reflectance (ρ_C) and total radiance factor (β_T) of a chromatic fluorescent and a white sample.

the color of a system containing fluorescent additives can be made only when spectrophotometric measurements of such materials represent both the contribution of reflectance and that of fluorescence. Such measurements cannot be obtained with conventional spectrophotometers in which the sample is placed between the monochromator and the photomultiplier. That is, in conventional spectrophotometry the sample is irradiated by monochromatic light and the reflected light is integrated and detected by a photomultiplier. This method is quite suitable for nonfluorescent materials. When a sample with fluorescence is introduced, however, a portion of the incident energy is absorbed and emitted at longer wavelengths. This emitted energy is detected as though it were reflected radiation of the same wavelength as the incident radiation, and erroneous data result. This is so because the fluorescent light, in such a measurement arrangement, is excited by the monochromatic incident light that reaches the photomultiplier undispersed. Obviously, the colorimetric parameters derived from such erroneous data do not agree with the visual perception of such a fluorescent color sample. Therefore, samples that fluoresce must be measured in an apparatus in which the sample is irradiated with nondispersed white light, and the radiation reflected and emitted by the sample is detected monochromatically. This type of measurement is called total spectral radiance factor. An example of such data is shown in Fig. 5. The assignment of the fluorescent energy to the wrong wavelength region is apparent in the conventional reflectance curve shown in the figure.

B. Definition of Total Spectral Radiance Factor

Total radiance factor is the sum of the reflected and the emitted flux from a fluorescent sample, relative to the reflected flux from a nonfluorescent reference, when both are irradiated by the same heterochromatic flux. This quantity has been defined by the CIE (1977) and was formerly identified in the literature as "relative radiance" or "spectral radiance factor" (Wyszecki, 1972; Allen, 1973; Grum and Wightman, 1960). Spectral total radiance factor can be expressed as

$$\beta_T(\lambda) = \beta_s(\lambda) + \beta_L(\lambda), \tag{1}$$

where $\beta_T(\lambda)$ is total radiance factor at wavelength λ, $\beta_s(\lambda)$ the reflected radiance factor at wavelength λ, $\beta_L(\lambda)$ the luminescence radiance factor at wavelength λ. A more detailed mathematic description of total radiance factor is

$$\beta_T(\lambda) = [E(\lambda)\rho_t + F(\lambda)]/E(\lambda)\rho_r(\lambda) = \rho_t(\lambda) + [F(\lambda)/E(\lambda)], \tag{2}$$

where $\rho_t(\lambda)$ is the true spectral reflectance of the sample, $\rho_r(\lambda)$ the spectral reflectance of the reference material (assumed to be unity), $F(\lambda)$ the true spectral fluorescence of the sample, $E(\lambda)$ the spectral irradiance of the light source used, and $F(\lambda)/E(\lambda)$ identical to $\beta_L(\lambda)$.

In the measurement of total radiance factor the specimen is illuminated diffusely (usually by an integrating sphere). The angle between the normal to the specimen and the axis of the viewing beam should not exceed 10°. Under these illuminating and viewing conditions, designated by the CIE (1931, 1971) as diffuse/normal (abbreviation $d/0$), reflectance $\rho_t(\lambda)$ becomes radiance factor $\beta_s(\lambda)$.

As stated earlier, $F(\lambda)$ is the true spectral fluorescence emittance with an irradiation source $E(\lambda)$ under which the sample has absorbed a total of N quanta. If the sample is irradiated with another source $E^*(\lambda)$ with which N^* quanta are absorbed, the true spectral fluorescence emittance becomes

$$F^*(\lambda) = (N^*/N)F(\lambda). \qquad (3)$$

Hence, the spectral distribution of fluorescence remains the same but the values have to be multiplied by the ratio of N^*/N [the number of quanta absorbed with source $E^*(\lambda)$ to the number of quanta absorbed with source $E(\lambda)$].

The fluorescent flux is produced on absorption of heterochromatic radiation by the sample and a subsequent emission of light at a longer wavelength due to the process of electronic relaxation.

The fluorescent flux affects the total spectral radiance data in two ways: by its spectral distribution and by its intensity. The spectral distribution of the fluorescent flux depends only on the emission spectrum of the fluorescent species. The intensity of the flux, however, depends on two separate parameters: (1) the fluorescent quantum yield of the material and (2) the total number of photons, absorbed by the material, capable of exciting the fluorescence. Since fluorescence occurs at wavelengths other than the illuminating wavelength, total radiance factor is inadequate for the evaluation of fluorescence. This topic will be discussed in further detail in Section XII.

XI. MEASUREMENTS OF TOTAL SPECTRAL RADIANCE FACTOR

Total spectral radiance factor of fluorescent materials can be measured in two ways. Direct measurement is made according to the definition, in which case the spectral-power distribution of the irradiating source must be standardized and carefully controlled. Alternatively, the components β_S and

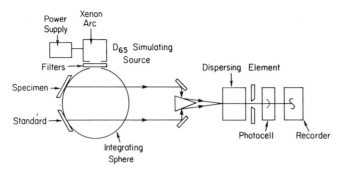

Fig. 6.   Schematic diagram of an optical system used for total radiance factor measurement.

β_L may be measured separately and combined numerically, using a tabulated spectral-power distribution.

There are several ways to measure the reflected radiance factor (β_S) of a fluorescent sample. Perhaps the best way is to use a dual-monochromator system in which the fluorescent sample is irradiated monochromatically and the energy reflected from the sample is detected also monochromatically at the same wavelength. The only requirement in such a system is that the bandwidth of the analyzing monochromator is kept at least as narrow as the bandwidth of the irradiating monochromator, to exclude the emitted fluorescence in such a system.

A typical setup for measuring total radiance factor directly is shown in Figs. 6 and 7. As shown in Fig. 6, the fluorescent specimen is illuminated with undispersed white light from a xenon lamp filtered to simulate D_{65} illuminant. [The CIE has recommended that the standard illuminants A and C be supplemented by the illuminant D representing daylight, described by Judd *et al.* (1964). This illuminant should have a correlated color temperature of 6500 K, and its relative spectral-power distribution should be similar to that of daylight at the same correlated color temperature.] The main application of such an illuminant is visual color matching of object color, particularly fluorescent materials. (This type of source can be used in total radiance

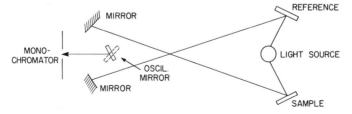

Fig. 7.   Optical diagram of a setup for measuring total radiance factor in 0°/45° geometry.

factor measurements of fluorescent materials.) The white light from the
xenon source is, by the use of the integrating sphere, diffusely illuminating
the specimen. The advantage of this type of illumination is that no surface
correction for the samples is needed and the orientation of the sample is
immaterial. There is another advantage and/or importance of using an
integrating sphere in these measurements. The integrating sphere coated with
$BaSO_4$ has a high depolarization coefficient because of multiple reflections
before detection. Grum and Costa (1974) reported 96% for the depolarization coefficient of an integrating sphere coated with $BaSO_4$. For the effects of
polarization on directional and bidirectional reflectance measurements, see
also the work of Grum and Spooner (1973).

The light emitted and reflected from the sample enters the monochromator,
where it is dispersed and then measured monochromatically. The amount of
radiation from the sample at each wavelength is measured directly or

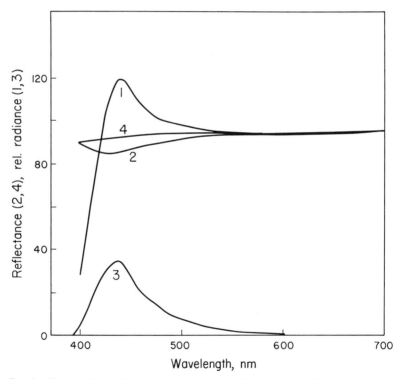

Fig. 8. Total radiance factor and conventional reflectance data of two paper samples.
Curve 1 represents total radiance factor data, curve 2 is the conventional reflectance of the same
sample, curve 3 is the relative fluorescence of that sample, and curve 4 represents both total
radiance factor and reflectance data of a nonfluorescent sample.

recorded on the instrument recorder. This radiation then is the sum of the fluorescence and reflectance from a sample.

Figure 7 shows an optical diagram of a setup for measuring total radiance factor in the 0°/45° geometry arrangement. The standard reference material is pressed $BaSO_4$, the type described by Grum and Luckey (1968; also Grum and Wightman, 1977). Commercially available instruments (spectrophotometers and colorimeters) suitable for this type of measurement have been well described in the literature (Grum and Wightman, 1960; Hunter, 1960; Stenius, 1965; Berger and Brockes, 1971).

A. Examples of Measurements

Figure 8 shows total radiance factor and reflectance spectra of two paper samples. Curve 1 represents total radiance factor of the sample treated with a fluorescent whitening agent, curve 2 is the conventional reflectance of the same sample, and curve 3 is the relative fluorescence of the first sample. Curve 4 represents the total radiance factor and the reflectance of a non-fluorescent white sample. Note that a single curve represents both β_T and reflectance (i.e., they are identical, since the sample is nonfluorescent).

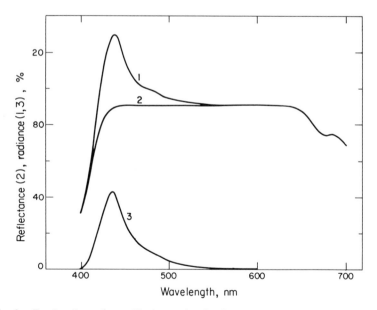

Fig. 9. Total radiance factor (1), conventional reflectance (2), and luminescent radiance factor (3) of a white fluorescent plastic sample.

TABLE I.   Colorimetric Data of White Samples

Data type	Fig. no.	(D$_{65}$ illuminant) Curve no.	x	y	$Y\%$	λ_D
$\beta_T(\lambda)$	8	1	0.3051	0.3221	94.8	479.5
$\beta_c(\lambda)$	8	2	0.3181	0.3377	93.9	572.8
$\beta_T(\lambda)$ & $\rho_c(\lambda)$	8	4	0.3152	0.3326	95.4	572.5
$\beta_T(\lambda)$	9	1	0.2995	0.3158	90.1	478.1
$\beta_c(\lambda)$	9	2	0.3141	0.3330	89.9	564.6

Another example showing the error in spectral reflectance data of a fluorescent plastic sample is shown in Fig. 9. The colorimetric data for the samples in Figs. 8 and 9 are given in Table I and in Fig. 10.

An examination of Table I and Fig. 10 clearly shows a huge shift in the dominant wavelength between the values computed from conventional reflectance and total radiance factor data.

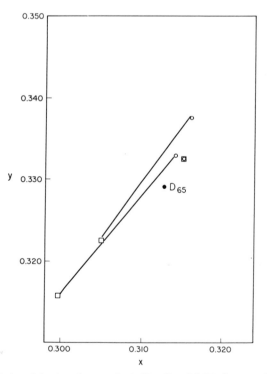

Fig. 10.   CIE plot of the data for samples in Figs. 8 and 9 (□: from total radiance factor, ○: from conventional reflectance).

TABLE II.   Colorimetric Parameters Computed from Curves in Fig. 11

Data type	Curve no.	x	y	$Y\%$	λ_D
$\beta_T(\lambda)$	1	0.5914	0.3099	24.0	623.0
$\beta_s(\lambda)$	2	0.4399	0.3160	23.9	633.3

Similar measurements made for a red fluorescent sample (Maxilon Brilliant Red 2B) are shown in Fig. 11. In this figure, curve 1 is the total radiance factor, curve 2 represents conventional reflectance, and curve 3 is the fluorescent radiance factor. Table II and Fig. 12 give the colorimetric parameters of the data from Fig. 11.

Figure 12 shows also colorimetric parameters determined for a fluorescent magenta sample. Two points should be made in regard to Fig. 12. First, note the large difference between the chromaticity points determined from the

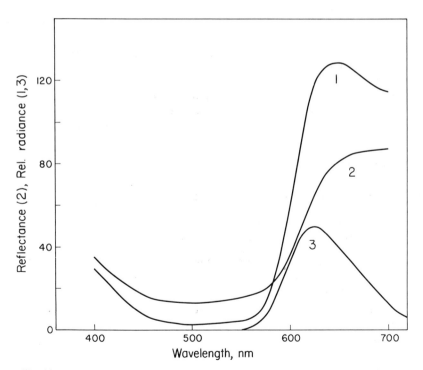

Fig. 11.   Spectral data for Maxilon Brilliant Red 2B sample. Total radiance factor (1), reflected radiance factor (2), luminescent radiance factor (3).

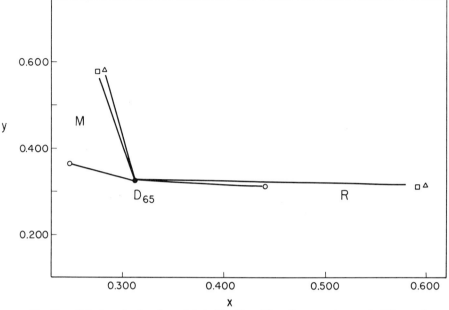

Fig. 12. Colorimetric plots from data in Fig. 11 and for a fluorescent magenta (*M*) sample (□: from β_T, ○: from ρ_C, △: from $\beta_S + \beta_L$).

conventional reflectance and total radiance factor data. Second, the chromaticity coordinates determined from total radiance factor data and those determined from ($\beta_S + \beta_L$) data are nearly identical, as expected.

XII. SEPARATION OF COMPONENTS

A. Dual-Monochromator Methods

As mentioned earlier, the total spectral radiance factor curve is a composite of two components, the reflected and the fluorescent radiance factor.

(1) The dual-monochromator systems mentioned earlier for measuring separated components have been described by Donaldson (1954) and by Grum (1971). The system used by Grum is schematically shown in Fig. 13 and consists basically of three components: an exciting monochromator, an optical relay system and sample-holding accessories, and an analyzing monochromator. The apparatus is completely automated and equipped with digital output for computer processing of the data. The apparatus can measure fluorescence emission and excitation (absorption) spectra and reflectance spectra with monochromatic irradiation and monochromatic

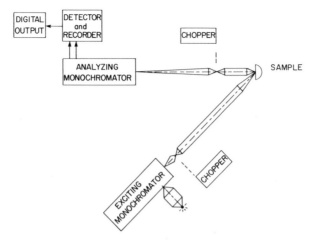

Fig. 13.   Optical diagram of a dual-monochromator system for measuring reflected radiance factor.

collection. A $0°/45°$ geometry is normally used; however, provisions are made for the use of integrating-sphere geometry also.

This apparatus is used for measuring true reflectance (reflected radiance factor) of any kind of fluorescent specimens. The energy from the exciting monochromator is calibrated in absolute terms and is constantly monitored. For fluorescence the losses due to scatter, self-absorption, and transmission are determined for each sample under test. Two sets of measurements, one from the sample, the other from the reference white, are required when true reflectance data (reflected radiance factor) are desired, because the equipment operates in a single-beam mode. The computer reduces the data and applies all the necessary parameters.

By use of the preceding apparatus, one can measure the following quantities: reflected radiance factor, true (source independent) fluorescence spectra, and true excitation spectra. An example of such data is shown in Fig. 14. These data are used to calculate such parameters as fluorescence quantum yield, tristimulus values for the reflected and fluorescent components, and any other colorimetric parameters (e.g., whiteness value for a white sample).

The colorimetric computations with separated components are done as follows:

$$T_i = K \int_{-\infty}^{\infty} \beta_s(\lambda)E(\lambda)t_i(\lambda)\, d\lambda + K \int_{-\infty}^{\infty} F(\lambda)t_i(\lambda)\, d\lambda, \qquad (4)$$

where T_i represents the total tristimulus values $\overline{X}, \overline{Y}, \overline{Z}$ (each computed separately); $\beta_s(\lambda)$ is the reflected radiance factor; $E(\lambda)$ is the spectral-power

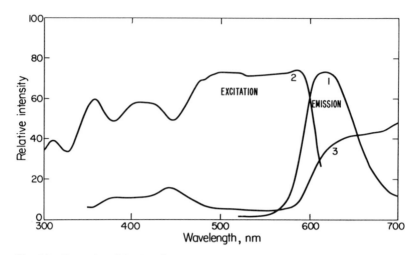

Fig. 14. Examples of the data from a dual-monochromator system. Emission (1), excitation (2), reflected radiance factor (3).

distribution of the illuminant; $t_i(\lambda)$ represents the CIE color mixture functions $\bar{x}(\lambda)$, $\bar{y}(\lambda)$, and $\bar{z}(\lambda)$; $F(\lambda)$ is the true fluorescence of the sample (source independent); and K is the normalization constant.

The technique of separation of components has several virtues, the most important of which is that colorimetric evaluations can be made for any illuminant without having to make the measurements with that source (illuminant). Hence, there is no need of having a standardized source that simulates a given illuminant, such as D_{65} illuminant. (The importance of a standardized daylight-simulating source will be discussed further in the sections dealing with standardization and with prediction methods.)

The chromaticity values obtained by the use of Eq. (4) and separated components should be identical with the chromaticity values obtained for the sample subjected to total spectral radiance factor measurement, provided that the source used in the measurement of total radiance factor simulates very closely the $E(\lambda)$ value used in Eq. (4).

When the colorimetric parameters are computed from the total spectral radiance factor data, the pertinent equation used is

$$T_i = K \int_{-\infty}^{\infty} \beta_T(\lambda)E(\lambda)t_i(\lambda)\, d\lambda, \qquad (5)$$

where $\beta_T(\lambda)$ is the total radiance factor, obtained with a source that closely approximates the illuminant $E(\lambda)$ for which the colorimetric data is computed. K, T_i, $E(\lambda)$, and $t_i(\lambda)$ were defined earlier.

The complexity of instrumentation involved in the dual-monochromator method for measuring a true reflected radiance factor is well documented in the literature (Donaldson, 1954; Grum, 1971).

(2) Donaldson, in his investigation on spectrophotometry of fluorescent pigments, measured complete spectral fluorescence distributions for each incident wavelength; thus, he obtained a series of fluorescence curves from which he determined fluorescence yield as a function of wavelength. Next he determined the absorption of the incident exciting light and computed the amount of fluorescence light from the sample, which he then added to the nonfluorescent reflected light to get the combined effect. Donaldson's approach is sound; however, the procedure is very elaborate and difficult to reduce to practice.

(3) Fukuda and Sugiyama (1961) described yet another method for calculating true tristimulus values for various illuminants from measurements with two monochromators. Their apparatus is extremely cumbersome and would be difficult to use in practice.

B. Abridged and Computational Methods

(1) Eitle and Ganz (1968) have proposed a modified measurement of spectral radiance factor for separating fluorescence and inertly reflected components. Their method is based on total radiance factor measurements and a subsequent mathematical separation of the inert reflectance from the fluorescence. Their method requires, besides knowledge of the spectral energy distribution of the light source for which colorimetric results are to be computed, knowledge of the reflectance of the nonfluorescent substrate of the sample. They measure total radiance factor with the nonfiltered source and with a series of cutoff filters to exclude the wavelengths of the incident light that excite fluorescence. Their data are used to compute the equivalent reflectance (reflectance of a nonfluorescent, idealized specimen that would give the same sensation of color as the fluorescent specimen under the same illuminant) for ordinary tristimulus calculations. This approach is useful and valid for white samples treated with fluorescent whitening agents but is not readily applicable to chromatic fluorescent samples.

The two-monochromator method described earlier is, however, applicable to fluorescent samples of any kind. However, this type of instrumentation is expensive and not accessible to every laboratory. It is for the reasons stated earlier that one has to turn to some abridged method for separating reflected and fluorescence components.

(2) Another method for separating the total spectral radiance factor curve of fluorescent substances into reflected and fluorescent components has been developed by Allen (1973). With his method the true reflectance

curve can be obtained by direct measurement below the emission region and also above the absorption region by the use of a fluorescence-killing filter. In the overlapping region (where absorption and emission overlap), he computes the true reflectance curve from two different total radiance factor curves. The basic idea of this method is similar to that of Eitle and Ganz and can be quite accurate if the measurements are made properly and if the fluorescence-killing filters are selected carefully.

The major steps of Allen's method are described in the following paragraphs with the aid of a schematic diagram.

In Fig. 15, curve 1 represents total radiance factor of a fluorescent sample, curve 2 is the reflected radiance factor of the sample, and curve 3 is the reflectance of the uncolored substrate. The shaded area, between the total radiance factor curve and the reflected radiance factor curve, is the area affected by fluorescence. The area under curve 3 is the effective absorption of the fluorescent species. Note an overlap of absorption and emission (designated C). Allen calls this overlapping area the crossover region.

Allen's first step is to introduce a sharp cutoff filter in the incident beam that completely eliminates all wavelengths below the maximum point on the total radiance factor curve. By doing so the fluorescence is not excited and the measured reflectance, at a wavelength longer than the maximum of the β_T curve, is the true reflectance (β_S). He calls this particular cutoff filter a "fluorescence-killing filter." Also, the β_T curve at wavelengths below the point where curve 1 (β_T) is the true reflectance curve.

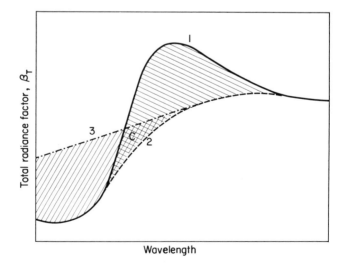

Fig. 15. Schematic presentation of Allen's method. Total radiance factor (1), reflected radiance factor (2), reflectance of uncolored substrate (3).

Next he treats the crossover region in the following manner. He measures first the total radiance factor with a source representing daylight; next he repeats the measurement with the same source in front of which he places a fluorescence-weakening filter (the proper choice of this filter is important). The filter should eliminate most of the fluorescence, but the cutoff wavelength should still be low enough so that the full total radiance factor curve is detectable in the crossover region. In practice, this means that the cutoff of the filter should occur at a wavelength about midway between β_T max and β_T min. However, it is preferable to select a fluorescence-weakening filter with cutoff wavelength near the β_T min.

Having measured the two total radiance factor curves, he computes the reflected radiance factor curve in the crossover region. He defines $\beta_T(\lambda)$ by

$$\beta_t(\lambda) = \rho_t(\lambda) + F(\lambda)/E(\lambda), \tag{6}$$

where $\rho_t(\lambda)$ is the true reflectance curve (identical with β_s), $F(\lambda)$ is the fluorescent power emerging from the sample, $E(\lambda)$ is the incident power on the sample.

The measured total spectral radiance factor for the sample irradiated with source 1, $E_1(\lambda)$ (unfiltered source), he identifies as

$$\beta_{1,T}(\lambda) = \rho_t(\lambda) + F(\lambda)/k_1 E_1(\lambda), \tag{7}$$

where k_1 is a wavelength-independent constant that corrects relative spectral-power distribution of source 1 to absolute values. A similar constant (k_2) is used with source 2.

An expression similar to Eq. (7) is set up for source 2 (the filtered source):

$$\beta_{2,T}(\lambda) = \rho_c(\lambda) + k_f F(\lambda)/k_2 E_2(\lambda). \tag{8}$$

Substituting $T(\lambda)$ for the ratio $E_2(\lambda)/E_1(\lambda)$ and noting that $E_2(\lambda) = T(\lambda)E_1(\lambda)$, he obtains

$$\rho_t(\lambda) = [\beta_{2,T}(\lambda)T(\lambda)k_2 - \beta_{1,T}(\lambda)k_1 k_f]/[k_1 T(\lambda) - k_1 k_f]. \tag{9}$$

The constant k_f [that converts the $F(\lambda)$ to the fluorescence emission curve (absolute) of sample irradiation with source 2] is directly proportional to k_2 (doubling the power of source 2 doubles both k_2 and k_f). Therefore:

$$k_f = kk_2/k_1, \tag{10}$$

where k is a new constant.

Substituting the expression in Eq. (10) into Eq. (9), one gets

$$\rho_t(\lambda) = [\beta_{2,T}(\lambda)T(\lambda) - \beta_{1,T}(\lambda)k]/[T(\lambda) - k]. \tag{11}$$

The constant k is evaluated from the lowest wavelength reading obtained

260

with the fluorescence-killing filter. Let λ_i represent this lowest wavelength reading. The β_T at this wavelength with source 1 and source 2 without fluorescence-killing filters are $\beta_{1,T}(\lambda_i)$ and $\beta_{2,T}(\lambda_i)$, respectively. The ratio E_2/E_1 (denoted T) is $T(\lambda_i)$; hence, writing Eq. (9) for wavelength λ_i and solving for k, one gets

$$k = T(\lambda_i)[\beta_{2T}(\lambda_i) - \rho_c(\lambda_i)]/[\beta_{1,T}(\lambda_i) - \rho_c(\lambda_i)]. \qquad (12)$$

Allen points out that relative power distribution curves $E_1(\lambda)$ and $E_2(\lambda)$ have been eliminated from Eqs. (11) and (12) and all that one needs to know is the ratio of the two sources, $T(\lambda)$.

An example showing a part of the results with the cut-off filter and also showing the type of cutoff filters required is given in Fig. 16.

We described this method in a fairly detailed manner because the 'method is quite accurate and also to show how complex and tedious the method may be.

We shall now turn to an abridged and greatly simplified (practical) method for separating components of fluorescent specimens.

3. The Two-Mode Method

The method developed by Eitle and Ganz (1968) (B.1 in this section) deals

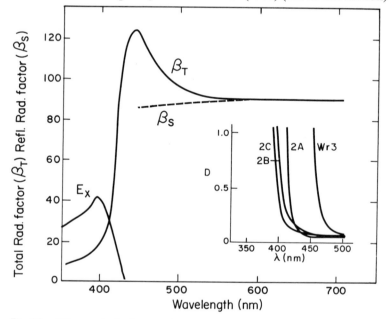

Fig. 16. Filter method of separating components; E_x is the excitation spectrum. The insert represents spectral densities of various cutoff filters.

with white fluorescent samples and is not readily applicable to colored fluorescent samples. Allen's (1973) method (XII.B.2) can be applied to chromatic fluorescent samples as well; however, the method is tedious and quite complex. Also, the use of yellow cutoff filters should be made judiciously, since many yellow filters exhibit fluorescence themselves.

The difficulty of characterizing color by separating fluorescence from reflectance in chromatic fluorescent samples is evident from the lack of publications on the subject. A summary of various available methods for determining the spectral reflected radiance factor of fluorescent specimens was published by Alman and Billmeyer (1976). They compared known one-monochromator methods for determining the reflectance of opaque fluorescent samples. From this work, they characterized various methods in order of decreasing accuracy. According to them, the double-monochromator method is the most accurate method for measuring reflected radiance factor and the two-mode method is the least accurate.

The two-mode method, published by Simon (1972), depends on making two separate measurements of a fluorescent sample. One measurement is made with monochromatic illumination and with the most conventional spectrophotometer; hence, it is called conventional reflectance. The second measurement is made with heterochromatic illumination, that is, with an instrumental setup in which the sample is illuminated with white light. It is preferable that both measurements be on the same instrument, if the instrument can be used in both modes of operation. When a nonfluorescent sample is subjected to these two types of measurements, the results obtained will be identical, that is, the conventional reflectance $\rho_c(\lambda)$ represents the reflected radiance factor $\beta_S(\lambda)$. For a fluorescent sample the situation is as follows.

The conventional reflectance $\rho_C(\lambda)$ represents the reflected radiance factor $\beta_S(\lambda)$ at wavelengths greater than the excitation wavelengths, whereas total radiance factor $\beta_T(\lambda)$ represents the reflected radiance factor $\beta_S(\lambda)$ for wavelengths shorter than the emission region.

With these two measurements there will be only the region of overlap (the crossover region) between the excitation and emission spectra where no reflected radiance factor is measured. In the two-mode method the reflected radiance factor in the overlap region is obtained by interpolation. The overlap region is usually about 50 nm for white fluorescent samples and normally somewhat larger for chromatic fluorescent samples. Figure 17 shows the excitation and emission spectra for a sample treated with a fluorescent whitening agent. The overlapping region is indicated by the shaded area. Figure 18 gives an example of the results obtained by dual-mode measurements; the interpolated portion of β_S is indicated by a dotted line. The interpolation can be done with a suitable mathematical model, such as cubic spline function.

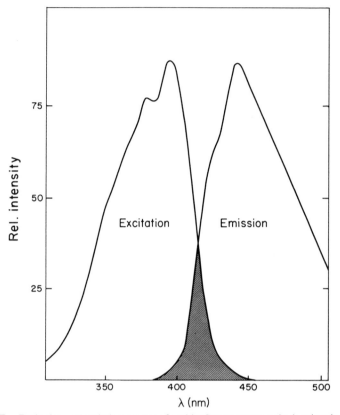

Fig. 17.   Excitation and emission spectra of a white fluorescent sample showing the overlap between the two.

The two-mode method is indeed a practical method, but it should be used with care for the interpolation to be valid; the method gives only an approximation of $\beta_S(\lambda)$ in the region of overlap. Also, the method becomes cumbersome and even less accurate with samples containing several fluorescent dyes. In such a mixture, various energy-transfer processes occur, such as the reabsorption of fluorescence, fluorescence quenching, and intramolecular transfer of energy. In spite of these problems the two-mode method is of practical importance in predicting formulations using basic dyes, as shown by Simon (1972). He showed that inasmuch as reflected radiance factor follows the Kubelka–Munk relationship, it provides a basis for color match predictions for fluorescent samples. Using the two-mode method for determining the reflected radiance factor, Simon showed a reasonably good

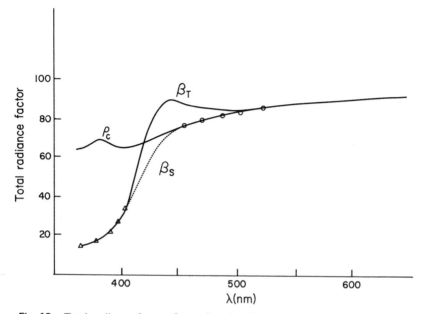

Fig. 18. Total radiance factor (β_T), reflected radiance factor (β_s), and conventional reflectance (ρ_c). The circles and the triangles denote the reflected radiance factor determined by the two-mode method.

formula prediction of a mixture of fluorescent colorants to match given patterns.

If the two-mode method, which is particularly suitable for measurement of a fluorescent sample containing a simple fluor, is used carefully, it does offer the complete information required for colorimetric specification of a fluorescent sample. The following quantities can be obtained by using the method. Total radiance factor is obtained by direct measurement, and the reflected radiance factor data are obtained partly from direct measurements of both types and partly by interpolation. Once the reflected radiance factor is known, the fluorescent radiance factor can be computed from the difference between the total spectral radiance factor curve and the reflected radiance factor curve, that is,

$$\beta_L(\lambda) = \beta_T(\lambda) - \beta_S(\lambda). \tag{13}$$

The fluorescent radiance factor determined this way is valid only for the irradiating source used in the measurement of total radiance factor. The actual spectral-power distribution of the source used in the measurement includes the entire instrumental illumination (the irradiance determined at the sample position).

Several commercial instruments can be readily adapted for the two-mode methods of measurement; these include the spectrophotometers made by Beckman Instruments, Cary, Diano, Hunter, Kollmorgen, Zeiss, and others.

XIII.   STANDARDIZATION AND MEASUREMENT PROBLEMS

A.   Light Source Requirements

Now that we have described instrumental arrangements required in color measurements of fluorescent materials and discussed various evaluation methods, it is essential that we review the standardization requirements of measuring apparatus and point out the problems associated with the measurements.

We have discussed the manner in which fluorescent samples must be measured for proper evaluation. We have shown that the color measurement of such samples must be made properly and that the interpretation of the data is based on a sound psychophysical relationship. The evaluation of color of fluorescent samples should be based either on total spectral radiance factor data or on the separated reflected and fluorescent radiance factor data.

In the measurements of total radiance factor it is the *conditio sine qua non* that the spectral-power distribution of the irradiating source be rigidly controlled if the results are to be reproducible and accurate. Furthermore, one must, in these measurements, know the spectral-power distribution of the complete irradiating system. This includes the light source and all the optical elements modifying the spectral-power distribution incident on the sample. One must keep in mind that the total spectral radiance factor $\beta_T(\lambda)$ represents the sum of the spectral reflected radiance factor $\beta_S(\lambda)$ and the spectral luminescent radiance factor $\beta_L(\lambda)$ of the sample. The spectral reflected radiance factor is independent of the spectral-power distribution of the irradiating system, but the spectral luminescent radiance factor is strongly influenced by this spectral-power distribution.

The problem unique to the fluorescent samples is the dependence of the spectral luminescent radiance factor on the irradiating system. With fluorescent white samples it is particularly important that the ultraviolet content of the irradiating source be adequate and that it simulate the corresponding viewing illuminant accurately.

The spectral-power distribution of the irradiating source will directly affect the value of $\beta_T(\lambda)$ obtained. This effect is illustrated in Fig. 19 for a

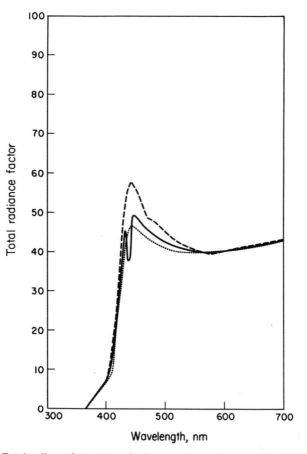

Fig. 19. Total radiance factor data of a fluorescent sample measured with three different irradiating sources (--- xenon, —— D_{65} fluorescent lamp, ... 3000 K tungsten lamp).

white fluorescent sample irradiated with a 3000 K tungsten–halogen source, with a filtered xenon arc simulating D_{65} illuminant, and with a fluorescent D_{65} lamp. The xenon lamp, which has the greatest amount of ultraviolet energy, produces the greatest fluorescence and, hence, the highest β_T values. (Fluorescent whitening agents are best excited with a wavelength just below 400 nm.) The fluorescent lamp, although it simulates the CIE D_{65} illuminant fairly well, is not ideal for instrumental use because of the strong line structure in the region of fluorescence. The tungsten provides the least amount of ultraviolet and therefore the least amount of excitation of the fluorescent component.

Variations up to about 10% may be expected in the total radiance factor data if the spectral-power distribution of the irradiating source is only slightly altered. An example of this variation is shown in Fig. 20 for a white fluorescent sample irradiated with a tungsten source at three different voltage settings, hence three different spectral-power distributions.

A very slight change in the quality of incident light that differs very slightly or not at all visually from the standard illuminant can cause large differences in the measured data. This is demonstrated in Fig. 21 for two fluorescent samples. The data in this figure clearly show that even small amounts of selectivity introduced by a spectrally selective neutral density filter produce large changes in the total spectral radiance factor measurement. The solid curve in this figure represents total radiance factor obtained for the samples irradiated with an unfiltered xenon source. When the intensity of the source was modulated (decreased) with a spectrally nonselective filter (neutral meshed screen), no change in total radiance factor is noted, as shown with dashed curves in the figure. However, if the same xenon source is filtered through a selective neutral modulator, which alters the spectral distribution of irradiating source, a huge variation in the results is noted, as shown in the

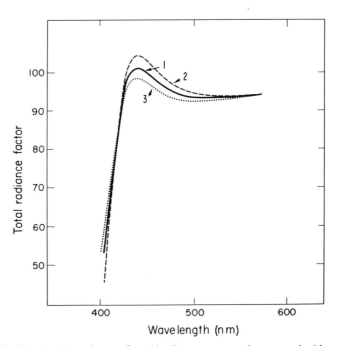

Fig. 20. Total radiance factor of a white fluorescent sample measured with a tungsten source at three different voltage settings (three different color temperatures: 1, 3000 K; 2, 3200 K; 3, 2650 K).

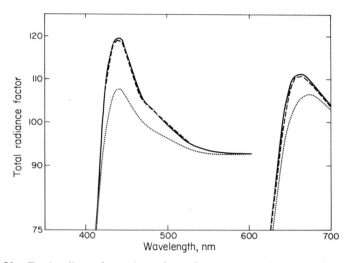

Fig. 21. Total radiance factor data of two fluorescent samples measured three ways: ——— with bare xenon lamp, – – – with xenon lamp modulated by a meshed screen, ... with xenon lamp modulated by a slightly selective neutral filter.

figure (dotted curves). Figure 22 shows spectral densities of the two neutral modulators used in the measurement given in Fig. 21.

Another comment should be made regarding the data in Fig. 21. The change of irradiating intensity of the xenon lamp (filtered through meshed screen) did not alter the $\beta_T(\lambda)$ distribution. This means that the intensity

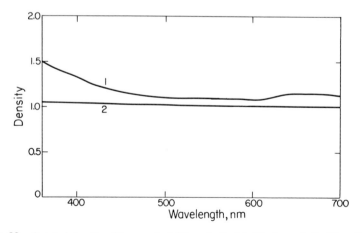

Fig. 22. Spectral density of two neutral filters: 1, Kodak Wratten gelatin filters No. 96; 2, mesh screen.

level of the irradiating source has little or no effect on the measurement, provided that (1) the intensity modulation does not alter the spectral-power distribution of the source and (2) the absorption is total (complete); hence, additional quanta supplied from higher intensity will not alter the emission of the fluorescent species.

B. Daylight-Simulating Sources

It has been mentioned earlier that the colors of fluorescent samples are usually appraised visually under daylight conditions; hence, the physical measurements of such samples should ideally also be made with the same source that was used for visual appraisals. CIE (1971) recommended the use of illuminant D for both visual appraisals and physical measurements. Hence, it is necessary to use in these measurements sources that simulate illuminant D closely for the spectral region 300–830 nm. Wyszecki (1970) published a collection of data representing the state of the art of reproducing CIE standard illuminants by means of artificial light sources. An example of spectral-power distribution of such a simulator is shown in Fig. 23.

Unfortunately, there is no accepted standard D_{65} simulator yet available,

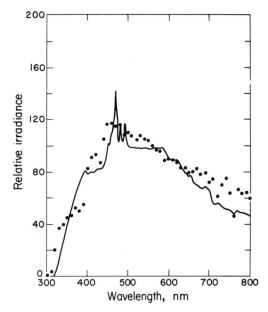

Fig. 23. Relative spectral power distribution of a D_{65} simulator using filtered xenon-arc source; dots represent D_{65} standard.

and different laboratories are using different D_{65} simulators with various degrees of goodness of fit. Also, there is no standard condition established yet for goodness of fit of spectral-power distribution of simulating sources to standard illuminants. [The CIE (1979a) TC-1.3 Subcommittee on Sources is now working on the methods to specify goodness of fit of daylight simulators and hopes to have a recommendation in 1979.] The best rule of thumb, however, would be to use the same light source in the measurement that is used in the visual assessment of fluorescent samples. This, however, is not practical. We shall discuss this matter further in a section on methods for predicting total radiance factor.

The appropriate and well-controlled light source is the single most important requirement for making accurate total radiance factor measurements. The intensity of the irradiating source, on the other hand, is of minor importance in physical measurements, provided that the variation of intensity does not produce changes in the spectral-power distribution of the source.

C. Standard of Reflectance and Geometry

Other important standardization parameters are the geometry of the measurement system, the efficiency of the integrating sphere (if one is used), and the standard of reflectance used. The use of an integrating sphere and $d/0$ (diffuse/normal) geometry is recommended. In this way the effects of surface structure and/or characteristics of the samples are minimized. The use of an integrating sphere is of practical importance, since most commercially available spectrophotometers utilize diffuse illumination obtained with an integrating sphere. One must remember, however, that the efficiency of the integrating sphere depends not only on the sphere coating and on the area of various ports but also on the nature of the material at its measurement port. The sample acts to modify the spectral-power distribution of the irradiating system in such a spectrophotometer, as shown by Alman and Billmeyer (1976).

The reference material must be stable, since in the total radiance factor measurements the standard and the sample are subjected to intense radiation by the source (normally of high UV content). Also, the reference material should be completely inert (free of any possible fluorescence). (Reagent-grade pressed $BaSO_4$ powder is commonly recommended for the reference standard.) The standard of reflectance, if not properly prepared, can cause objectionable variation in $\beta_T(\lambda)$ measurements, as is the case in nonfluorescent reflectance measurements. Figure 24 shows an example of the error produced by using a poor reference material.

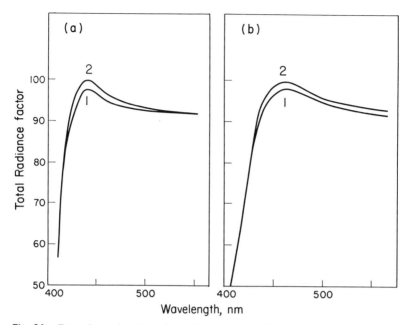

Fig. 24. Errors in total radiance factor data produced with variation in color temperature of the irradiating source (a), and errors due to poor reference standard (b).

Attention must also be paid to possible variation of the reflected radiance factor and fluorescent radiance factor of the sample as a function of sample temperature (thermochromic effects).

D. Calibration of the Measuring Spectrophotometer

As in conventional spectrophotometry, it is essential that the measuring apparatus be in good calibration before making total radiance factor measurements. The 100% line of the instrument, the photometric scale, and the wavelength scale must be well standardized with standard reference materials. A calibrated Vitrolite (white) tile or other standard reference whites can be used for checking the merit of the reference sample ($BaSO_4$). Calibrated colored or gray tiles can be used for checking the photometric scale of the instrument. (For further details see ASTM standard method for absolute calibration of reflectance standards ASTM-1966.) The reader is also referred to the literature on properties and measurement of white reference materials used as standard of reflectance (see Grum and Luckey,

1968; Budde, 1958, 1970, 1976; Goebel *et al.*, 1966; Erb, 1979; Grum and Wightman, 1977).

The total radiance factor values of standard reference materials mentioned earlier should be identical with their reflectance values, for which they are calibrated, since these reference materials are free from fluorescence.

Standard reference materials such as didymium glass or holmium oxide glass can be used for checking the wavelength scale of the instrument. Line sources with well-defined emission lines can also be used for this purpose.

XIV. METHODS FOR PREDICTING SPECTRAL RADIANCE FACTORS OF FLUORESCENT SAMPLES UNDER STANDARD ILLUMINANTS FROM MEASUREMENTS UNDER AN ARBITRARY SOURCE

It was stated earlier that measurements of total spectral radiance factor are made in a manner similar to measurements of spectral reflectance factor. The major difference between the two is that in total spectral radiance factor the specimen is illuminated with heterochromatic illumination and both reflected and fluorescent fluxes are collected monochromatically. This adds a new measure of complexity, since the contribution of fluorescence depends greatly on the spectral-power distribution of the irradiating source. Since there is no standard source readily available to simulate various phases of daylight, it is essential to establish a procedure through which total spectral radiance factor can be predicted accurately for any desired illuminant from measurements using an irradiation that only approximates that illuminant in spectral-power distribution.

Several predicting methods have been developed recently for this purpose. We shall first describe a simplified predicting method developed by Grum and Costa (1977); then we will review other known methods and compare them. However, before we describe these methods, we should again discuss the reason why predicting methods are not only desirable but also badly needed.

A. State of the Art of Measurement

It has been pointed out throughout this chapter how complex are the measurements of fluorescent materials, and many errors associated with such measurements were described in detail. It seems appropriate that we summarize the instrumental problems before discussing the most recent work in this field.

The instrumentation available for color measurement of fluorescent samples often falls short of achieving the desired objectives. Special instrumental setups have been described by Donaldson (1954), Fukuda and Sugiyama (1961), Grum (1971), and Baba *et al.* (1974–1975) that attempt to overcome these deficiencies by the use of a monochromator in each of the irradiating and viewing beams. Measurement can be made of spectral reflected radiance factor, the excitation spectrum, and the luminescence emission spectrum, with a possibility of calculating the total spectral radiance factor of the fluorescent material for any desired illuminant. The complexity and general unavailability of such instruments, however, precludes their consideration as a practical solution to the problem.

The following are areas where error can occur in the measurement of total spectral radiance factor, in order of increasing importance:

(1) The irradiating system in the spectrophotometer must be fully defined and standardized. The lack of uniformity among instruments of irradiation on the samples is perhaps the most important source of error to be considered.

(2) The samples must be stable under the measurement conditions. With direct sample irradiation by a heterochromatic source there may be significant irreversible fading of the sample during the measurement, or reversible change due to heat if the sample is thermochromic. The precautions required to test for and avoid this source of error are obvious.

(3) Appropriate selection of illuminating and viewing geometry must be made. For practical reasons, our discussion is based on the use of reflectance spectrophotometers equipped with integrating-sphere geometry. However, in this geometry the specimen acts to modify somewhat the spectral-power distribution of the irradiating system, as pointed out by Alman and Billmeyer (1976) and by Gundlach and Hammer (1976).

The use of integrating-sphere spectrophotometers is widespread for industrial color measurement. Only a few reflectance spectrophotometers with $45°/0°$ or $0°/45°$ geometry are available that are suitable for total spectral radiance factor measurements. Hence, we shall limit our discussion to the integrating-sphere geometry systems.

Of more compelling importance is the influence of the surface of the specimen on the measurement of total spectral radiance factor when $45°/0°$ geometry is used. Systematic errors resulting from this influence can be as large as those associated with integrating-sphere efficiency. Moreover, they cannot be corrected readily, whereas the integrating-sphere error can be reduced to minor, if not negligible, importance by calculation.

The reproducibility among integrating-sphere spectrophotometers of different types for the measurement of the total radiance factor of fluorescent materials was tested by Chong and Billmeyer (1977) using six chromatic paint samples. Four spectrophotometers were used in their comparative study; each spectrophotometer was equipped with two sources: an incandescent source, nominally CIE source A, and a daylight simulator, nominally a simulation of CIE illuminant D_{65}. The daylight simulators were filtered tungsten except for one filtered-xenon simulator.

Each instrument could be operated with the specular component included or excluded in the integrating sphere. If the samples have matte surfaces, the major effect of including or excluding the specular component is to alter the integrating-sphere efficiency. One must remember that in the measurement of total spectral radiance factor the sample is irradiated diffusely by the use of integrating sphere and in such an illumination there is no specular component.

Figure 25 illustrates the spread in total spectral radiance factor obtained in a typical experiment using the daylight simulators and a yellow fluorescent sample. This example clearly demonstrates the problem at hand. This variability among instruments can be attributed to the following factors:

(1) Differences in the goodness of fit with which the various daylight simulators simulate the desired standard illuminant—in short, the lack of a standard source that simulates D_{65} illuminant.

(2) Lack of control of the source's power supplies.

(3) Differences in the integrating-sphere efficiency.

(4) Differences, although minor, in instrument illuminating and viewing geometry.

Subjecting the data from Fig. 25 to colorimetric analysis gives the results shown in Table III. Mean results for the sample are expressed as CIE 1976 $L*a*b*$ (CIELAB)° color differences calculated as the color difference

TABLE III. Average Color Differences ΔE between the Mean and Individual Measurements of the Data in Fig. 25 (D_{65})

Specular component	ΔE
Included	4.1
Excluded	6.7

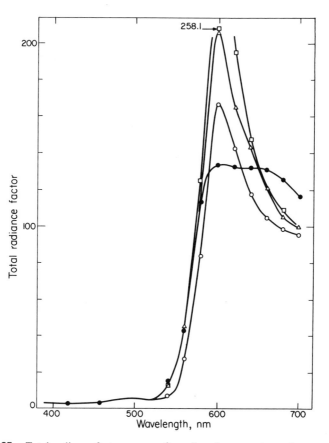

Fig. 25. Total radiance factor curves of a yellow fluorescent sample measured on four different instruments with D_{65} simulating sources.

between results with a given sample for specified measurement conditions and those for the average of all measurements of that sample.

The total spectral radiance factor of the same yellow sample measured with the same four instruments but using the incandescent source of irradiation produced much less variability in the data. The average ΔE in that case was 2.0 CIELAB units.

The performance of a variety of instruments in the measurement of nonfluorescent samples under typical industrial conditions was studied by analyzing published results of a Collaborative Reference Program (1977). For 20 samples of different colors the average color difference between the "grant mean" and the results with the various instruments was 1.3 CIELAB units. Thus, there is a much smaller variability in the measurement of nonfluorescent samples, as expected.

In view of the state of the art, just described, for measuring total radiance factor of fluorescent materials, one must seek an alternative approach, that is, calculating desired values of total spectral radiance factor under standard illuminants from measurements obtained under arbitrary irradiation.

1. A SIMPLIFIED PREDICTION METHOD

The method developed by Grum and Costa (1977) is based on work in which they used a dozen white samples containing fluorescent whitening agents and six color samples with fluorescent additives. All samples were subjected to the following tests:

(1) Total radiance factor $\beta_T(\lambda)$ measurements for four selected sources,
(2) Conventional reflectance measurement $\rho_C(\lambda)$,
(3) Measurement of reflected radiance factor $\beta_S(\lambda)$.

The four sources used in the experiment were greatly different from one another in spectral-power distribution. They were comprised of a tungsten–halogen source, operated at 3000 K; a nonfiltered 150 W xenon source, whose correlated color temperature was about 5900 K; a D_{65} simulator (150 W xenon source plus filters); a Macbeth D_{65} fluorescent lamp.

They computed the luminescent radiance factor $\beta_L(\lambda)$ for all samples and sources using the relationship shown in Eq. (13).

Their approach is based on an analytical expression for specifying the number of relative absorbed quanta under a given irradiating source, that is,

$$Q = \int_{\lambda_1}^{\lambda_2} E_i(\lambda)\alpha(\lambda) \, d\lambda, \tag{14}$$

where Q represents the relative number of absorbed quanta, $E_i(\lambda)$ is the spectral irradiance of any desired source, $\alpha(\lambda)$ is the spectral absorptance of the fluorescent sample under test. The terms λ_1 and λ_2 represent the region of excitation. This range is shown for a green fluorescent sample in Fig. 26.

As is evident from Fig. 26, the upper limit of the excitation region λ_2 is the point of intercept of the true reflectance and conventional reflectance. The lower limit λ_1 is the point of closest proximity of β_T and ρ_c.

Since it is difficult to determine in practice the absorptance of fluorescent species $\alpha(\lambda)$, Grum and Costa made the following assumptions and/or approximations.

(1) In view of the high concentration of fluorescent species they assumed

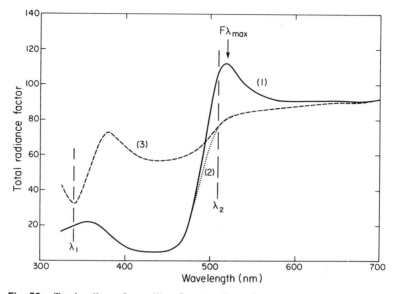

Fig. 26. Total radiance factor (1), reflected radiance factor (2), and conventional reflectance (3) of a green fluorescent sample. Vertical lines denote the excitation region.

that the absorptance can be considered nearly constant ($\alpha = 1$) in the "excitation range."

From this assumption, $\alpha(\lambda)$ in Eq. (14) can be placed outside the integral sign.

(2) If the preceding is valid, then one can compute the ratio of the relative number of quanta absorbed by two different sources, that is,

$$Q_i/Q_m = \alpha \int_{\lambda_1}^{\lambda_2} E_i(\lambda) \, d\lambda \bigg/ \alpha \int_{\lambda_1}^{\lambda_2} E_m(\lambda) \, d\lambda, \qquad (15)$$

where Q_m and E_m are the quantities related to the source actually used in the measurements. Since, according to assumption (1), the absorptance is constant and hence cancels out in Eq. (15), the ratio of the number of relative absorbed quanta for the two sources is directly proportional to the ratio of integrated irradiances of the two sources (expressed in units of quanta) over the excitation region.

With the preceding assumptions, one can then relate total spectral radiance factor from the source used in the measurement to any other desired and defined illuminant. First one has to compute the luminescent radiance factor for the desired source; this is done by the use of the following relationship:

$$\beta_{L,i}(\lambda) = \beta_{L,m}(\lambda)[E_m(\lambda)/Q_m][Q_i/E_i(\lambda)], \qquad (16)$$

where $\beta_{L,i}(\lambda)$ is the luminescent radiance factor for any desired irradiating source in the region of emission and $E_m(\lambda)$ is the spectral power distribution of the source used in the measurement in the emission region (expressed in units of quanta); $E_i(\lambda)$, Q_m, and Q_i have been defined earlier.

Once $\beta_{L,i}(\lambda)$ is computed one can determine $\beta_{T,i}(\lambda)$ using Eq. (1), that is, $\beta_{T,i}(\lambda) = \beta_s(\lambda) + \beta_{L,i}(\lambda)$.

The method just described may not be the most accurate for predicting $\beta_T(\lambda)$ because it is based on the assumption that the absorptance is constant in the excitation region. However, since the intensity of the irradiating source is not critical in these measurements, the assumption of constant ($\alpha = 1$) absorption is valid if the measurements are made with a source only slightly dissimilar from the standard illuminant. For example, an unfiltered (bare) xenon source can be used in the measurement and the predictions computed for D_{65} illuminant, which is very similar in spectral-power distribution to a xenon-arc source. The goodness of prediction by this method can be measured from colorimetric parameters. The CIELAB color differences computed from the D_{65} illuminant point ranged from 0.5 to 2.0 CIELAB units for the white samples. The largest color differences were observed when the predictions for D_{65} were made from the measurement with the 3000 K source. The color differences for chromatic fluorescent samples ranged from 1.5 to 9.0 CIELAB units. Again the largest color difference was observed when measurements of total spectral radiance factor were made with the 3000 K source and predictions for D_{65} were made.

The method described is relatively simple and adequate for most practical applications.

In this method, one measures (1) the total spectral radiance factor with a stable and easy-to-obtain source and (2) reflected radiance factor (determined by any methods described earlier). The luminescent radiance factor, the only quantity that is affected by the variation in the spectral-power distribution of the source, is computed by Eq. (16). As in any other method, one must know the spectral-power distribution of both the irradiating source system and the illuminant for which the predictions are made.

2. CIE WORK ON THIS PROJECT

The color measurement and specification of fluorescent materials are of great practical importance, and since the color measurements are complex and because there is no standard source readily available that simulates a desired daylight condition, it is imperative to develop practical predicting methods of total spectral radiance factor. It is because of this urgent need that the CIE Technical Committee on Materials (TC-2.3), chaired by Grum, undertook a study in comparing various predicting methods and recommending a suitable procedure for measuring and evaluating the color of fluorescent materials. A task force on "Colorimetric Aspects of Luminescence" has been formed by this committee and is now preparing a CIE (1979b) Technical Report on the subject. The objectives of the study are (1) to report on the state of the art of measurement of the total radiance factor of fluorescent samples, (2) to give details of methods for calculating values of total spectral radiance factor under standard illuminants from measurements made under an arbitrary source by applying various corrections for known systematic errors in both the measurement and the calculation steps, and (3) to recommend tentatively satisfactory procedures for correcting for these systematic errors.

This task force (CIE, 1979b) made a comparative study of various known methods for predicting total spectral radiance factor under any desired illuminant. The methods compared are those of Eitle and Ganz (1968), Alman and Billmeyer (1977), Allen and Sanders (1979), and Grum and Costa (1977). This comparative study is based on experimental work in which six fluorescent samples (white and colored) were subjected to total spectral radiance factor measurements using an integrating-sphere geometry. These measurements were made with three different sources: incandescent (designated A), daylight (designated D), and modified daylight (designated D^1). The relative spectral-power distribution of the irradiating systems of the spectrophotometer used in the measurement were measured with samples in place.

The measurement procedure recommended for total spectral radiance factor is as follows:

(1) Irradiate the entrance port of the sphere with the desired stable simulator and view the specimen at the sample port using near-normal geometry with a monochromator as part of the detection system.

(2) Operate the instrument with the specular cup in the position, to increase sphere efficiency.

(3) Calibrate the instrument using appropriate procedures and standards (see also Carter and Billmeyer, 1978).

(4) Using an appropriate nonfluorescent specimen in double beam instruments, measure a nonfluorescent white working standard that has a nearly nonselective reflectance factor near unity.

(5) Replace the working standard with the fluorescent specimen to be measured, thus providing diffuse irradiation with the spectral-power distribution of the irradiating system, including all effects of the integrating sphere.

3. METHOD OF EITLE AND GANZ (METHOD A)

This method is a basis of various calculation methods and utilizes the total radiance factor of a sample under source 1 of spectral irradiance $S_1(\lambda)$, the total number of quanta absorbed by the sample under source 1 (N_1), and total number (N_2) of quanta absorbed by source 2 with spectral irradiance S_2 to calculate $\beta_{T,2}(\lambda)$. Their basic equation is

$$\beta_{T,2}(\lambda) = \beta_s(\lambda) + \beta_{L,i}(\lambda)[S_1(\lambda)N_2/S_2(\lambda)N_1]. \tag{17}$$

They compute the total number of quanta absorbed using

$$N_i = \int_{\lambda^1} [1 - \beta_s(\lambda')]A(\lambda')S_i(\lambda')\lambda' \, d\lambda', \tag{18}$$

where λ' denotes the excitation region, $[1 - \beta_s(\lambda')]$ is the absorptance, and $A(\lambda')$ is the factor determined by Kubelka–Munk analysis, which is not always applicable and requires several assumptions, according to Alman and Billmeyer (1977).

4. MODIFICATION USING EXCITATION SPECTRUM FROM A TWO-MONOCHROMATOR METHOD (METHOD B)

In this method the excitation spectrum $\overline{X}(\lambda')$, defined in Section VII of the sample determined by the two-monochromator method, is used in place of the quantity $[1 - \beta_s(\lambda')]A(\lambda')$.

5. MODIFICATION USING EXCITATION SPECTRUM FROM A ONE-MONOCHROMATOR METHOD (METHOD C)

Alman and Billmeyer (1977) modified method A by calculating the excitation $\overline{X}(\lambda')$ from the conventional reflectometer value and the total radiance factor value of the sample.

6. MODIFICATION REPLACING EXCITATION SPECTRUM BY ABSORPTANCE (METHOD D)

Allen and Sanders (1979) approximated the geometry $[1 - \beta_s(\lambda')]A(\lambda')$ by setting $A(\lambda') = 1$. By doing so, Eqs. (17) and (18) may be written as follows:

$$\beta_{T,2}(\lambda) = \beta_s(\lambda) + \beta_{L,1}(\lambda)\frac{S_1(\lambda)\int_{\lambda^1} S_2(\lambda')[1 - \beta_s(\lambda')]\lambda' \, d\lambda}{S_2(\lambda)\int_{\lambda^1} S_1(\lambda')[1 - \beta_s(\lambda')]\lambda' \, d\lambda}. \tag{19}$$

The report also compares the method of Grum and Costa (described earlier). Their method is designated "method E."

The results of the comparative study are summarized in Table IV in terms of mean absolute spectral differences and mean CIELAB color differences. The differences in calculating methods B, D, and E are small. Hence, one may well conclude that the use of the excitation spectrum, is attractive.

TABLE IV.   Mean Spectral and Colorimetric Errors of All the Samples for
All Cases Used for the Prediction Methods

| Method based on: | Method | Based on modified daylight illumination | | Based on tungsten illumination | |
		Mean spectral difference	CIELAB ΔE	Mean spectral difference	CIELAB ΔE
Irradiating system[a]	B	0.0098	0.88	0.0405	3.61
	E	0.0144	1.33	0.0462	4.55
	D	0.0153	1.42	0.0368	3.36
Calculated irradiating system[b]	B	0.0193	1.77	0.0405	3.59
	E	0.0259	2.50	0.0455	4.52
Irradiating system of white sample[c]	B	0.0504	4.84	0.0638	5.99
	E	0.0551	5.34	0.0658	6.57
Simulator[d]	B	0.0598	6.03	0.0613	6.03
	E	0.0655	6.60	0.0639	6.59

[a] Simulator plus all optical elements modifying the spectral-power distribution incident on the sample.

[b] Spectral-power distribution of the irradiating system is computed by integration-sphere theory, that is, by multiplying the spectral-power distribution of the simulator by the relative spectral efficiency of the integrating sphere.

[c] Spectral-power distribution of the irradiating system is measured with a white sample at the sample port.

[d] Spectral-power distribution of a filtered source that simulates the desired illuminant.

Methods B and C are very similar; in both, one has to obtain excitation spectrum. In B the excitation is computed from measurements on a dual-monochromator system, whereas in C the excitation is obtained from measurements on a one-monochromator instrument using

$$\overline{X}(\lambda') - [R_C(\lambda') - \beta_S(\lambda')] \cdot [r(\lambda')/r_f], \qquad (20)$$

where $r(\lambda')$ is the detector responsivity.

The quantity r_f is the responsivity averaged over the emission band and is an instrument constant.

In view of this and because dual-monochromator instruments are seldom available in practice, the simpler method C is recommended over method B. These two methods are compared in Table V. Colorimetric results, obtained for modified daylight source D^1 and for the incandescent source A using orange (O), yellow (Y), and red (R) samples, are compared between the two methods. The results are equivalent.

TABLE V. Comparison of Spectral and Colorimetric Data Computed by Methods B and C

| | Method B | | | | Method C | | | |
| | Source D′ | | Source A | | Source D′ | | Source A | |
Sample	ΔE	Mean spectral difference	ΔE	Mean spectral difference	ΔE	Mean spectral difference	ΔE	Mean spectral difference
Y	0.67	0.006	1.94	0.029	0.66	0.006	2.02	0.029
O	0.41	0.006	4.06	0.059	0.28	0.006	3.51	0.053
R	1.07	0.017	4.49	0.069	1.10	0.018	4.89	0.074
Average	0.72	0.010	3.50	0.052	0.68	0.010	3.47	0.052

7. CONCLUDING REMARKS ON PREDICTION METHODS

From the discussion of the CIE study on evaluation of methods for predicting the spectral radiance factors of fluorescent samples under standard illuminants from measurements made under an arbitrary source, one can deduce the following conclusions:

(1) Total radiance factor can be predicted accurately from measurements made on a one-monochromator instrument equipped with an integrating sphere, provided that such an instrument can be used in two modes of operation, that is, can measure conventional reflectance as well as total radiance factor.

(2) The irradiating system for measurement of total radiance factor should have a spectral-power distribution close to that of the standard illuminant. If the measurements are made with an incandescent source, such as source A, and the predictions of total radiance factor are made for D_{65}, the error in prediction will be quite large, as shown in Tables IV and V. It is quite appropriate to use an unfiltered xenon-arc lamp in the measurement for computation of values of total spectral radiance factor under CIE illuminant D_{65}.

(3) A computational method (method C) based on the use of excitation spectrum will produce very accurate results.

(4) Determine by spectroradiometry the spectral-power distribution of the irradiating system with the sample being tested at the measurement port.

(5)   If the necessary calibrations for the application of method C cannot be made (i.e., measurement of conventional reflectance and spectral calibration of the detection system), then the much simpler methods D or E can be used.

(6)   In any case, provisions should be made to determine the spectral-power distribution of the irradiation falling on the sample.

(7)   Total spectral radiance factor under standard illuminants measured with arbitrary heterochromatic irradiation can be predicted with an average color difference of less than 1.5 CIELAB units, using any of several calculation methods provided that the appropriate calibration is made over the variables involved. The most important variables are (1) proximity of the spectral-power distribution of the irradiating system to that of the illuminant for which the values are predicted and (2) the knowledge of the spectral-power distribution of the irradiating system with the sample at the sample port of the instrument.

XV.   APPLICATIONS

In the beginning of this chapter we pointed out that fluorescent additives are everywhere; hence, it is impreative to develop and to use physical methods that will correctly represent the color of such materials. The desirability of achieving good correlation between the psychophysical parameters determined instrumentally and the perceived appearance of fluorescent materials is obvious. Yet instrumental irradiating systems are almost invariably different from sources available for visual observations. It is particularly desirable that the psychophysical parameters of fluorescent materials be those for the samples irradiated by CIE standard illuminants and evaluated by a CIE standard observer. The CIE illuminants, in particular D_{65}, are recommended for color measurement and evaluation because they represent various phases of natural daylight. The daylight appearance of materials containing fluorescent whitening agents could not be adequately described under standard source C or B owing to their low ultraviolet content compared with natural daylight.

Samples containing fluorescent additives cannot be properly evaluated by conventional colorimetric methods. Thus, physical measurements on fluorescent materials must be made similar to the way they are viewed visually for the results of such measurements to agree with visual assessment.

Total radiance factor measurements are the proper measurements for fluorescent materials and have been widely used in the evaluation of whiteness of samples containing fluorescent whitening agents. However, very few data are available concerning the use of total radiance factor measurements as the basis for color specification of highly chromatic fluorescent samples. The reason for this is the complexity of such measurements and the lack of suitable instrumentation. Only recently articles appeared in the literature dealing with color evaluation of chromatic color samples; the articles by Clarke (1975) and Baba and Segoku (1973) are typical examples.

The use of total spectral radiance measurements in studying the purely physical effect of illuminants on halftone color tints was introduced by Grum and Clapper (1969). Clapper and Grum (1975) also showed the effect of the samples' fluorescence in color evaluation and color reproduction studies. Now it is well recognized that fluorescent materials require measurement of the total spectral radiance factor for their colorimetric characterization. But most of all, these types of measurements have been used widely in the characterization of fluorescent whites.

A.   Fluorescent Whiteness

The color of white material is a subject that has been discussed and investigated for a long time. A sample is considered white if its spectral reflectance in the visible region of the spectrum is high and nearly flat. In practice, however, most white materials exhibit a sharp decrease in the spectral reflectance in the blue spectral region. The larger this decrease, the yellower the sample and the poorer it is as a white. For this reason it was common and successful practice to describe the whiteness of white and near-white paper in terms of its blue reflectance (TAPPI Standard Method T452m-58). When blue dyes are added that supposedly counteract the yellowish appearance of the undyed paper or coating, the method for specifying whiteness in terms of blue reflectance is no longer as successful as that for the undyed material. The addition of blue dyes reduces the reflectance of the red region of the spectrum to make it closer to the reflectance in the blue. The net effect is an overall decrease of reflectance. More recently, chemical manufacturers have made available optical or fluorescent whitening agents to be used with materials for improving their whiteness appearance. Nowadays these whitening agents are used widely in paper, textiles, printing, detergents, and many other items.

Fluorescent whitening agents, which in the past 15 years have been commonly added to paper to improve its white appearance, complicate the spectrophotometric problem. Since these agents are rarely white themselves, they cause tints that are at least as noticeable as those caused by blue dyes. The amounts of fluorescent whitening agents that can be used, therefore, are limited in a way similar to the limitation of the amount of blue dyes that can be used to produce white or to improve whiteness. Fluorescent whitening agents can cause noticeable tints in illumination that differs only slightly or not at all, visually, from the standard light for which the whiteness has been optimized. Such peculiar sensitivity to visually trivial variations of quality of illumination is often as objectionable as the actual tints that result from nonstandard lighting.

As is evident from the avalanche of literature, fluorescent whiteness has been investigated extensively in the past 15 or more years. A good literature review on the instrumentation for whiteness measurement is given by Grum (1971). Since it is not our objective to deal in detail with this subject, the reader is referred to a paper by Ganz (1976). In this paper, entitled "Whiteness: Photometric Specification and Colorimetric Evaluation," Ganz surveys the present state of the requirements of industrial whiteness measurement and gives generic formulas with adjustable parameters for the evaluation of whiteness and tint. In his description of measurement requirements for fluorescent white samples, he deals with problems associated with sample irradiation requirements, standardization of sample irradiation, and conversion of experimental data to standard illuminants. For a review on the mode of action and the chemistry of fluorescent whitening agents, assessment of whiteness, and whitening textile materials, see Ciba–Geigy (1973).

Grum *et al.* (1974) discussed the boundaries between the white space and color space and the application of the tristimulus weighting functions to low-purity colors. They concluded that the relationship between instrumental and visual whiteness evaluation can best be established by using the samples that have chromaticity coordinates within the whiteness space. They also concluded that white fluorescent samples should, ideally, be measured with the same light source used for visual observations (see also Swenholt *et al.*, 1978).

When fluorescent whitening agents are used to improve the whiteness of the samples, the fluorescent quantum yield must be (Grum, 1971) related to whitening power and hence to the visual stimuli that give rise to the sensation of whiteness. Having determined the fluorescent quantum yields and luminous reflectance, based on the reflected radiance factor data, one may hypothesize that the whiteness may be linearly related to fluorescent quantum yield and luminous reflectance.

The basic problems of the perception of whiteness are yet to be resolved. A CIE (1979c) task force is presently looking into this problem with an attempt to narrow the number of whiteness formulas to the two or three most practical. Presently, there are as many whiteness formulas as people working with this problem.

Like whiteness, justice is not susceptible to degrees. When they are qualified by "more or less," they owe this qualification to the things with which they are mixed. [Dante, *De Monarchia*] (See: Encyclopedia Britanica Vol. 7 (1969) Chicago).

XVI. CONCLUSIONS

The accuracy of spectral measurements of fluorescent materials is heavily dependent on close control of such factors as the spectral-power distribution of the irradiating source, the geometry of the system, and the reference standards used in the measurement.

The visual stimuli that give rise to the sensation of color can be determined by physical measurements, provided these measurements are made properly and the interpretation of the data is based on sound psychophysical relationships.

There are several alternative methods for deriving colorimetric parameters necessary for colorimetric evaluation of fluorescent material.

The measurement of total spectral radiance factor is the most direct method for color measurements of such materials. In this type of measurement the sample is diffusely irradiated, by the use of an integrating sphere, with undispersed white light, and the combined effects of fluorescence and reflectance from the sample are detected monochromatically. In the measurement of total spectral radiance factor the irradiating system must be well standardized and controlled.

Methods of separating components may be used as an indirect way to specify colors of fluorescent specimens. Several methods of separating the reflected and fluorescent components have been reviewed. The most accurate of these is the dual-monochromator method, in which the sample is irradiated monochromatically and the reflected flux is also detected monochromatically.

In the filter reduction method, which is also precise, the reflected spectral radiance factor is estimated by measuring total radiance factor with a series of sharp cutoff filters in the irradiating beam to reduce and/or eliminate the excitation energy. At wavelengths longer than the upper limit of the excitation region, all the excitation is cut off by the filters; hence, reflected radiance factor is measured directly. In the region where both excitation and emission spectra overlap, one must calculate the reflected radiance factor based on the data with reduced excitation.

The two-mode method is the simplest method for measuring reflected radiance factor. The measurements of conventional reflectance and of total radiance factor are made on the same instrument, if the instrument can be operated in both modes. In this method the total radiance factor below the emission region represents the reflected radiance factor, and in the region above excitation wavelengths, the conventional reflectance represents the reflected radiance factor. In the region of overlap an interpolation for reflected radiance factor is necessary. The validity of this interpolation depends on how wide the region is and which mathematical function is used in the interpolation. This method of measuring the reflected radiance factor may be adequate for quality control operations because it is simple to use and can be quite reproducible. However, for very accurate analysis, and if the data are to be compared with those of other laboratories, this method may fall short.

Since there is no standard light source available for instrumental use that simulates various phases of daylight, alternative methods must be used for color measurements of fluorescent materials; that is, there must be a predicting method.

Several methods have been described in this chapter for predicting spectral radiance factor under standard illuminants from the measurements made under an arbitrary source. All of these methods require accurate calibration of the irradiating system, preferably with the sample under test at the sample port.

The most accurate predicting method is the one that requires the calculation of the number of fluorescent quanta absorbed by the test source and by the source for which prediction is to be made. These calculations are not simple and cannot be made readily. For this reason one must resort to alternative methods, some of which are just as valid and accurate as the one based on determination of the number of absorbed quanta by the fluorescent species.

The method based on excitation spectrum calculation is certainly acceptable. The excitation spectrum is calculated from measurements of the conventional reflectance of the sample, from total radiance factor measurements, and from the calibrated detection system.

A simpler and a little less accurate method assumes a constant absorptance in the excitation region. In this method all that needs to be done is to compute the luminescent radiance factor for the desired illuminant by multiplying the luminescent radiance factor obtained with the source used in the measurement by the ratio of the integrated irradiances of the two sources in the excitation region.

Other simplified computational alternatives may be derived from the basic concepts. However, the irradiating source should always simulate closely the illuminant for which the prediction is made. Large errors in colorimetric parameters will result if the spectral-power distribution of the source used in the measurement and the desired illuminant are too dissimilar. For example, it is not recommended that the measurement of total radiance factor be made with a tungsten source if the predictions are to be made for a daylight illuminant. On the other hand, it is quite appropriate to make the measurements of total spectral radiance factor with an unfiltered xenon source and predict the values for D_{65} illuminant, since the spectral-power distributions of the two are not greatly dissimilar.

It was shown in this chapter that color measurement of fluorescent material is a complex measurement involving complex systems and should be treated as such. Since the basic measurement is total radiance factor, specially arranged instrumentation is required in such measurements. Commercially available instruments suitable for these measurements include those from Zeiss, Cary, Beckman, Hitachi, Diano, Kollmorgen, Hunter, and others.

We cannot stress enough the importance of accurate standardization of sample-irradiating systems associated with these measurements. With instrumentation well calibrated and the irradiating system well defined, the total radiance factors of fluorescent samples can be measured as accurately as those of nonfluorescent samples measured by conventional means.

REFERENCES

Allen, E. (1957). *J. Opt. Soc. Am.* **47**, 933.
Allen, E. (1973). *Appl. Opt.* **12**, 289.
Allen, E., and Sanders, C. L. (1979). *Color Res. Appl.* (in preparation).
Alman, D. H., and Billmeyer, F. W., Jr. (1976). *Color Res. Appl.* **1**, 141.
Alman, D. H., and Billmeyer, F. W., Jr. (1977). *Color Res. Appl.* **2**, 19. ASTM (1966). *Book ASTM Stand.* Part 30.
Baba, G., and Sengoku, M. (1973). *Colour 73, 2nd Congr. Int. Colour Assoc., York* p. 335. Hilger, London.
Baba, G., Sengoku, M., and Minegishi, H. (1974–1975). *Acta Chromat.* **2**, 225.
Berger, A., and Brockes, A. (1971). *Bayer Farben Rev., Sonderh.* 3/1.
Budde, W. (1958). *Farbe* **7**, 295.
Budde, W. (1970). *Farbe* **19**, 94.
Budde, W. (1976). *J. Res. Natl. Bur. Stand., Sect. A* **80**, 585.
Carr, W. (1955). *Paper Maker* (*London*) **129**, 41.
Carter, E. C., and Billmeyer, F. W., Jr. (1978). ISCC Tech. Rep. 78-2.
Chong, T. F., and Billmeyer, F. W., Jr. (1977). Unpubl. Rep., Rensselaer Color Meas. Lab., Rensselaer Polytech. Inst., Troy, New York.
Ciba–Geigy, Ltd. (1973). *Ciba–Geigy Rev.* 6, 4.
CIE (Commission Internationale de l'Eclairage (1931). *C. R., 8ᵉ Sess., Cambridge, Eng.* p. 19.
CIE (1970). Publ. No. 17 (E-1.1). CIE, Paris.
CIE (1971). Publ. No. 15 (E-1.3.1), p. 14. CIE, Paris.
CIE (1977). Publ. No. 38. CIE, Paris.
CIE (1978). "Terminology," TC-2.3. CIE, Paris.
CIE (1979a). "Task Force on Sources," TC-1.3. CIE, Paris.
CIE (1979b). "Task Force on Colorimetric Aspects of Luminescence," Tech. Rep. CIE, Paris. In preparation.
CIE (1979c) TC-1.3 Task Force on Whiteness (in preparation).
Clapper, F. R., and Grum, F. (1975). *Proc. Int. Kongr. Reprogr. Inf.*, 261; Hannover, Germany.
Clarke, F. J. J. (1975). NPL Rep. MOMR, Natl. Phys. Lab., Teddington, Middlesex, England.

Collaborative Reference Program (1977). "Color and Appearance (Color and Color Differences)." U.S. Dep. Commer., Natl. Bur. Stands., Washington, D.C.

Day-Glow Color Corp. (1971). Tech. Booklet No. 1176-A.

Donaldson, R. (1954). *Br. J. Appl. Phys.* **5**, 210.

Eitle, D., and Ganz, E. (1968). *Textilveredlung* **3**, 389.

Erb, W. (1975). PTB-Bericht, PTB-OPT-3, Braunschweig, Germany.

Fukuda, T., and Sugiyama, Y. (1961). *Farbe* **10**, 73.

Ganz, E. (1976). *Appl. Opt.* **15**, 2039.

Gaunt, T. N. (1958). U.S. Patent No. 2,851,423.

Goebel, D. G., Caldwell, B. P., and Hammond, H. K. (1966). *J. Opt. Soc. Am.* **56**, 783.

Grum, F. (1971). *C. R.*, *17^e Sess. CIE, Barcelona* Pap. No. P.71.22.

Grum, F., and Clapper, F. R. (1969). *Tappi* **52**, 1352.

Grum, F., and Costa, L. F. (1974). *Appl. Opt.* **13**, 2228.

Grum, F., and Costa, L. F. (1977). *Tappi* **60**, 119.

Grum, F., and Luckey, G. W. (1968). *Appl. Opt.* **7**, 2289.

Grum, F., and Spooner, D. (1973). *J. Color Appearance* **2**, 6.

Grum, F., and Wightman, T. E. (1960). *Tappi* **43**, 400.

Grum, F., and Wightman, T. E. (1977). *Appl. Opt.* **16**, 2775.

Grum, F., Witzel, R. F., and Stensby, P. (1974). *J. Opt. Soc. Am.* **64**, 210.

Gundlach, D., and Mallwitz, E. (1976). Farbe **25**, 113.

Harkavy, C. M. (1958). *Tappi* **41**, 199A.

Hunter, R. S. (1960). *J. Opt. Soc. Am.* **50**, 44.

Judd, D. B., MacAdam, D. L., and Wyszecki, G. (1964). *J. Opt. Soc. Am.* **54**, 103.

Kazenas, Z. (1957). U.S. Patent No. 2,809,954.

Kazenas, Z. (1960). U.S. Patent No. 2,938,873.

Maycumber, S. G. (1964). *Color Eng.* **2**, 12.

Morton, T. H. (1963). *J. Soc. Dyers Colour.* **79**, 238.

Preucil, F. (1959). *In* "Printing Inks and Color" (. W. H. Banks, ed.), p. 9. Pergamon, New York.

Pringsheim, P. (1949). "Fluorescence and Phosphoresence," Chap. 5. Wiley (Interscience), New York.

Simon, F. T. (1972). *J. Color Appearance* **1**, 5.

Stenius, A. S. (1965). *Tappi* **48**, 45A.

Stensby, P. S. (1967a). *Soap Chem. Spec.* **43**, 41.

Stensby, P. S. (1967b). *Soap Chem. Spec.* **43**, 84.

Swenholt, B. K., Grum, F., and Witzel, R. F. (1978). *Color Res. Appl.* **3**, 141.

Switzer, J. L., and Switzer, R. C. (1950). U.S. Patent No. 2,498,592.

Switzer, J. L., and Switzer, R. C. (1958). U.S. Patent No. 2,851,424.

Wilson, C. M. (1967). *Print. Technol.* **2**, 121.

Wyszecki, G. (1970). *Farbe* **19**, 43.

Wyszecki, G. (1972). *J. Color Appearance* **1**, 18.

7

Colorant Formulation
and Shading

EUGENE ALLEN

Center for Surface and Coatings Research
Lehigh University
Bethlehem, Pennsylvania

289

I. INTRODUCTION

A. What We Cover in This Chapter

This chapter deals with one of the most important industrial applications of the principles of color. The recent great increase of activity in the twin fields of colorant formulation and shading marked the coming of age of color science as a tool for industry. Although color difference measurements have been used in quality control for many years, it is only with the advent of computers that the real utility of color science for manufacturing became appreciated. Formulation and shading computer programs made it possible for the first time to do something about color rather than merely to measure after the fact how far off a color is.

We will begin by defining colorant formulation and shading, and we will discuss how these functions are handled visually. We will then show in a general way how computer programs treat these problems. Next will come a presentation of the basic theory and mathematical methods used in the most prevalent formulation and shading programs, including methods of calibrating colorants. Following is the application of these basic concepts to the writing of colorant formulation and shading computer programs. Finally we will touch on some special problems and the use of advanced methods.

B. What Do Colorant Formulation and Shading Mean?

The best way to explain colorant formulation and shading is by example. We will imagine that an architect is designing a condominium complex and he wishes to paint all the walls of the apartments a certain shade that has struck his fancy. He has a plastic chip sample of the color he wants and has given it to a certain paint company with a large order for a specified number of gallons.

The first task of the paint company is to work out a formula for the paint. This formula must recognize that certain physical properties are required in the paint itself, such as nonsettling, brushability, and the like. Also, the final paint film must have certain qualities, such as wash resistance, light-fastness, and continued film integrity. But, relevant to our present problem, the paint film must have the proper color. A pigment formula must be found that (1) matches the desired color, (2) is nonmetameric against the desired color, and (3) is low in cost. Finding such a formula is an example of the activity known as *colorant formulation*.

We are now at the stage where the paint is being manufactured. The pigments are ground into the vehicle by suitable milling equipment to produce the paint. The standard formula, however, is never made up immediately, because there are enough differences in the milling conditions and in successive shipments of the pigments themselves to make it very doubtful that the required color will be obtained by merely weighing up the standard formula. Instead, the correct color is reached by successive approximation. A typical procedure is to make up, say, 80% of the required formula by grinding the pigment into the vehicle. The reason for making up only 80% is that it is easy to add more colored pigment if we have too little but impossible to remove it if we have too much. Then the color is examined and a decision is made as to how much of each pigment must be added to match the shade. Somewhat less than these amounts is added, usually in paste or preground form, and another check is made. This procedure is repeated until the quality control laboratory is satisfied that a match within tolerance has been achieved. With visual methods as many as 10 shots may be required to reach this point; with computers and color measurement, as little as one or two shots may be enough. This operation is known as *shading*.

C.   How Colorant Formulation Is Handled Visually

Let us return to the colorant formulation problem and see how the visual colorist handles it. He first selects from all the colorants at his disposal those that have the required application and fastness properties plus any other special properties that may be dictated by the requirements of the job. If cotton is to be dyed, cotton dyes must be used. Furthermore, if washfastness is important, only washfast cotton dyes must be chosen. Usually about 7–15 colorants are picked as being suitable for the job at hand; this group is so selected as to have all the basic colors available. A typical group of 11 colorants would comprise three reds, one orange, two yellows, one violet, two blues, one green, and one black. (If pigments are used, an opaque white is usually included.) Next, the colorist must decide which colorants from this group to select for the match and how much of each to use. He knows that there usually are many possible combinations of these colorants he can use, each of which will provide a match. He cannot possibly try all these combinations, but by experience and intuition he selects a few of the most promising and finally decides on one. He makes several trial samples, varying the amounts of the colorants he has selected, until he has matched the color to his satisfaction under some standard light source. If the match does not hold up under some other light source, it is a metameric match and the pigment combination will have to be changed. When he is through, he has produced a match formula that, he hopes, has close to the lowest possible metamerism index and cost.

It is quite obvious that much experience is needed to work out such a match in a reasonable time. If the colorants in the sample being matched are known and if the same colorants can be used in the match, the problem is merely one of finding the right amounts. Otherwise, the colorist must settle on the best colorant combination first. If he is only using visual methods, he has no guarantee that he has really found the best combination. We can see that the better the match, the more time it takes. Some help from the computer would be very welcome.

D. How Batch Shading Is Handled Visually

With batch shading, the problem of finding the right combination of colorants does not exist. The colorants are already known, and the task is to get the batch out of the plant in the shortest time. But the difficulty is that each add to the batch must be followed by an inspection of the color, and this inspection always involves a preliminary application step. In the case of a paint batch, for example, a drawdown must be made and the paint must dry before a judgment can be made. If 5–10 adds are necessary before the paint batch is on shade, the batch may have a residence time in the plant of one or more days. This production bottleneck is often the rate-determining step in paint production. Here also help from the computer is greatly appreciated.

II. TYPES OF COMPUTER PROGRAM FOR FORMULATION AND SHADING

A. Degrees of Freedom in Color Matching

We now look at the kinds of computer programs that have been developed to solve the problems just presented. We will leave the details for later. Our first concern is with the number of colorants necessary for a match, which brings in a consideration of degrees of freedom.

The concept of degrees of freedom is all-important in color matching. If we wish to match the tristimulus values of a certain sample exactly (three quantities), we have to be able to vary the color three ways. In textile and paper dyeing, as well as in the coloration of transparent and translucent plastics, we accomplish this by manipulating the concentrations of three colorants relative to the weight of the substrate. Changing the concentration of each colorant separately affects all three of the tristimulus values, but only by having three different colorants to work with can we expect, *in the general case*, to get a certain desired set of tristimulus values. We thus say that we need three degrees of freedom to get a tristimulus match.

We are really solving a set of three simultaneous equations and we need three unknowns to get a complete solution. Of course, it may happen that a match is possible with two colorants or even with one colorant, just as in algebra certain sets of three simultaneous equations can be solved with only two, or sometimes only one, unknown. These are exceptions, however, to the general rule.

For opaque paints and plastics the presence of four pigments is necessary to obtain the required three degrees of freedom. No change in color is possible in principle by changing the pigment concentration in a paint containing only a single pigment, provided that there is at least enough pigment to give opacity and provided that we neglect second-order PVC effects. With two pigments in the formula, we obtain one degree of freedom based on the amounts of one in relation to the other. Accordingly, for a tristimulus match in an opaque system, the quantities to be manipulated are the percentages of three of the pigments in the total pigment formula, with the percentage of the fourth being given by difference from 100.

B.   Combinatorial Programs

1.   MATCH TO TRISTIMULUS VALUES

a.   *The Combinatorial Approach*   The most widely used type of computer program for colorant formulation is the combinatorial program, so called because it tries combination after combination of colorants. It works as follows: The color to be matched has three tristimulus values under a standard illuminant, say illuminant D_{65}. The matching combination of colorants must have the same tristimulus values or else there is no match. The combinatorial program goes systematically through every possible combination of the colorants originally selected as appropriate and calculates a tristimulus matching formula for each combination if possible.

As shown in Section A, in order to match tristimulus values we need combinations of colorants taken three at a time for textile and paper dyeing and for the coloration of transparent or translucent plastics. We need four-colorant combinations for opaque paints and plastics. If n is the number of colorants from which the computer is to calculate a match, the number of three-colorant combinations that can be formed from the n colorants is given by

$$\binom{n}{3} = \frac{n(n-1)(n-2)}{6}, \tag{1}$$

where $\binom{n}{3}$ means the number of combinations of n things taken 3 at a time.

The number of four-colorant combinations is given by

$$\binom{n}{4} = \frac{n(n-1)(n-2)(n-3)}{24}. \tag{2}$$

It thus becomes very important to choose the list of n colorants carefully, with enough colorants to ensure that a sufficient choice is available, yet not so many that the running time for all the combinations will be excessive. For example, $\binom{10}{3} = 120$ but $\binom{12}{3} = 220$, showing that lengthening the colorant list from 10 to 12 results in almost doubling the computer time for calculating matching formulas.

When we say that a tristimulus match is calculated, we mean specifically that we determine the concentrations of the three colorants needed to give a match that has the same tristimulus values as the standard. Also, we say, "If possible," because some of these combinations cannot possibly provide a match; for example, you cannot match a blue sample with a combination consisting of three reds. But enough of the combinations are usually left to provide a choice of match formulas.

b. *Match Criteria—Metamerism and Cost* The formulation program must now decide which of these matches is the best. The choice is made on the basis of the criteria mentioned previously. The first of these, that the formula should provide a match under the standard light source, is already taken care of, since if we have matched the tristimulus values under D_{65}, we have matched the color under D_{65}.

The second criterion, that the match should be nonmetameric, requires that the computer program calculate some sort of metamerism index for each match and then choose one with a low index. A commonly used index of metamerism is given by considering the formula that has given a match under D_{65} and calculating the color difference between the formula and the standard under another quite different light source, usually source A, some fluorescent lamp source, or both. The greater this color difference, the more metameric the match by this convention. The CIEs index of illuminant metamerism is of the same type (CIE, 1972).

Another type of metamerism index that might be used was suggested by Nimeroff and Yurow (1965). It is based on the fact that the reflectance curves of match and standard may differ from each other even though the tristimulus values are the same. A general index of metamerism, which would not be tied down to specific light sources or in fact to any light sources, would then be given by calculating the mean square deviation between the two reflectance curves. If desired, these deviations might be weighted by the color mixture functions in accordance with Nimeroff and Yurow's suggestions.

The cost of each matching formula is very easy to calculate if we know the price of each colorant. The formulation program would then give a metamerism index and a cost for each of the matches, and the programming could be done in such a way as to print out the matches in any desired order based on these two criteria.

2. MATCH TO FOUR STIMULUS VALUES

It may happen that none of the combinations providing three degrees of freedom yields a match with sufficiently low metamerism. We may then find it advantageous to consider combinations with four degrees of freedom. With the extra degree of freedom we can match one of the tristimulus values under source A in addition to the three under D_{65}. Which source A tristimulus value shall we choose?

Figure 1 shows a plot, in source A color space, of all the possible dye matches for a certain textile sample. The matches were calculated under illuminant D_{65}, and a similar plot in the color space of this latter illuminant would show only a single point, since all the formulas match the standard. In source A color space, however, we have a spread of points, and the farther away any point is from the standard (indicated by a full circle), the more metameric the match. The plot shows only tristimulus values X and Z. We can see immediately from Fig. 1 that there is much more difference among the matches along the X direction than along the Z direction. It turns out, as we can see by the greatly expanded plot shown in Fig. 2, that the Y tristimulus value under source A shows about the same degree of variation among matches as does the Z tristimulus value. Accordingly, we are more likely to strike a nonmetameric four-dye combination by using the X tristimulus value for the extra degree of freedom than we are by using the Y or Z tristimulus value.

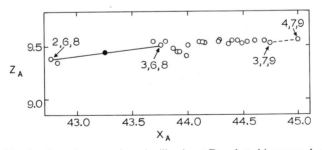

Fig. 1. Matches for a given sample under illuminant D_{65} plotted in source A color space with tristimulus value coordinates X and Z. The matches were calculated by computer from all possible combinations of nine colorants. The full circle represents the sample that was matched. The full line connects two combinations that together would form a four-colorant combination that can match X_A; not so for the four-colorant combination indicated by the broken line.

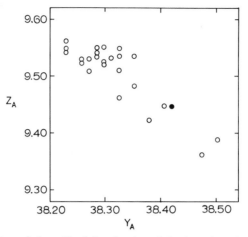

Fig. 2. Same type of plot as Fig. 1, but showing tristimulus values Y and Z under source A as coordinates. The scale is magnified $5\times$ compared to Fig. 1.

If we are using 11 dyes, we would have to go through 330 possible combinations of these dyes taken four at a time. A large amount of computation time would be needed if it were not for a short-cut procedure that is very useful. We assume that every four-dye combination that matches the tristimulus values under D_{65} consists of two three-dye combinations, each of which matches the standard under D_{65}, these two three-dye combinations differing by only one dye and having the other two dyes in common. But in order to be able to match the X tristimulus value under source A, these two three-dye combinations must lie on either side of the X_A value. Figure 1 shows two such three-dye combinations connected by a full line; the numbers are dye identifiers. These two combinations would form a four-dye combination that would be capable of matching not only the three tristimulus values under D_{65} but also the X tristimulus value under source A. Two other three-dye combinations differing in only one dye are shown connected by a broken line; these two combinations could never produce a four-dye combination that would be capable of matching the X_A value. We therefore program the computer to search through all the possible three-dye matches and select pairs of these that differ by only one dye and that lie on either side of the X_A value. These are the four-dye combinations for which the computer calculates matches.

Combinatorial programs will be discussed under the section dealing with algorithms.

C. Other Programs

There is another class of formulation program that works on a completely different principle. This kind of program is not combinatorial; that is, it does not go through all possible combinations of colorants. Instead, it calculates a formula using all the colorants at once that will give the closest possible approximation of the reflectance curve to that of the standard. This calculation can be made either by linear regression (Gugerli, 1963; McGinnis, 1967) or by a more sophisticated algorithm (Cairns, 1975) devised by Marquardt (1963). Note the difference in principle between this kind of formulation program and the other; we are now matching the entire spectrophotometric curve as closely as possible instead of the tristimulus values under a specific light source, and we are using all the colorants at once instead of taking them three or four at a time.

Computer shading programs differ from the formulation programs in several important respects. There are two samples to consider: a trial batch sample that is to be shaded over to the standard and the standard itself. The problem is how much of each of the colorants to add to the batch in order to match the standard. We are not concerned here with alternative formulations from which we must select the best, which means that we are also not concerned with metamerism. The reason for this last statement is that in formula shading the standard is not the original sample supplied by the customer but rather a production sample that has been set aside as standard—perhaps the first batch made. For this reason the matching is entirely nonmetameric.

We will have more to say about both linear regression formulation programs and shading programs in the algorithm section.

III. THEORY OF TRANSPARENT AND TURBID MEDIA

A. Light Absorption and Light Scattering

Our discussion in the preceding section has taken for granted that there is a basic relationship between colorant and color. Let us now put this relationship on a quantitative basis.

Figure 3 shows three colored films, where we use the word "film" to mean any thin slice of substance that transmits, reflects, or absorbs light. Each of the three diagrams represents a well-known model for the interaction of light and matter. We shall see, however, that the center diagram

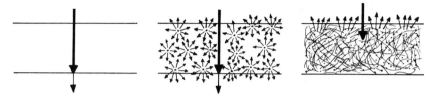

Fig. 3. Three well-known models for the passage of light through a film: the general case in the center and the two extreme cases on the left and right. See text for explanation.

portrays the general case and that the left and right diagrams represent extremes of this general case in opposite directions.

In the diagram at the left, all the colorant is dissolved in the film. Light strikes the film and starts to pass through. But because the film is colored, some of the light is absorbed at certain wavelengths. Let us imagine that the light striking the film is monochromatic (contains radiation at only a single wavelength) and that this wavelength is one that is absorbed. The light that emerges is weakened by its passage through the film, and this weakening is expressed quantitatively by the *absorption coefficient* of the film for that wavelength of light, as we shall see.

The middle diagram again shows monochromatic light striking a film and starting to pass through. This time, however, some discrete colorant particles are present, so that in addition to being absorbed, some of the light is scattered. Scattering is essentially a change in direction, so that light falling on a particle in a single downward direction leaves the particle in all possible directions in a *scattering pattern*, certain directions being favored over others. Light emerges from both the top and the bottom of such a film, the total amount of light that comes out being weakened by absorption. Some of the light comes right out the bottom without having been scattered by the colorant particles, but the rest of the light coming out the bottom and all the light coming out the top has been scattered. The amount of light scattering is expressed quantitatively by the *scattering coefficient* of the film for that wavelength of light.

In the diagram at the right, a much higher concentration of colorant particles is present, and scattering has increased to the point where no light can get through the film. All the light striking the film is eventually either absorbed or scattered back out the top. If the colorant particles do not absorb but only scatter light of the particular wavelength that strikes the film, all of the light is sent back out the top, and such a sample is said to have a reflectance of 100%. Otherwise, the reflectance is reduced by the amount of absorption that takes place.

The film on the left is said to be *transparent*, the film in the center to be *translucent*, and the film on the right to be *opaque* to light of the particular wavelength in question.

B. Transparent Media—Bouguer–Beer Law

The equations worked out in this section apply to colored materials in solution in liquids, as well as to colored transparent plastic or glass materials such as are used for signal lights.

1. DERIVATION OF EQUATIONS

If money is kept in a savings account, it undergoes an increase in unit time that is proportional to the amount of money on deposit. This compound interest law is true in reverse for light passing through a film and is at the very basis of all colorant formulation calculations.

Consider a transparent film such as the one shown on the left of Fig. 3 with total thickness X. The intensity of a beam of monochromatic light before passing through the film is I_0; after passing through it is I. Now consider an infinitesimal thickness dx inside the film. Light of intensity i passing through this element suffers a weakening of intensity, di, *that is proportional to its intensity*. We therefore have

$$di/dx = -Ki, \tag{3}$$

where K is the absorption coefficient of the film. The reason for the minus sign is that absorption coefficients are positive by convention, and we must express the fact that the intensity decreases.

Integration of this differential equation over the entire thickness of the film gives

$$I/I_0 = T_i = e^{-KX}, \tag{4}$$

which can also be expressed as

$$-\log T_i = \log(1/T_i) = KX, \tag{5}$$

where T_i is the internal transmittance of the film. These relationships hold only for monochromatic light, and indeed this restriction holds for all the relationships worked out for colorant formulation theory. We must therefore understand that not only are I_0, I, and T_i functions of wavelength, but so is K. If we wish color information, as we always do for colorant formulation and shading work, we must use our equations at equally spaced wavelengths across the spectrum and then integrate with the appropriate color mixture functions to obtain tristimulus values. Methods for doing this kind of calculation will be considered in Section IV.

Now a very important relationship involving the absorption coefficient

K when only one colorant is present is that if we disregard any color that might have originally been present in the substrate, K is proportional to the concentration of colorant. An exception to this rule may occur when a change in concentration affects the chemistry of the system, but otherwise the rule is valid. Thus, for no substrate color, we can say that $K = ck$ and therefore

$$\log(1/T_i) = ckX, \tag{6}$$

where c is the concentration of colorant and k the unit absorption coefficient. Equation (6) is known as the Bouguer–Beer law and is widely used in analytical work in solution. The quantity $\log(1/T_i)$ is commonly called the optical density, or the absorbance. It should be noted that logarithms for this application are commonly taken to the base 10 and the unit absorption coefficients must be defined accordingly.

When we have n colorants present, and if we no longer disregard the color of the substrate, we can write

$$K = k_t + c_1 k_1 + c_2 k_2 + \cdots + c_n k_n, \tag{7}$$

where k_t is the absorption coefficient of the substrate without colorant, $c_1, \ldots, c_n$ represent the concentrations of the various colorants, and $k_1, \ldots, k_n$ are their respective unit absorption coefficients. The latter are characteristic of these particular colorants in this particular substrate; we must also remember that they are functions of wavelength.

2. CORRECTION FOR REFRACTIVE INDEX DISCONTINUITY

Before we can go any further in colorant formulation theory, we must face up to one annoying fact: There is usually a refractive index change between air and the colored film. This discontinuity complicates the calculation but must be allowed for if we want accurate results.

Consider a collimated beam of light from a spectrophotometer striking the surface of a transparent film. Some of the light is immediately reflected from the boundary as a result of the refractive index change; if the refractive index of the film is 1.5 and the light is perpendicular to the film, the reflectance is 4%. The rest of the light starts on its way through the film and a fraction T_i (the internal transmittance) reaches the other side. Most of the light emerges from the bottom, but some of it is reflected back at the bottom boundary. This light goes back through the film, and a fraction T_i reaches the top, where the cycle is repeated. Table I shows the relationships involved, where the fraction of light reflected at either boundary is denoted by K_1. Note that the reflectance at the boundary K_1 is the same whether the light goes from the medium of lower to the medium of higher refractive index or vice versa. We assume that one unit of light originally strikes the film.

TABLE I. Derivation of Refractive Index Correction for Transparent Samples

Cycle	Amount of light					
	leaving top boundary and going up	leaving top boundary and going down	arriving at bottom boundary	emerging from bottom boundary	leaving bottom boundary and going up	arriving at top boundary from below
1	K_1	$1 - K_1$	$(1 - K_1)T_i$	$(1 - K_1)^2 T_i$	$(1 - K_1)K_1 T_i$	$(1 - K_1)K_1 T_i^2$
2	$(1 - K_1)^2 K_1 T_i^2$	$(1 - K_1)K_1^2 T_i^2$	$(1 - K_1)K_1^2 T_i^3$	$(1 - K_1)^2 K_1^2 T_i^3$	$(1 - K_1)K_1^3 T_i^3$	$(1 - K_1)K_1^3 T_i^4$
3	$(1 - K_1)^2 K_1^3 T_i^4$	$(1 - K_1)K_1^4 T_i^4$	$(1 - K_1)K_1^4 T_i^5$	$(1 - K_1)^2 K_1^4 T_i^5$	$(1 - K_1)K_1^5 T_i^5$	$(1 - K_1)K_1^5 T_i^6$
etc.						

Summing up the light emerging from the bottom boundary, which is really the light transmitted by the film, we obtain

$$T = (1 - K_1)^2 T_i(1 + K_1^2 T_i^2 + K_1^4 T_i^4 + \cdots) = (1 - K_1)^2 T_i/(1 - K_1^2 T_i^2), \qquad (8)$$

where T represents total (as opposed to internal) transmittance.

3. CALCULATION OF TRANSMITTANCE VALUES

We are now in a position to calculate the transmittance of a transparent sample of any thickness containing any desired mixture of colorants, provided that we know the unit absorption coefficients of these colorants as well as the absorption coefficient of the substrate. (We will explain how to determine absorption coefficients in the next section.) If we have n colorants with concentrations $c_1, c_2, \ldots, c_n$, we first use Eq. (7) to get the overall absorption coefficient of the film K. We then substitute this K value as well as the desired thickness X into Eq. (4) in order to get the internal transmittance T_i. Finally we use Eq. (8) to convert the internal transmittance to the total transmittance T. The constant K_1 can be calculated from the Fresnel equation

$$K_1 = (\eta - 1)^2/(\eta + 1)^2, \qquad (9)$$

where η is the refractive index of the medium.

If we know the unit absorption coefficients of all the colorants and of the substrate at equally spaced wavelengths across the spectrum, we can perform the calculations just described at all these wavelengths to obtain a transmittance curve, which can then be integrated to obtain tristimulus values. We will postpone a discussion of this topic until Section IV.

4. DETERMINATION OF ABSORPTION COEFFICIENTS

We determine absorption coefficients in transparent media by working backward. Whereas in the previous section we started with (presumably) known absorption coefficients and ended up with a calculated transmittance value, in this section we start with a measured transmittance value and end up with an absorption coefficient.

To determine the absorption coefficient of the substrate, we measure the transmittance of the substrate without colorant. We must first convert this measured transmittance to internal transmittance by the inverse of Eq. (8). Solving Eq. (8) for T_i gives

$$T_i = \frac{-(1 - K_1)^2 + [(1 - K_1)^4 + 4K_1^2 T^2]^{1/2}}{2K_1^2 T}. \qquad (10)$$

We then substitute this value of T_i into Eq. (5) and get KX. If we know the thickness of the sample X, we can calculate K, the absorption coefficient of the substrate, which we will denote as k_t.

To determine the absorption coefficient of each colorant, we prepare a sample of the colorant alone at known concentration in the substrate. We measure the transmittance of the sample and convert this value to internal transmittance by Eq. (10). We then use Eq. (5) and the thickness to get K as described in the preceding paragraph. But this value of K now includes the colorant as well as the substrate. Equation (7) rewritten for a single colorant and solved for k_1 gives

$$k_1 = (K - k_t)/c_1. \tag{11}$$

Substitution of the known values of k_t and c_1 gives the unit absorption coefficient of the colorant k_1.

C. Translucent and Opaque Media—Kubelka–Munk Law

Opaque samples comprise the most important group for colorant formulation work, including, as they do, paints, textiles, paper, and most plastics. The general method of approach with such samples is similar to that used with transparent samples, but the mathematical model is somewhat more complicated. Translucent samples are less important commercially and require equations that are still more complex; nevertheless we start with the translucent case as being the most general. The mathematical treatment that is almost universally used is known as Kubelka–Munk theory (Kubelka and Munk, 1931).

1. MATHEMATICAL MODEL

a. *Derivation of General Equation* Consider a film that both absorbs and scatters light but through which some light can pass. The thickness of the film is X. The film is placed in optical contact with a background of reflectance R_g. The depth parameter x is considered to be zero at the background and X at the illuminated side of the film.

The picture is essentially similar to that shown in the center portion of Fig. 3. We have made several changes, however, because we are primarily interested in reflecting samples and also because we wish to conform to the specific model used by Kubelka and Munk. The revised picture is shown in Fig. 4.

The first point of difference is the presence of the background in the latter figure. The second point of difference lies in the directional characteristics of the light. In Fig. 3 the light is shown as striking the film as a collimated beam and then being scattered from the individual pigment

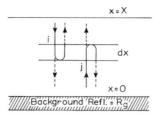

Fig. 4. An element dx in a film of thickness X, indicating how the Kubelka–Munk equations were derived. There are two channels, up and down. The light in both channels, in the film as well as above and below, is considered to be diffuse. Both absorption and scattering occur in dx, the scattered light being reversed in direction.

particles in all directions. But in the Kubelka–Munk model shown in Fig. 4 the light is considered to be scattered in only two directions, down and up. Thus, there are only two channels, one downward and one upward, with the downward channel also containing the original light before it is scattered. This original light, however, as well as all the scattered light (in other words, the light in both channels) is considered to be diffuse rather than collimated. Since no known spectrophotometer both emits and receives diffuse light, the model is not realistic, but it is simple and it works.

Just as for the transparent case, we take an element of thickness dx at an arbitrary depth. But this time we have two channels to manage instead of one. We will call the intensity of light in the downward channel i and in the upward channel j. Now light in the downward channel passing through the thickness element suffers a weakening of intensity di that is proportional to its intensity. But the proportionality constant now consists of an absorption coefficient K plus a scattering coefficient S. The light emerging from the bottom of the element is *strengthened*, however, by the light scattered backward out of the upward channel. We therefore write

$$di/(-dx) = -(K + S)i + Sj, \tag{12}$$

where the minus sign before dx refers to the fact that we are proceeding in a negative direction through the element when we are going downward. Exactly the same thing happens in the reverse direction, except that this time dx is positive. We have

$$dj/dx = -(K + S)j + Si. \tag{13}$$

To solve this pair of differential equations, we first define ρ as being the ratio j/i and then we convert the pair of equations into a single differential equation in ρ. From the quotient rule for differentiation,

$$\frac{d\rho}{dx} = \frac{d(j/i)}{dx} = \frac{i(dj/dx) - j(di/dx)}{i^2} \tag{14}$$

We substitute Eqs. (12) and (13) into this equation and change j/i to ρ wherever it occurs. We obtain

$$d\rho/dx = S - 2(K + S)\rho + S\rho^2, \tag{15}$$

a first-order differential equation in which we can separate the variables. To integrate we apply the following boundary conditions: since at any depth x, ρ represents the light flux streaming up divided by the light flux streaming down, ρ must equal R_g (the reflectance of the background) when $x = 0$, and ρ must equal R (the reflectance of the film) when $x = X$. We therefore have

$$\int_0^X dx = \int_{R_g}^R \frac{d\rho}{S - 2(K + S)\rho + S\rho^2}. \tag{16}$$

Performing the indicated integrations and solving the resulting equation for R, we find that

$$R = \frac{1 - R_g(a - b \coth bSX)}{a - R_g + b \coth bSX}, \tag{17}$$

where $a = 1 + (K/S)$ and $b = (a^2 - 1)^{1/2}$. The symbol "coth" refers to the hyperbolic cotangent function and is defined by $\coth bSX = [\exp(bSX) + \exp(-bSX)]/[\exp(bSX) - \exp(-bSX)]$.

This is a basic form of the Kubelka–Munk equation. It expresses the reflectance of a translucent film as a function of four parameters: the absorption coefficient K, the scattering coefficient S, the film thickness X, and the reflectance of the background R_g.

Equation (17) is seldom used as such. It is interesting primarily because it is the ancestor of the much more familiar and much simpler equation used for opaque samples. But before going in the direction of opacity, let us go in the opposite direction and see what happens when the scattering coefficient S is allowed to approach zero in Eq. (17). We get

$$\lim_{S \to 0} R = R_g e^{-2KX}, \tag{18}$$

which is just what we would expect from the Beer–Bouguer law, considering that the light must first traverse the film inward, then be reflected from the background, and finally traverse the film outward again. (Remember, however, that the K in this equation is not the same as the K used in the Beer–Bouguer law, since the former is defined for diffuse light and the latter for collimated light.) We therefore see that the Kubelka–Munk law reduces to a form compatible with the Beer–Bouguer law for transparent samples.

b. *Simplification for Opaque Samples* In Eq. (17) if we let either the scattering coefficient S or the thickness X increase gradually, we quickly reach a point where $\exp(-bSX)$ becomes negligible compared to $\exp(bSX)$, and therefore the coth function becomes unity. The resulting equation simplifies to the following form:

$$R_\infty = 1 + (K/S) - [(K/S)^2 + 2(K/S)]^{1/2}, \tag{19}$$

where R_∞ means the reflectance at infinite thickness, that is, a thickness such that any further increase in thickness has no effect on the reflectance of the sample. Solving this equation for K/S in terms of R_∞ gives

$$K/S = (1 - R_\infty)^2/2R_\infty. \tag{20}$$

Equations (19) and (20) are the universally used equations for opaque samples. The latter equation is presented in many texts as *the* Kubelka–Munk equation. Note that neither the film thickness X nor the reflectance of the background R_g appears in either equation. We also see that K and S appear only as the ratio K/S.

c. *Additivity of Absorption and Scattering Coefficients—Two-Constant Theory* We now consider how to use Eqs. (19) and (20). Our first concern is the building up of the overall absorption and scattering coefficients from those of the individual colorants and the substrate.

Equation (7) still holds for Kubelka–Munk calculations, provided that we are sure not to try to use Beer's law Ks and ks for such work. We must rederive these constants for use with Kubelka–Munk theory, as will be explained. We also have a corresponding equation for the scattering coefficient S:

$$S = s_t + c_1s_1 + c_2s_2 + \cdots + c_ns_n \tag{21}$$

with analogous symbols. With the exceptions mentioned in the next section, any work with Kubelka–Munk theory requires the use of both the unit absorption coefficients and the unit scattering coefficients of all the colorants involved as well as the absorption and scattering coefficient (if any) of the substrate. We call this approach Kubelka–Munk *two-constant theory.*

d. *Single-Constant Simplification* There are cases where the scattering coefficient is essentially constant no matter what changes in the colorant formula occur. For example, for dyed textile or paper the light scattering is effected by the textile or paper fibers. As a good first approximation, any dye added to the textile or paper can be imagined to dissolve in the fiber and not to contribute to the scattering capability of the substrate. Another example is paint in which most of the pigment is a highly scattering white such as titanium dioxide (creating a so-called pastel shade). Adding small amounts of colored pigment to such a paint does not materially affect its scattering capability. In both of these examples the scattering power of each colorant can be considered to be equal to the same quantity. This quantity is the scattering power of the substrate, if we consider for the moment the white pigment in the paint example to be a kind of substrate.

Dividing Eq. (7) by Eq. (21) gives

$$\frac{K}{S} = \frac{k_t + c_1 k_1 + c_2 k_2 + \cdots + c_n k_n}{s_t + c_1 s_1 + c_2 s_2 + \cdots + c_n s_n}, \tag{22}$$

which is the equation we will have to use to calculate K/S of a mixture by two-constant theory. But if the scattering coefficient is constant and equal to that of the substrate, this equation simplifies to

$$\frac{K}{S} = \frac{k_t + c_1 k_1 + c_2 k_2 + \cdots + c_n k_n}{s_t}.$$

If we let

$$\left(\frac{k}{s}\right)_t = \frac{k_t}{s_t}, \left(\frac{k}{s}\right)_1 = \frac{k_1}{s_t}, \left(\frac{k}{s}\right)_2 = \frac{k_2}{s_t}, \text{etc.},$$

we can write the mathematically equivalent expression

$$\frac{K}{S} = \left(\frac{k}{s}\right)_t + c_1 \left(\frac{k}{s}\right)_1 + c_2 \left(\frac{k}{s}\right)_2 + \cdots + c_n \left(\frac{k}{s}\right)_n. \tag{23}$$

The reason this is a simplification is that we need only one parameter per wavelength, (k/s), to characterize a colorant instead of two, k and s. We therefore call this approach Kubelka–Munk *single-constant theory*.

2. CORRECTION FOR REFRACTIVE INDEX DISCONTINUITY (SAUNDERSON CORRECTION)

We need a correction for refractive index discontinuity for work with opaque reflecting samples that is similar to that derived in Section III.B.2. To derive this correction we assume, as we did for transparent samples, that collimated light strikes the film and a fraction K_1 [usually calculated by Eq. (9) as before] is reflected from the surface. The rest of the light enters the film and is reflected diffusely upward. When the reflected light encounters the film boundary, however, a fraction K_2 is reflected back into the film to undergo another cycle. Since Fresnel's equations show that the greater the angle of incidence of a light beam on a boundary between substances of different refractive index, the higher the reflectance, and since

TABLE II. Derivation of Saunderson Correction

Cycle	leaving top boundary and going up	Amount of light leaving top boundary and going down	arriving at top boundary from below
1	K_1	$1 - K_1$	$(1 - K_1)R_\infty$
2	$(1 - K_1)(1 - K_2)R_\infty$	$(1 - K_1)K_2R_\infty$	$(1 - K_1)K_2R_\infty^2$
3	$(1 - K_1)(1 - K_2)K_2R_\infty^2$	$(1 - K_1)K_2^2R_\infty^2$	$(1 - K_1)K_2^2R_\infty^3$
4	$(1 - K_1)(1 - K_2)K_2^2R_\infty^3$	$(1 - K_1)K_2^3R_\infty^3$	$(1 - K_1)K_2^3R_\infty^4$
etc.			

the light becomes diffuse as soon as it enters the film, K_2 is much larger than K_1. For perfectly diffuse light the theoretical value of K_2 is 0.6.

We derive the correction in a way similar to the derivation for transparent samples. (Table II gives the details.)

Summing up the light leaving the top boundary and going up, which is really the light reflected by the film, we obtain

$$R_m = K_1 + (1 - K_1)(1 - K_2)R_\infty(1 + K_2R_\infty + K_2^2R_\infty^2 + \cdots)$$

$$= K_1 + \frac{(1 - K_1)(1 - K_2)R_\infty}{1 - K_2R_\infty}, \tag{24}$$

where R_m represents the reflectance value that would be measured in a spectrophotometer and R_∞ represents the reflectance calculated by the Kubelka–Munk theory for an opaque sample. Equation (24) is widely known as the Saunderson correction (Saunderson, 1942).

3. Calculation of Reflectance Values

We are now in a position to calculate the reflectance of an opaque sample containing any desired mixture of colorants, provided that we know the unit Kubelka–Munk coefficients of each colorant as well as those of the substrate. We will consider the methods for determining these constants in a subsequent section. Basic units used are usually percentage of dye on the weight of the fiber for textiles and percentage of pigment on the weight of total pigment for paints.

a. *Single-Constant Theory* We need to know the unit k/s ratio for each colorant at each wavelength and for the substrate. If we are working with textiles or paper, the substrate is the undyed textile or paper. For pastel paints, single-constant theory considers the substrate to be the paint pigmented with the white pigment alone.

We first use Eq. (23) to get the K/S value of the mixture of colorants on the substrate. We then convert K/S to R_∞ by Eq. (19). We now use the Saunderson correction only if we are working with paints; if so, we convert R_∞ to R_m by Eq. (24). For textile and paper samples, on the other hand, the fibers that are scattering the light are immersed in air, and we therefore do not have a refractive index discontinuity. Accordingly, we usually use $R_m = R_\infty$ for these cases (but see Section III.C.4). A discussion of the appropriate values of K_1 and K_2 to use with the Saunderson correction will be reserved for a subsequent section.

b. *Two-Constant Theory* For opaque samples for which we can expect the scattering coefficient to vary with the colorant formula, we need the two-constant theory. We need to know the unit values of both k and s for each colorant. The procedure is the same as for single-constant theory, except that we calculate K/S of the mixture by Eq. (22) instead of Eq. (23). Several things should be noted: First, the cs add to 100%. Second, s_t and k_t represent the scattering and absorption coefficients of the vehicle in the case of paints. The quantity s_t is almost always zero, and k_t is usually small enough to be neglected. This is not so for the $(k/s)_t$ of single-constant theory, since the typical fiber (textiles or paper samples) or white pigment (paint samples) may absorb an appreciable amount of light.

4. DETERMINATION OF CONSTANTS

a. *k/s Ratio in Single-Constant Theory—Textile and Paper Samples* We first determine $(k/s)_t$, the k/s value of the substrate, by preparing a "mock dyeing," that is, putting a sample through the complete dyeing procedure but omitting the dye. We measure the reflectance of the sample and convert to K/S [in this case $(k/s)_t$] by Eq. (20).

We then determine the unit k/s value for each dye by preparing a series of dyeings of this dye alone at increasing concentrations on the substrate. Each dyeing is measured spectrophotometrically, and the reflectance values are converted to K/S values by Eq. (20). We choose several wavelengths of strong absorption for each dye and plot K/S against concentration of dye originally taken.

Rewriting Eq. (23) for the presence of only one colorant, we obtain

$$K/S = (k/s)_t + c(k/s), \tag{25}$$

from which we would expect that a plot of K/S against concentration would give a straight line with slope equal to the unit k/s value of the dye. In practice we are likely to get a curve that is concave downward instead of a straight line (see Fig. 5). Two possible reasons for this behavior suggest themselves. The first is that there exists a surface reflection of the fiber

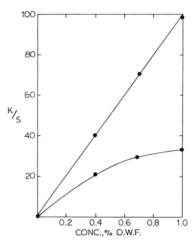

Fig. 5. The lower curve is a plot of K/S against concentration for carbon black in dope-colored polypropylene. The upper curve shows the same data calculated after subtracting a surface reflection correction of 1.0% from the reflection values.

such that even when we have added so much dye that we have increased the light-absorptive capacity of the fiber to its utmost degree, some light is still reflected from the surface. We see that this is analogous to the initial reflection from a paint film discussed earlier under the Saunderson correction. This would indicate that we should subtract a small constant reflectance value from the measured reflectance to allow for this surface reflection, and, in fact, if we do so we can quite often transform the curve to a straight line, as is shown in Fig. 5.

An alternative technique is to invoke the Saunderson correction even though it is seldom applied to textiles and to use both K_1 and K_2. The method of finding the appropriate K_1 and K_2 values to use would be to select a pair of these values that seem reasonable and then convert all the measured reflectance values (R_m) for all the dyes at all wavelengths of strong absorption to R_∞ by the use of the following equation:

$$R_\infty = (R_m - K_1)/(1 - K_1 - K_2 + K_2 R_m), \tag{26}$$

which was obtained from Eq. (24) by solving it for R_∞. Then all the R_∞ values would be converted to K/S values by Eq. (20). Finally the best straight lines, for each dye at each wavelength, would be calculated by considering the K/S values to be a linear function of concentration. The overall root mean square deviation between the K/S points and the lines would be determined. A new pair of K_1 and K_2 values would then be chosen by some systematic iterative procedure, and the calculation would be repeated until that pair of K_1 and K_2 values would be found that gives the minimum root mean square deviation between straight lines and experimental points.

The other possible reason for obtaining a curve that is concave downward is that the dye may not completely exhaust onto the fiber in the preparation of the calibration dyeings. As the concentration of dye increases and as the saturation value of the fiber is approached, it is possible that more and more dye might remain behind in the dye bath. We can determine if this is the case by suitable analysis of the dye-bath liquor. We can correct for this effect by suitable mathematical handling of the curve of K/S against concentration; a typical technique is to write, in place of Eq. (25),

$$(K/S)^p = (k/s)_t + c(k/s) \tag{27}$$

where p represents a power slightly greater than 1 (Gall, 1973). Another possibility is to use a polynomial:

$$K/S = a_0 + a_1 c + a_2 c^2 + a_3 c^3, \tag{28}$$

in which the constant a_1 represents approximately the unit k/s value, and the other constants a_2 and a_3 serve to correct for the concavity of the curve. It is probably not necessary to use a higher-degree polynomial than the third. The constants in Eq. (27) or Eq. (28) can be fitted by suitable regression programs.

b.   *k/s Ratio in Single-Constant Theory—Paint Samples*   With modern computer techniques there is usually no valid reason for using the single-constant simplification for paints. It was used in the past when the available computational aids (such as COMIC I or !og log K/S plots) could not handle Kubelka–Munk two-constant theory. The $(k/s)_t$ values were determined by measuring a drawdown of the paint containing white pigment but no colored pigment. The Saunderson correction was always used. Otherwise the techniques were similar to those used for fiber samples.

c.   *k and s Separately in Two-Constant Theory*   Our first concern in two-constant theory is with the scattering coefficient of the white pigment (usually titanium dioxide), because all other absorption and scattering coefficients are expressed relative to this one. There are two commonly used schemes. The first is to consider that $s_w = 1$ at all wavelengths, where the subscript w refers to the white pigment. This scheme is simple because it does not require the absolute determination of s_w, and it gives satisfactory results provided that we always work with samples that are opaque at all wavelengths. If we are not sure that this will be the case, we would be better advised to use the second scheme, which requires an absolute determination of s_w.

The dimensions of both K and S can be deduced from a consideration of Eqs. (12) and (13) to be reciprocal length (actually reciprocal thickness). If we had a method of measuring film thickness directly that is precise to $1-2\%$, we would use centimeters as the unit of thickness and express K and S as centimeters^{-1}. But we find that the only precise way is to weigh a measured area of film, which leads to grams per centimeter squared (or grams per meter squared) as the most natural unit of film thickness. The dimensions of K and S are therefore usually taken as centimeters squared per gram (or meters squared per gram).

To determine K and S of the white pigment on an absolute basis, we prepare a drawdown of the white pigment alone in the appropriate vehicle at the correct pigment-volume concentration. We use a measured film thickness thin enough to allow the background to show through and use both a black and a white background. Before making the drawdown, we measure the reflectance of both the black and the white background and apply Eq. (26) to these values in order to correct for the fact that after the paint film is applied there will (practically) no longer be a refractive index boundary between the film and the background. (We leave the question of what values of K_1 and K_2 to use open for the moment.) After the drawdown has dried, we measure its reflectance over both the black and the white background and apply Eq. (26) to the reflectance values. We now solve Eq. (17) simultaneously for K and S, using the two pairs of values for R and R_g and the measured value for X. We need a nonlinear iterative algorithm for this solution.

An alternative method is to use the film over the black background plus an opaque film. If we call the reflectance of the thin film over the black background R, the reflectance of the black background R_g, and the reflectance of the opaque film R_∞, we can calculate the scattering coefficient of the white pigment directly by the equation

$$S = \frac{R_\infty}{X(1 - R_\infty^2)} \log_e \left[\frac{(1 - RR_\infty)(R_\infty - R_g)}{(1 - R_g R_\infty)(R_\infty - R)} \right]. \tag{29}$$

This equation can be derived from Eq. (17). Note that all measured reflectance values must be put through the Saunderson correction in reverse, Eq. (26).

We have possibly already determined the k of the white pigment if we have used the black and white background method. If not, we do so through a knowledge of K/S [obtained from the reflectance of an opaque film plus Eq. (20)] and the s value.

In determining k and s of a colored pigment, we will be guided by the basic principles laid down by Duncan (1949). But we will present the most popular methods in use today. We start with the tint ladder method used by Cairns *et al.* (1976). Mixtures of the colored pigment to be characterized and titanium dioxide are made at different ratios of colored pigment to white. About five to eight such mixtures are made, including a masstone (no titanium dioxide). The reflectance values for each of these mixture drawdowns are measured and converted to K/S by Eqs. (26) and (20) in tandem. We therefore have, for each pigment and at each wavelength, a series of $(K/S)_i$ values and a corresponding series of c_i values, the latter being the fraction of colored pigment in the mixture of colored pigment and white. The subscript i refers to the ith mixture of colored pigment and titanium dioxide.

Let us now rewrite Eq. (22) for the present situation. We will let the subscript 1 refer to the colored pigment and subscript 2 to the white pigment, these being the only two pigments present. We assume that there is no substrate to speak of; any absorptive or scattering properties of the vehicle can be neglected. We thus have

$$(K/S)_i = [c_i k_1 + (1 - c_i)k_2]/[c_i s_1 + (1 - c_i)s_2]. \tag{30}$$

If we multiply throughout by the denominator of the right-hand side and rearrange, we get

$$c_i k_1 - c_i(K/S)_i s_1 = (1 - c_i)[(K/S)_i s_2 - k_2]. \tag{31}$$

This equation has two unknowns, k_1 and s_1; everything else is known. If we had only two such equations, we could solve them simultaneously for these two unknowns. But since we have five to eight mixtures, we have as many equations, and we therefore have an overdetermined system that we can solve by least squares. The solution gives us k_1 and s_1, the unit absorption and scattering coefficients of the colored pigment.

The tint-ladder method gives good results except in regions of the spectrum where the colored pigment has a high reflectivity approaching that of the white. For example, chrome yellow absorbs as little light as titanium dioxide does in the red region of the spectrum and is also a good scatterer, as is titanium dioxide. Therefore, the spectrophotometer cannot easily distinguish between chrome yellow and titanium dioxide at these wavelengths, and the simultaneous equations give inaccurate answers for that reason. Despite this difficulty, workable results are obtained by the tint-ladder method and pigments standardized in this way give good results in computerized color formulation.

Other investigators, however, prefer to eliminate the kind of inaccuracy just described by using a black reduction as well as a white reduction. The thinking is that if you mix the colored pigment with a black instead of a white, using enough black to reduce the reflectance to about 50%, you have much more accuracy where the colored pigment does not absorb. You can still use the white pigment to work on the regions of the spectrum where the reflectivity is low; the general idea is to mix unlike pigments so as to obtain maximum accuracy. The procedure, then, would be to prepare three drawdowns: a masstone, a mixture with white, and a mixture with black. The masstone and the white reduction would be used to determine the absorption and scattering coefficients of the colored pigment in the strongly absorbing regions of the spectrum; the masstone and the black reduction would be used in regions where the colored pigment hardly absorbs any light. In both cases Eq. (31) would be used and solved simultaneously, two equations at a time. For the black reduction the subscript 2 now refers to the black pigment instead of the white pigment.

In order to make the black reduction method work, a carefully controlled and standardized black pigment is necessary. Many black pigments suffer from agglomeration difficulties, and some investigators find that a pre-formulated black paint works best for this purpose. The absorption and scattering coefficients of the black must be determined from a masstone and a reduction with white in the usual manner.

d. *Saunderson K_1 and K_2* An important part of the standardization process is the setting up of workable Saunderson constants K_1 and K_2. As mentioned previously, K_1 has been usually calculated by Eq. (9), and therefore special attention has been given to K_2.

We saw previously that $K_2 = 0.60$ would be correct if the light inside the film were completely diffuse. It is not, however, and many workers recommend lower values for K_2. Saunderson himself recommended 0.4 for pigmented plastics (Saunderson, 1942), and Mudgett and Richards (1973) have worked out a theoretical basis for preferring a K_2 value of about 0.4 to higher values. Many workers use K_2 as another constant to be fitted to the experimental results. A good example is the work of Andrade (1976), who found that he was able to write equations that expressed K_2 as a function of pigment volume concentration and Kubelka–Munk scattering coefficient.

An interesting approach to the determination of K_2 was used by Orchard (1977). He considered K_2 to vary with the reflectance R_∞ by the equation

$$K_2 = (a + bR_\infty)/(1 + cR_\infty), \qquad (32)$$

where a, b, and c are adjustable parameters. Orchard showed that the form of this equation is reasonable on the basis of theory. He used a nonlinear optimization program with a 15-step tint ladder to obtain the best least-squares values of k for the white; k and s for the colored pigment; and a, b, c in Eq. (32), all at one time. The procedure was to calculate, for every step in the tint ladder and with the current set of these six parameters, R_m by the following sequence of equations: $(30) \rightarrow (19) \rightarrow (32) \rightarrow (24)$. This calculated value of R_m was compared with the measured value for each step in the tint ladder, and the values of the six parameters that gave the best least-squares fit were found.

If we wish to go the reverse route and calculate K/S from a measured R_m value by the use of Orchard's K_2 value (having determined a, b, and c), we would need an equation that expresses K_2 as a function of R_m rather than R_∞. We can get such an equation, as Orchard suggests, by eliminating R_∞ from Eqs. (32) and (24).

IV. ALGORITHMS FOR COLORANT FORMULATION AND SHADING

A. General Calculation Scheme

All the mathematical apparatus that we have marshalled in Section III applies to a single wavelength at a time. As we have already mentioned, if we wish to match tristimulus values we need an algorithm that integrates across wavelengths. In this section we will consider such algorithms, and we will start by solving the following problem: Let us suppose that the computer is running a combinatorial program, in which combination after combination of colorants is tried. For any specific combination of colorants, how can we program the computer to calculate the amounts of the colorants necessary to give the same tristimulus values as those of the standard being matched?

The basic mathematics for making this calculation have already been presented. We could guess at the concentrations of the colorants needed, and use the equations recommended in Section III.C.3 to calculate the reflectance at each wavelength separately. This gives us the reflectance curve of our postulated mixture. We can integrate this reflectance curve and see how close the tristimulus values of the mixture come to those of the standard. If our guess has been fortunate, they will match. If not, we must make a correction to each of the three concentrations and start over again.

Now there are algorithms available that make it unnecessary to guess at the three starting concentrations for a tristimulus match. These algorithms do not give a direct answer, and the three starting concentrations usually do not produce an exact match to the tristimulus values. But the match is much closer than would be obtained by merely guessing. Furthermore, the same algorithm tells how to correct the three concentrations to get a much closer match. The program works by iteration and stops when the tristimulus values of the match equal those of the standard to within a preassigned tolerance. To understand how to set up such an algorithm, we will have to familiarize ourselves with matrix notation.

B. Matrices in Colorimetry

Computers are built to work with large amounts of data. These data naturally fall into groups of related items, such as the 39 values comprising the reflectance curve of a sample at 380, 390, ..., 760 nm. It seems natural to assign a single symbol to the entire array of 39 numbers rather than to assign a separate symbol to each. The computer could then be instructed to read in the entire reflectance curve at once from punched cards by the FORTRAN command

$$READ \ldots R,$$

which is preceded at some point in the program by the explanatory statement

$$DIMENSION \ R \ (39).$$

The dimension statement merely instructs the computer to reserve 39 memory spaces for "R." The symbol "R" would thus represent an array, or matrix, containing 39 items.

Certain arrays are better represented in two-dimensional form. An example is the array of color mixture functions, which may be given the symbol T. If we wish to read these values into the computer, we would write

$$READ \ldots T,$$

preceded by

$$DIMENSION \ T \ (3, 39).$$

The symbol "T" would then represent an array, or matrix, of 117 items, but conceptually arranged in a format of three rows and 39 columns (see below).

The use of matrices is not confined to shortening the process of transferral of data into and out of the computer. Computations involving masses of related data are more easily handled by manipulating the data in groups by the use of certain symbolic rules and conventions.

In presenting the algorithms for use in color matching, we will use matrix notation. We will first define the various matrices to be employed and then write the matrix equations without proof. The algorithms used are those published by the writer (Allen, 1966, 1974a), and we refer the interested reader to these papers for details. The matrix operations of addition, subtraction, multiplication, and inversion are in the repertoire of all computers with software documentation, and so it should be possible for the reader to write the computer programs using these algorithms even though he or she does not understand how to carry out the various matrix operations.

C. Match to Tristimulus Values

1. SINGLE-CONSTANT THEORY (COMBINATIONS OF THREE COLORANTS)

a. *Initial calculation*   We define the following matrices and vectors:

$$\mathbf{T} = \begin{bmatrix} \bar{x}_{380} & \bar{x}_{390} & \cdots & \bar{x}_{760} \\ \bar{y}_{380} & \bar{y}_{390} & \cdots & \bar{y}_{760} \\ \bar{z}_{380} & \bar{z}_{390} & \cdots & \bar{z}_{760} \end{bmatrix},$$

where $\bar{x}$, $\bar{y}$, and $\bar{z}$ are color mixture functions. We have assumed 10 nm intervals from 380 to 760 nm in this formulation, but any other desired intervals or limits may be used.

$$\mathbf{E} = \begin{bmatrix} E_{380} & 0 & \cdots & 0 \\ 0 & E_{390} & \cdots & 0 \\ \vdots & \vdots & & \vdots \\ 0 & 0 & \cdots & E_{760} \end{bmatrix},$$

where E represents the relative spectral power distribution of the light source at the wavelength in nanometers indicated by the subscript.

$$\mathbf{f}^{(a)} = \begin{bmatrix} f(R)^{(a)}_{380} \\ f(R)^{(a)}_{390} \\ \vdots \\ f(R)^{(a)}_{760} \end{bmatrix}; \quad \mathbf{f}^{(t)} = \begin{bmatrix} f(R)^{(t)}_{380} \\ f(R)^{(t)}_{390} \\ \vdots \\ f(R)^{(t)}_{760} \end{bmatrix}.$$

In these definitions R refers either to spectral reflectance for opaque samples or to spectral transmittance for clear samples. The superscript (a) refers to the sample being matched, (t) to the substrate on which the match is to be made. The function $f(R)$ is specifically $(1 - R)^2/(2R)$ for opaque samples [see Eq. (20)] and $\log(1/R)$ for clear samples [see Eq. (5)].

$$\mathbf{D} = \begin{bmatrix} d_{380} & 0 & \cdots & 0 \\ 0 & d_{390} & \cdots & 0 \\ \vdots & \vdots & & \vdots \\ 0 & 0 & \cdots & d_{760} \end{bmatrix},$$

where $d_i = [dR/df(R)]_i$. Specifically, $d_i = -2R_i^2/(1 - R_i^2)$ for opaque samples, and $-2.3026 R_i$ for clear samples.

$$\mathbf{\Phi} = \begin{bmatrix} \phi_{380}^{(1)} & \phi_{380}^{(2)} & \phi_{380}^{(3)} \\ \phi_{390}^{(1)} & \phi_{390}^{(2)} & \phi_{390}^{(3)} \\ \vdots & \vdots & \vdots \\ \phi_{760}^{(1)} & \phi_{760}^{(2)} & \phi_{760}^{(3)} \end{bmatrix},$$

where the ϕ values represent unit k/s ratios for the colorants used with opaque samples, and unit absorption coefficients for the colorants used with clear samples. The subscripts refer to wavelengths and the superscripts to colorant number. (The three colorants that will be used in the match are arbitrarily assigned numbers 1, 2, 3.)

$$\mathbf{c} = \begin{bmatrix} c_1 \\ c_2 \\ c_3 \end{bmatrix},$$

where c_1, c_2, and c_3 refer to concentrations of the three colorants.

A colorant formula, expressed by the vector $\mathbf{c}$, that will provide a fairly close but not exact correspondence to the tristimulus values of the sample being matched is given by the following matrix equation:

$$\mathbf{c} = (\mathbf{TED\Phi})^{-1}\mathbf{TED}[\mathbf{f}^{(a)} - \mathbf{f}^{(t)}]. \tag{33}$$

The matrix $\mathbf{TED\Phi}$ is a 3×3 matrix that is inverted and multiplied into the 3×1 vector $\mathbf{TED}[\mathbf{f}^{(a)} - \mathbf{f}^{(t)}]$. The result is the 3×1 vector of the concentrations of the three colorants that is the computer's first guess as to the matching formula.

b. *Iterative Improvement*   As shown in preceding sections, we can calculate the tristimulus values corresponding to the formula just derived and determine how far off they are from those of the sample being matched. We then start iterating toward a closer match, and stop the iterations when the tristimulus differences have all become smaller than some preassigned value. Each iteration is conducted by first defining

$$\Delta\mathbf{t} = \begin{bmatrix} \Delta X \\ \Delta Y \\ \Delta Z \end{bmatrix}, \qquad \Delta\mathbf{c} = \begin{bmatrix} \Delta c_1 \\ \Delta c_2 \\ \Delta c_3 \end{bmatrix},$$

where ΔX, ΔY, and ΔZ represent the differences in tristimulus values between standard and rough match, and Δc_1, Δc_2, and Δc_3 refer to an iterative correction in the concentrations. We then use

$$\Delta c = (\mathbf{TED\Phi})^{-1} \Delta t. \tag{34}$$

Note that the inverted 3×3 matrix used for the iterations is the same as the one used for calculating the first rough match. We calculate a new concentration vector by $c_{new} = c_{old} + \Delta c$ and again determine how close the tristimulus values are. In most cases no more than four or five iterations will be needed.

2. Two-Constant Theory (Combinations of Four Colorants)

a. *Initial Calculation* In this presentation we will neglect any absorptive or scattering effects caused by the vehicle. The concentration values refer to the fraction of each pigment in the pigment mixture; $c_1 + c_2 + c_3 + c_4 = 1$. We define the following additional matrices and vectors:

$$\mathbf{k}^{(a)} = \begin{bmatrix} K^{(a)}_{380} \\ K^{(a)}_{390} \\ \vdots \\ K^{(a)}_{760} \end{bmatrix}, \qquad \mathbf{s}^{(a)} = \begin{bmatrix} S^{(a)}_{380} \\ S^{(a)}_{390} \\ \vdots \\ S^{(a)}_{760} \end{bmatrix},$$

where the $K^{(a)}$ refer to the absorption coefficients and the $S^{(a)}$ to the scattering coefficients of the sample being matched. Admittedly, we do not know these quantities separately; we only know $K^{(a)}/S^{(a)}$, which we calculate from the $R^{(a)}$ values by Eq. (20). But see the following for the way to avoid this difficulty.

$$\mathbf{D}_k = \begin{bmatrix} d_{k,380} & 0 & \cdots & 0 \\ 0 & d_{k,390} & \cdots & 0 \\ \vdots & \vdots & & \vdots \\ 0 & 0 & \cdots & d_{k,760} \end{bmatrix},$$

$$\mathbf{D}_s = \begin{bmatrix} d_{s,380} & 0 & \cdots & 0 \\ 0 & d_{s,390} & \cdots & 0 \\ \vdots & \vdots & & \vdots \\ 0 & 0 & \cdots & d_{s,760} \end{bmatrix},$$

where

$$d_{k,i} = \frac{\partial R^{(a)}_i}{\partial K^{(a)}_i} = \frac{-2R^{(a)2}_i}{S^{(a)}_i(1 - R^{(a)2}_i)}, \tag{35}$$

$$d_{s,i} = \frac{\partial R_i^{(a)}}{\partial S_i^{(a)}} = \frac{R_i^{(a)}[1 - R_i^{(a)}]}{S_i^{(a)}[1 + R_i^{(a)}]}. \tag{36}$$

These equations have been derived from Eq. (19) by differentiating with respect to K and S, respectively. The subscript i refers to a specific wavelength.

$$\Phi_k = \begin{bmatrix} k_{380}^{(1)} & k_{380}^{(2)} & k_{380}^{(3)} \\ k_{390}^{(1)} & k_{390}^{(2)} & k_{390}^{(3)} \\ \vdots & \vdots & \vdots \\ k_{760}^{(1)} & k_{760}^{(2)} & k_{760}^{(3)} \end{bmatrix}, \qquad \mathbf{k}^{(4)} = \begin{bmatrix} k_{380}^{(4)} \\ k_{390}^{(4)} \\ \vdots \\ k_{760}^{(4)} \end{bmatrix},$$

$$\Phi_s = \begin{bmatrix} s_{380}^{(1)} & s_{380}^{(2)} & s_{380}^{(3)} \\ s_{390}^{(1)} & s_{390}^{(2)} & s_{390}^{(3)} \\ \vdots & \vdots & \vdots \\ s_{760}^{(1)} & s_{760}^{(2)} & s_{760}^{(3)} \end{bmatrix}, \qquad \mathbf{s}^{(4)} = \begin{bmatrix} s_{380}^{(4)} \\ s_{390}^{(4)} \\ \vdots \\ s_{760}^{(4)} \end{bmatrix},$$

where k and s are unit absorption and scattering coefficients and the superscripts refer to colorants.

$$\mathbf{u} = \begin{bmatrix} 1 & 1 & 1 \end{bmatrix}.$$

We start the calculation by converting the reflectance curve of the sample to be matched, already reverse Saunderson corrected by Eq. (26), to K/S by Eq. (20). We now assume, temporarily only, that $S^{(a)} = 1$ at all wavelengths. This gives us the $K^{(a)}$ values at all wavelengths as being equal to the K/S values. We next substitute the values for $R^{(a)}$ and $S^{(a)}$ into Eqs. (35) and (36) in order to calculate the d values, and we form the $\mathbf{k}^{(a)}$ vector from the $K^{(a)}$ values just obtained.

We now have everything we need to calculate the computer's first guess— a formula that will provide a fairly close but not exact correspondence to the tristimulus values of the sample being matched. The equation is the following:

$$\mathbf{c} = (\mathbf{TE}\{\mathbf{D}_k[\Phi_k - \mathbf{k}^{(4)}\mathbf{u}] + \mathbf{D}_s[\Phi_s - \mathbf{s}^{(4)}\mathbf{u}]\})^{-1}$$
$$\cdot \mathbf{TE}\{\mathbf{D}_k[\mathbf{k}^{(a)} - \mathbf{k}^{(4)}] + \mathbf{D}_s[\mathbf{s}^{(a)} - \mathbf{s}^{(4)}]\}. \tag{37}$$

Just as for single-constant theory, the matrix within large parentheses is a 3×3 matrix that is to be inverted. The rest of the right member of this equation is a 3×1 vector. The answer, $\mathbf{c}$, is a 3×1 vector containing as elements the fractions of the first three colorants, c_1, c_2, and c_3. The fraction of the fourth colorant is found by difference: $c_4 = 1 - c_1 - c_2 - c_3$.

b. *Iterative Improvement* The first step in determining a matrix to be used for the iterative part of the program is to compute a better value for $S^{(a)}$ at each wavelength. We do this by assuming that $S_i^{(a)} = S_i^{(m)}$, the scattering coefficient of the colorant mixture that constitutes the first computer guess. We use

$$S_i^{(a)} = S_i^{(m)} = c_1 s_i^{(1)} + c_2 s_i^{(2)} + c_3 s_i^{(3)} + (1 - c_1 - c_2 - c_3)s_i^{(4)}, \quad (38)$$

where the subscript i refers to a specific wavelength. Next we recompute the **D** matrices by substituting our new values of $S_i^{(a)}$ together with the $R_i^{(a)}$ values into Eqs. (35) and (36).

We are now ready for the iterative calculation leading to a closer match. Each iteration uses

$$\Delta c = (TE\{D_k[\Phi_k - k^{(4)}u] + D_s[\Phi_s - s^{(4)}u]\})^{-1} \Delta t. \quad (39)$$

It might seem that, just as is true for single-constant theory, the same inverted matrix that was used to obtain the first guess is used for the iteration. But this is only true in a formal sense. If we attempt to use the same numerical matrix we will not converge to the final answer. We must recompute and reinvert the matrix in large parentheses, using the new **D** matrices derived as just explained. We only have to do this once, however, right after the calculation of the first guess. We do not have to recalculate the matrix after each iteration.

D. Least-Squares Match

In Section II.C we discussed a match by linear regression using all the colorants simultaneously. We now present a matrix algorithm for performing such a match by the use of single-constant theory. This type of calculation was used by Gugerli (1963) and was written up in matrix form by McGinnis (1967).

We assume that we have n colorants at our disposal and define

$$P = \begin{bmatrix} \phi_{380}^{(1)} & \phi_{380}^{(2)} & \cdots & \phi_{380}^{(n)} \\ \phi_{390}^{(1)} & \phi_{390}^{(2)} & \cdots & \phi_{390}^{(n)} \\ \vdots & \vdots & & \vdots \\ \phi_{760}^{(1)} & \phi_{760}^{(2)} & \cdots & \phi_{760}^{(n)} \end{bmatrix},$$

where the symbol ϕ has the same meaning as before. We expand the **c** vector to

$$\mathbf{c} = \begin{bmatrix} c_1 \\ c_2 \\ \vdots \\ c_n \end{bmatrix}$$

and use the $\mathbf{f}^{(a)}$ and $\mathbf{f}^{(t)}$ vectors as previously defined.

It can be shown that the $\mathbf{c}$ vector that provides a least-squares match to the K/S values of the standard is given by

$$\mathbf{c} = (\mathbf{P'P})^{-1}\mathbf{P'}[\mathbf{f}^{(a)} - \mathbf{f}^{(t)}], \tag{40}$$

where $\mathbf{P'}$ means the transpose of $\mathbf{P}$, obtained by interchanging rows and columns. It is quite likely that some of the c values will be negative as a result of such a calculation. One possible way of proceeding is to eliminate the colorants with negative cs and repeat the algorithm.

If the colorist prefers to get a least-squares match to reflectance values instead of K/S values, or if it is necessary to use two-constant theory, the simple least-squares algorithm just presented cannot be used. The Marquardt algorithm referred to in Section II.C might then be tried.

One useful application of this type of calculation is to find matches with fewer colorants than are required to give three degrees of freedom. For example, it is always worthwhile to see if a textile sample could be matched with two dyes, or even one dye, instead of three. We have found that a regression calculation gives much better results than matching only one or two of the tristimulus values. It is worthwhile to write down the specific equations for these types of match in ordinary notation. Remember that the equations refer to single-constant theory only.

One-dye matches by least-squares K/S fit are given by

$$C = \sum_i \{\phi_i[f(R)_i^{(a)} - f(R)_i^{(t)}]\}/\sum_i (\phi_i^2), \tag{41}$$

where ϕ_i is the unit k/s ratio at wavelength i of the dye being calculated, and $f(R)_i^{(a)}$ and $f(R)_i^{(t)}$ have their previously assigned meanings. Two-dye matches by least-squares K/S fit are given by the following pair of simultaneous equations, which should be solved for c_1 and c_2:

$$\begin{aligned} \left(\sum_i \phi_i^{(1)^2}\right)c_1 + \left(\sum_i \phi_i^{(1)}\phi_i^{(2)}\right)c_2 &= \sum_i \phi_i^{(1)}[f(R)_i^{(a)} - f(R)_i^{(t)}], \\ \left(\sum_i \phi_i^{(1)}\phi_i^{(2)}\right)c_1 + \left(\sum_i \phi_i^{(2)^2}\right)c_2 &= \sum_i \phi_i^{(2)}[f(R)_i^{(a)} - f(R)_i^{(t)}], \end{aligned} \tag{42}$$

where the superscripts and subscripts 1 and 2 refer to the two dyes being calculated.

When devising a combinatorial program testing three-dye combinations to obtain the best match, it is good practice first to test each dye in the group to see if it alone can provide a good match and then to test all possible two-dye combinations to see if one of these can possibly serve to give a good match. The technique would be to solve Eq. (41) for the single dyes and Eq. (42) for the two-dye combinations, to synthesize the reflectance curve for each of these solutions, and then to see if perhaps the color difference between this curve and the curve of the standard would be small enough to accept the formula as a workable match.

Another situation where an exact tristimulus match is impossible occurs in plant shading. As mentioned earlier, it is common manufacturing practice to start with less than the expected formula so as not to overshoot. These corrections are made cautiously, always guarding against adding too much of any one pigment or dye. Despite these precautions it often happens that too much red, say, has been added, and no matter how much other colorants are used the tristimulus values of the standard can no longer possibly be matched. It then becomes important to determine how much colorant to add so as to produce the closest possible visual match, granted that an exact match cannot be produced. This problem was first posed by Park and Stearns (1944) in their pioneering paper on colorant-formulation theory, and finally solved by Nelson and Stearns (1976) 32 years later. The latter paper contains FORTRAN source-program listings for tristimulus value matching as well as for the problem under discussion.

E. Correction Factors for Shading Programs

Shading programs are essentially like the iterative portion of formulation programs. We are fairly close to a match and we wish an exact match. We must therefore calculate the proper Δc vector to overcome the remaining Δt vector. The only difference is that we are not dealing with a computer guess that we wish to refine, but rather with an actual physical batch that we wish to bring on exact shade. Now there is one technique that we can use in shading programs but are unable to use in the iterative part of formulation programs, namely, *correction factors*.

When a batch of colored product is made up, it often happens that one or more of the colorants that go into the batch may not give the color value that you would expect from its calibration values. Possibly a pigment shipment is strong or weak. (Remember that each individual lot of pigment or dye is not calibrated.) Or possibly two specific dyes in a formula interact in such a way as to change the unit k/s ratios of one or both.

It is therefore advisable in the first part of the shading program to disregard the standard temporarily and instead to call for a match for the batch itself. The computer will calculate this match on the basis of the calibration values of the colorants that went into the batch, and the result of this calculation will be a formula that assumes that we have a collection of colorants of exactly standard strength interacting with each other in approved Kubelka–Munk fashion. But in the real world we do not have such colorants, and we will see a difference between the formula given by the computer and the formula that was actually used to make the batch. The ratio : quantity used to make batch/quantity given by computer, for each colorant, can be used as a correction factor to be applied to all further computer pronouncements involving this colorant in this batch.

After this calculation the standard to be matched is brought into play. The shading program determines the Δt vector representing the tristimulus difference between the standard and the batch. It then calculates the inverted matrix shown in Eq. (34) or Eq. (39), and multiplies this matrix into the Δt vector to obtain Δc. Before printing out Δc, which is the required addition to the batch, however, the program multiplies each individual Δc value by the corresponding correction factor in order to get the real effect rather than the theoretical effect from that colorant.

We should point out that often in paint manufacture the bulk of the batch is first made up from dry pigment, but the shading of the batch is then done with preground pigment pastes made from different pigment lots. In that case the determination of correction factors based on the bulk of dry pigment would be inappropriate, and the program would have to be modified accordingly.

V. SURVEY OF ADVANCED METHODS

A. Shall We Go Beyond Kubelka–Munk?

Successful as Kubelka–Munk theory is, workers in the field have long recognized that there are some cases for which the results are not satisfactory. In general, thin films, dark shades, and metallics lead to trouble. The reason is easy to understand if one remembers the basic assumption used in deriving the theory, namely, that both the light entering the film and the light emerging from the film are completely diffuse. In most spectrophotometers the light used for measurement is not diffuse but collimated (it enters the film as a beam of parallel rays) in a direction perpendicular to the surface of the film. For thick films without very much light absorption, considerable scattering occurs before the light penetrates too far into the film, and the entering light quickly becomes diffuse and obeys the precepts of Kubelka–Munk theory. But if the film is thin so that the light does not have a chance to scatter or if considerable absorption occurs before scattering takes place (as would be the case with dark samples), the fact that the entering beam is not diffuse makes a great difference in the results. Kubelka–Munk theory might also be expected to fail in the case of metallic finishes, where the presence of aluminum flakes parallel to the surface minimizes the diffusion of the entering light and favors specular reflectivity at film depths less than the actual film thickness.

Another source of difficulty in using Kubelka–Munk theory is the need for reliance on calibration. It is well known that best results are obtained if the absorption and scattering coefficients of a series of colorants are determined under circumstances as close as possible to those that exist in actual use of those colorants. The reason is that Kubelka–Munk theory is a phenomenological theory in the sense that it depends for its successful operation on some very specific measurements that have only a limited application. For example, the Kubelka–Munk absorption and scattering coefficients of a certain pigment made by manufacturer A might be measured and used for color formulation work. If we wish to use instead another lot of pigment made by manufacturer B with a different particle size, we would have to redetermine the absorption and scattering coefficients. We do not know, from Kubelka–Munk theory alone, how to correct for the change in particle size. In the case of textiles, we do not know how to calculate absorption and scattering coefficients for a dye on a certain substrate if we know the values for the same dye on some other substrate.

In view of these limitations of present-day colorant formulation methods, it is natural to see if some more fundamental theoretical considerations might be applied. Mie theory (1908) appears to be the first logical approach to consider.

B. Mie theory—Single Scattering

Mie theory describes what happens to light impinging on and emerging from a single pigment particle (in paints, inks, and plastics) or a single dyed fiber element (in textiles). For economy of language we will refer to such a scattering element as a particle, no matter what the actual element may be. We are talking about single scattering, which means that once the light is scattered by a particle it does not strike another particle. Expositions of Mie theory are found in books by van de Hulst (1957), Kerker (1969), Born and Wolf (1970), and Stratton (1941).

To use Mie theory we need to know only two ratios. We need first the ratio of the diameter of the scattering particle (which is assumed to be a sphere) to the wavelength of the light that strikes the particle. Next, we must know the ratio of the complex refractive index of the particle to the refractive index of the medium in which the particle is immersed. The complex refractive index consists of a real part, which refers to the velocity of light through the particle, and an imaginary part, which refers to the capacity to absorb light. (In Mie theory the medium is not considered to have any light-absorbing power, so that its refractive index consists only of a real part.)

We get from Mie theory, first, two quantities that represent the power of the single particle to absorb light and the power to scatter light. These are referred to as the absorption and scattering cross sections of the particle. We also get the scattering pattern of the particle, or the relative intensity of the scattered light as a function of the scattering angle. The latter is defined as the angle between the direction of incidence of the impinging light and the direction of emergence of the scattered light.

Scattering patterns calculated by Mie theory are usually very complex and structured. A typical Mie scattering pattern is shown in polar plot form in Fig. 6. Note the many maxima and minima in the curve. The mathematical expression of the scattering pattern of a particle is called the phase function of the particle.

Mie theory alone, without the aid of any other theoretical approach to supplement it, has certain very definite limitations. It applies only to the scattering from a single particle isolated in space and says nothing about what happens when an assembly of such particles must be considered, as is the case in a paint film or a textile fabric. In actual practice, light scattered from one particle impinges on and is rescattered by another particle; this process is repeated, resulting in an endless web of scattered and rescattered light. Any attempt to work out an adequate theory on the basis of successive Mie scattering encounters leads to frustration because of the unbelievable complexity of the mathematics. Obviously a supplementary theoretical approach must be used to provide the necessary information about what happens in multiple scattering. Such approaches are known, and a good start toward using them has been made, as we will see later.

Another deficiency of Mie theory is perhaps more serious. As particles are crowded closer and closer together, the very assumptions on which Mie theory was built cease to hold. The scattered light from one particle begins to interfere with the scattered light from a neighboring particle, and with enough crowding the calculations gradually become unreliable. If Mie theory is to hold strictly, the separation between particles must be greater than about three particle diameters—a situation that generally does not exist in practice. Scattering that is characterized by this kind of interference

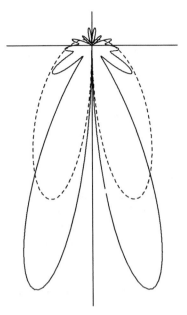

Fig. 6.   Polar plots of the distribution of scattered light from a particle with Mie parameters $\alpha = 6.0, n = 1.84, k = 0$. The full line was calculated by Mie theory; the broken line represents the Henyey–Greenstein phase function having the same average cosine of scattering angle as the Mie phase function. The phase functions have been weighted by the sine of the scattering angle to allow for the solid-angle content in a given element of scattering angle. The direction of the incident light is downward toward the reader.

is called dependent scattering. The existence of dependent scattering has so far proved to be a great obstacle in the way of widespread use of Mie theory for practical problems.

The differences between dependent scattering and multiple scattering deserve more explanation. Both are brought about by interaction between neighboring particles. Both are obstacles to the use of Mie theory alone to get practical results. Multiple scattering refers to scattered light from one particle rescattered by another particle and can be handled by methods now available. But dependent scattering refers to the perturbation and twisting of the very scattering pattern itself because of the presence of neighboring particles. This is a harder problem and to our knowledge has not been satisfactorily solved, although one possible way of attacking it will be mentioned. A well-known example of this effect is the loss in hiding efficiency of titanium dioxide as the concentration in a paint film is increased. This effect has been described by Ross (1971). An interesting review paper on pigment optics by Orchard (1968) has additional references on the subject of dependent scattering.

Other restrictions of Mie theory are somewhat less important. The fact that the theory strictly holds only for spheres has been shown to be not as serious as it may first seem. With the advent of high-speed digital computers, the great complexity of the calculations is no longer a problem.

Summarizing, we see that Mie theory starts with fundamental information about the particle and the surrounding medium and ends with the calculation of the absorption and scattering cross sections (and also the scattering pattern). Kubelka–Munk theory, on the other hand, starts by accepting absorption and scattering coefficients and ends with the calculation of such useful quantities as reflectance, hiding power, contrast ratio, and the like. Since Kubelka–Munk theory seems to begin where Mie theory ends, it would seem logical to take the results calculated from Mie theory, plug them into Kubelka–Munk theory, and in this way create a clear path from the fundamental colorant constants to the ultimate reflectance of a paint film or textile fabric. Unfortunately, however, it is not that simple. The absorption and scattering cross sections calculated by Mie theory apply to isolated particles, not to a paint film or textile fabric where we have multiple scattering effects. We must find some way to allow for multiple scattering before we can use the results of Mie theory as data for Kubelka–Munk calculations. In fact, Kubelka–Munk theory does not even recognize the existence of individual scattering centers. Some kind of common ground must be established.

An alternative approach would be to circumvent Kubelka–Munk theory entirely and go directly from Mie theory to a more rigorous treatment not involving the approximations of Kubelka–Munk. Let us look at such an approach.

C. The Multichannel Technique

In their study of the transmission of light through stellar and planetary atmsopheres the astrophysicists have developed a comprehensive theory of the passage of light through absorbing and scattering media that is known as radiative transfer theory. The classic exposition of this theory is found in a book by Chandrasekhar (1960). Kubelka–Munk theory is itself an extremely simplified version of radiative transfer theory and can be derived from it.

An important aspect of this theory is that we can use the results of Mie-type calculations directly to obtain answers meaningful to everyday use, such as reflectance and hiding power. Furthermore, we are not restricted to the assumptions of Kubelka–Munk theory but can incorporate whatever assumptions we wish (collimated illumination, for example). In their original form the radiative transfer equations are quite complex, and one or another simplifying assumption must be made in order to render the equations manageable. We will describe one method of using radiative transfer theory that is particularly suited to our problem, namely, the multichannel technique.

Although some previous attempts were made to overcome the limitations of complete diffusion of light imposed by the Kubelka–Munk theory (Ryde, 1931; Duntley, 1942), it was not until fairly recently in the work of Völz (1962, 1964) and of Beasley *et al.* (1967) that the number of channels was increased from two to four. In a recent series of papers, Richards (1970) and Mudgett and Richards (1971, 1972) have shown how this concept can be extended to any number of channels. These authors' work has paralleled the work of some other authors in the field of heat transfer. Their papers contain references to this work.

In the general approach the light passing through the film is divided up into as many channels as desired (26 channels were used by Mudgett and Richards for one series of calculations). Each channel covers a different range of angles from the perpendicular to the horizontal; among them the channels cover all of space. Light in half of the channels is traveling downward and is traveling upward in the other half.

The procedure is to start with light of any geometry (a collimated beam, for example). By means of available theory the light that passes into any channel can be calculated. Scattering coefficients are then calculated with the aid of Mie theory, showing how much light is scattered from any one channel into any other channel. A series of absorption coefficients is also calculated by Mie theory, showing how much light is absorbed in each channel. Appropriate boundary conditions relating to the reflectance of the background and the interior reflectance of the film are then applied. The uncertainties of the Saunderson correction are eliminated by this approach, because each channel has its own interior reflectance value that can be calculated from the Fresnel equations. Finally, by the solution of a series of equations, one can determine how much light emerges from any channel and the overall reflectivity of the film.

We have just explained that we need Mie theory to calculate the various inter- and intrachannel absorption and scattering coefficients required by the calculations. More specifically, we need the results of Mie theory calculations; that is, we need the single-particle absorption cross section, single-particle scattering cross section, and scattering pattern for each pigment used. The scattering pattern, or phase function, can be expressed as an expansion in Legendre polynomials, and we need as many coefficients as are necessary for convergence of the series. For certain pigments one may need some 30 or even 40 of these coefficients before convergence is obtained. These coefficients are, however, calculable from Mie theory by the use of special methods (Chu and Churchill, 1955; Clark *et al.*, 1957; Allen, 1974b).

Now where does all this leave us? It leaves us with the ability to start with fundamental pigment properties, particle size and complex refractive index, and to use this information to calculate, from Mie theory, single-particle absorption and scattering cross sections and a phase function (in the form of a series of numerical coefficients). We can then plug these numbers into multichannel theory and calculate reflectance values. We can thus, by a combination of Mie theory and multichannel theory, calculate, for example, the effect of particle size on color, the effect of refractive index on color, and so on. We have solved the problem of using Mie theory even though we have multiple scattering.

Note, however, that we have not solved the problem of dependent scattering. Our results are only valid if we can assume that our scattering particles are separated from each other by more than three particle diameters. This restriction is almost never found in practice; in paints, for example, the pigment particles are crowded to the point where they are almost in contact. One may well ask, therefore, if we have done anything really worthwhile by bringing in the Mie and multichannel theories. The answer is yes, as we will now see.

D.   A Possible Improved Approach for the Future

Suppose that instead of calculating the absorption and scattering cross sections and phase-function coefficients by Mie theory, which does not hold because of dependent scattering, we work backward from the reflectance values, just as we do with Kubelka–Munk. We will standardize a pigment or dye by starting with its reflectance curves under various conditions or in various mixtures and then finding values of the various coefficients that, when inserted into the multichannel calculations, will give us back these reflectance curves. In this way we will get values that work. Also, in view of the much more powerful multichannel apparatus we should get much better results than with Kubelka–Munk over a broader range of conditions.

The first reaction to this proposal may be one of dismay. If we are dealing with a pigment that requires, say, 30 Legendre coefficients to express fully the phase function at a certain wavelength, we would have to determine 32 numbers to characterize the pigment at that wavelength: the 30 phase-function coefficients plus the absorption and scattering cross sections. That means that we would have to have 32 drawdowns to work from in order to solve that many simultaneous equations. This is a bit more complicated than just the two coefficients, K and S, used with Kubelka–Munk!

The answer to this objection is that we do not need a phase function as elaborate as that shown in Fig. 6 for Mie theory. A much simpler phase function is available. Henyey and Greenstein (1941) have proposed a phase function with only one adjustable parameter, g, the average cosine of the scattering angle. If $g = 0$, the scattering pattern is isotropic, and there is as much backward as there is forward scattering. If $g = 0.5$, the scattering has a strong forward thrust, as is the case with a typical titanium dioxide particle. If $g = 0.9$, the scattering is almost entirely forward. The highest possible value for g is 1.0. Figure 7 shows some Henyey–Greenstein scattering patterns with various values of g. We see that these patterns do not have the complicated structure of the Mie patterns, but these latter complicated structures undoubtedly get washed out anyway in the process of multiple scattering. Figure 6 shows, together with the Mie scattering pattern, the Henyey–Greenstein pattern with the same average cosine of the scattering angle as that of the Mie pattern.

Now the Henyey–Greenstein phase function was proposed for astrophysical work, but there is no reason not to use it for colorants (Allen,

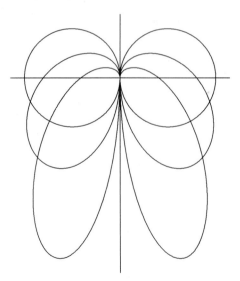

Fig. 7.   Three Henyey–Greenstein phase function plots of the same type as those shown in Fig. 6. The two circles at the top on either side of the vertical axis represent the function with $g = 0$. The two ovoid plots in the middle are for $g = 0.3$, and the ones at the bottom are for $g = 0.6$.

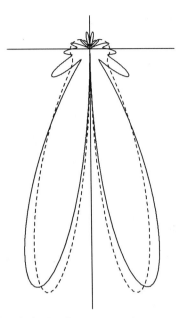

Fig. 8. Same as Fig. 6 but the broken line represents the new two-parameter phase function, with both the average cosine and the average cosine squared of the scattering angle equal to the corresponding quantities for the Mie phase function.

1975). If we use the Henyey–Greenstein scattering pattern, we need only three parameters to characterize a pigment: the absorption cross section, the scattering cross section, and g. This is only one more parameter than Kubelka–Munk uses. Also, we do not have to worry about Saunderson's K_1 and K_2, as the refractive index discontinuity takes care of itself in the multichannel calculations.

The writer has done some work with a new phase function (Allen, 1975) that comes a little closer to the overall shape of the Mie phase function than does Henyey–Greenstein. It has two adjustable parameters instead of one. Figure 8 shows the approximation to the Mie phase function achieved by the two-parameter function. This function may give better results.

The multichannel technique requires extensive calculation time and would probably cost too much to use for routine colorant formulation and shading calculations today. But it is inevitable that computation costs will be reduced, and this technique may well be the standby of the future. Meanwhile, advanced tools of this type can be used to obtain a better insight into the cases where Kubelka–Munk fails and may also show how to correct these failures and obtain greatly improved results.

VI.   SPECIAL PROBLEMS

A.   Fluorescent Samples

When an investigator measures a nonfluorescent sample in a spectrophotometer, he obtains a set of values that enables him to find out everything he needs to know about the color of the sample. He can calculate tristimulus values under any light source he chooses. On the other hand, when he measures a fluorescent sample in a spectrophotometer, he obtains a set of values (the spectral radiance factor curve) that provides only a limited knowledge of the color. If he knows the exact spectral power distribution of the light source that was used for the measurement, he can calculate the tristimulus values specifically under that light source. He cannot, however, calculate the tristimulus values under any other light source without additional data. This means, for example, that since illuminant D_{65} does not exist in the physical world, tristimulus values under D_{65} cannot be obtained from the spectral radiance factor curve alone.

Furthermore, from only the spectral radiance factor curve no Kubelka–Munk-type calculations are possible. These difficulties have stood in the way of the development of a reliable method for the colorant formulation of fluorescent samples. Recent advances in the measurement of fluorescent materials, however, lead to a suggestion for a simple method for programming the formulation and shading of materials containing fluorescent colorants.

The spectral radiance factor curves of all fluorescent materials are composed of two components: the true reflectance portion and the fluorescent portion (Judd and Wyszecki, 1975). The true reflectance portion follows Kubelka–Munk rules and can be handled by the equations given in the preceding sections of this chapter. The fluorescent portion is more complicated, being, in effect, proportional in strength to the amount of light absorbed by the colorant that does the fluorescing (Allen, 1957, 1964, 1972; Eitle and Ganz, 1968). Reliable methods are now available whereby the true reflectance curve can be separated out from the total spectral radiance factor curve (Eitle and Ganz, 1968; Allen, 1973).

It would therefore appear that the colorant formulation calculations could be confined to the true reflectance curve, which in common with all reflectance curves does not depend on the light source used and which obeys the Kubelka–Munk law. Colorant calibrations would be made as usual by multiple dyeings at increasing concentrations or multiple reductions of pigment with white to form a tint ladder. The true reflectance curves of all the calibration samples would be isolated from the composite spectral radiance factor curves, and the Kubelka–Munk constants would then be determined by the procedures before outlined. A sample to be matched would be treated in the same way, with only the true reflectance portion used for getting the match.

It is quite probable that such a method would be best suited to matches made with the identical colorants used in the sample being matched, and on the same substrate. Quenching effects on fluorescent colorants, in which the fluorescence of the colorant is diminished because of interaction with other substances present, are always possible and rather unpredictable. It therefore may happen that even though the nonfluorescent portions of the spectral radiance factor curves may match, the fluorescent portions may differ to such a degree that the match would be destroyed. This, of course, would not happen if the fluorescent colorant were placed in the same environment in the match as in the sample, which would be the case if the same colorants and substrate were used. The correct colorant combination would be easy to find in a combinatorial program, because it would appear as a match of close to zero metamerism (within the limits of accuracy of the measurements). Batch correction programs based on these principles would automatically work well, since metamerism is absent.

In the event that the fluorescent sample to be matched is based on a substrate different from the one on which the match is to be made, this procedure may still be tried. Trial colorations will quickly determine whether the fluorescence intensity is the same for match and standard.

B. Metallic Paints and Fabrics

Modern automotive finishes and upholstery are likely to contain aluminum flakes as well as pigments or dyes. The presence of such flakes imparts certain color effects not possible with pigments alone. One of the most pronounced of these effects is a change of color when the observer changes from a straight-on to a grazing viewing angle.

The Kubelka–Munk theory fails to a greater or lesser degree with metallic samples. A paper by Davidson (1965) describes some of the anomalies obtained and suggests approximate methods of matching. A patent by Armstrong *et al.* (1972) makes use of equipment that can measure the sample at two different angles and proposes a method that matches both resulting sets of tristimulus values.

Recent work by Lau and Allen (1979) has investigated three possible methods for matching metallic samples. The first makes use of a variable Saunderson K_2 value, with K_2 ranging from 0.04 (equal to K_1) for a film with only aluminum flake and no pigment to 0.6 for a film with only pigment and no aluminum flake. The K_2 value also depends on an additional parameter, gamma, determined by a best-fit procedure for each individual system. The second method uses the multichannel approach, with the Henyey–Greenstein g assuming negative values (strong backward scattering) for aluminum. The third method, being investigated primarily for batch shading, makes use of a high front surface reflectance value (K_1) for the colorant film.

REFERENCES

Allen, E. (1957). *J. Opt. Soc. Am.* **47**, 933–943.
Allen, E. (1964). *J. Opt. Soc. Am.* **54**, 506–515.
Allen, E. (1966). *J. Opt. Soc. Am.* **56**, 1256–1259.
Allen, E. (1972). *Color Appearance* **1**(5), 28–32.
Allen, E. (1973). *Appl. Opt.* **12**, 289–293.
Allen, E. (1974a). *J. Opt. Soc. Am.* **64**, 991–993.
Allen, E. (1974b). *Appl. Opt.* **13**, 2752–2753.
Allen, E. (1975). *J. Opt. Soc. Am.* **65**, 839–841.
Andrade, D. (1976). Personal communication.
Armstrong, W. S., Jr., Edwards, W. H., Laird, J. P., and Vining, R. H. (1972). U.S. Patent No. 3,690,771.
Beasley, J. K., Atkins, J. T., and Billmeyer, F. W., Jr. (1967). *In* "Electromagnetic Scattering" (R. L. Rowell and R. S. Stein, eds.), pp. 765–785. Gordon & Breach, New York.
Born, M., and Wolf, E. (1970). "Principles of Optics," 4th Ed. Pergamon, Oxford.
Cairns, E. L. (1975). Personal communication.
Cairns, E. L., Holtzen, D. A., and Spooner, D. L. (1976). *Color Res. Appl.* **1**, 174–180.
Chandrasekhar, S. (1960). "Radiative Transfer." Dover, New York.
Chu, C. M., and Churchill, S. W. (1955). *J. Opt. Soc. Am.* **45**, 958–962.
CIE (1972). "Special Metamerism Index: Change in Illuminant," Suppl. No. 1 to "Colorimetry," CIE Publ. No. 15 (E-1.3.1) (1971).
Clark, G. C., Chu, C. M., and Churchill, S. W. (1957). *J. Opt. Soc. Am.* **47**, 81–84.
Davidson, H. R. (1965). *Color Eng.* **3**(1), 22–32.
Duncan, D. R. (1949). *J. Oil Colour Chem. Assoc.* **32**, 296–321.
Duntley, S. Q. (1942). *J. Opt. Soc. Am.* **32**, 61–70.
Eitle, D., and Ganz, E. (1968). *Textilveredlung* **3**, 389–392.
Gall, L. (1973). *Colour 73, 2nd Congr. Int. Colour Assoc., York*, pp. 153–178. Hilger, London.
Gugerli, U. (1963). *Text.-Rundsch.* **18**, 252–267.
Henyey, L. G., and Greenstein, J. L. (1941). *Astrophys. J.* **93**, 70–83.
Judd, D. B., and Wyszecki, G. (1975). "Color in Business, Science and Industry," 3rd Ed. Wiley, New York.
Kerker, M. (1969). "The Scattering of Light and Other Electromagnetic Radiation." Academic Press, New York.
Kubelka, P., and Munk, F. (1931). *Z. Tech. Phys.* **12**, 593–601.
Lau, K. C., and Allen, E. (1979). Paper in preparation for Color Res. & Appl.
McGinnis, P. H., Jr. (1967). *Color Eng.* **5**(6), 22–27.
Marquardt, D. W. (1963). *J. Soc. Ind. Appl. Math.* **11**, 431–441.
Mie, G. (1908). *Ann. Phys. (Leipzig)* **25**, 377–445.
Mudgett, P. S., and Richards, L. W. (1971). *Appl. Opt.* **10**, 1485–1502.
Mudgett, P. S., and Richards, L. W. (1972). *J. Colloid Interface Sci.* **39**, 551–567.
Mudgett, P. S., and Richards, L. W. (1973). *J. Paint Technol.* **45**, 43–53.

Nelson, R., and Stearns, E. I. (1976). *Ind. Manage. Text. Sci.* **15**, 55–82.

Nimeroff, I., and Yurow, J. A. (1965). *J. Opt. Soc. Am.* **55**, 185–190.

Orchard, S. E. (1968). *J. Oil Colour Chemists' Assoc.* **51**, 49–60.

Orchard, S. E. (1977). *Color Research and Application.* **2**, 26–31.

Park, R. H., and Stearns, E. I. (1944). *J. Opt. Soc. Am.* **34**, 112–113.

Richards, L. W. (1970). *J. Paint Technol.* **42**, 276–286.

Ross, W. D. (1971). *J. Paint Technol.* **43**, 49–66.

Ryde, J. W. (1931). *Proc. Roy. Soc.* **A131**, 451–475.

Saunderson, J. L. (1942). *J. Opt. Soc. Am.* **32**, 727–736.

Stratton, J. (1941). "Electromagnetic Theory." McGraw-Hill, New York.

van de Hulst, H. C. (1957). "Light Scattering by Small Particles." Wiley, New York.

Völz, H. G. (1962). *FATIPEC Congr. (Fed. Assoc. Tech. Ind. Peint., Vernis, Emaux Encres Impr. Eur. Cont.)* **VI**, 98–103.

Völz, H. G. (1964). *FATIPEC Congr. (Fed. Assoc. Tech. Ind. Peint., Vernis, Emaux Encres Impr. Eur. Cont.)* **VII**, 194–201.

8

Modern Color-Measuring Instruments

M. PEARSON

Graphic Arts Research Center
Rochester Institute of Technology
Rochester, New York

I. INTRODUCTION

The human eye is, of course, the oldest means of color measurement and specification, but because of its design characteristics of adaptation, individuality of spectral response, and lack of color memory, it is a poor instrument for analytical specification of color.

The eye is fundamentally a radiation-sensitive system consisting of rods and cones that are selectively responsive to radiation between wavelength limits normally of about 380 and 760 nm. Radiation in this range is called light, and color is defined here as that characteristic of light by which an observer may distinguish differences between two structure-free fields of view of the same size and shape, such as may be caused by differences in the

337

spectral composition of the radiant energy. This is termed *psychophysical color* and is specified by the tristimulus values of the radiant energy entering the eye.

An instrument that is designed to measure color must, therefore, evaluate the sample by illuminating the sample with the light of the same spectral characteristics and have sensors with the same spectral-response characteristics of the "color-normal" human observer.

Color-measuring instruments are grouped into two classes, spectrophotometers and colorimeters. The spectrophotometer is the most fundamental instrument for color measurement. Spectrophotometers, however, do not directly measure the color itself; instead, they measure the physical attributes of a material concerning its reflection or transmission of light. This information through tristimulus integration results in the numerical expression of color as

$$X = k \int S(\lambda)\rho(\lambda)\bar{x}(\lambda)\, d\lambda,$$
$$Y = k \int S(\lambda)\rho(\lambda)\bar{y}(\lambda)\, d\lambda, \qquad (1)$$
$$Z = k \int S(\lambda)\rho(\lambda)\bar{z}(\lambda)\, d\lambda,$$

where S is the source, ρ is the modulator, and $\bar{x}$, $\bar{y}$, $\bar{z}$ are the detectors.

The spectral power of the source $S(\lambda)$, the reflectance of the object $\rho(\lambda)$, and the response of the observer $\bar{x}(\lambda)$, $\bar{y}(\lambda)$, $\bar{z}(\lambda)$ are all needed to produce the resulting tristimulus values. The colorimeter responds more nearly like the human visual system. It achieves the same numerical expression for color as that obtained by tristimulus integration of spectrophotometric data, but it achieves this by performing an analog integration optically within the instrument. This is accomplished by having a source that conforms to the spectral-power distribution of the source being used to view the sample and a response that simulates that of the standard observer. The sample modulates the radiation between the source and the detectors and the resulting response is an expression of the color characteristics of the sample.

A third type of instrument, the densitometer, is not a standard color-measuring instrument in that its response does not relate to that of the standard observer. It does, however, have red, green, and blue response, and since the human visual system has red, green, and blue response sensors there is a relationship between the response of a color densitometer and that of a human observer. The densitometer can under some circumstances then give an approximation to color measurement. It is, however, capable of detecting quite accurately changes in color or color differences regardless of how they may be perceived by a human observer; as such, it is an extremely useful instrument for the quality control of color and color processes.

II. SPECTROPHOTOMETERS

The spectrophotometer is the most fundamental instrument in color measurement. As indicated earlier, however, it does not measure color directly; it measures the light reflected or transmitted by a material. This information after tristimulus integration is transformed into numerical expressions of color. The word spectrophotometer means a measurement of the spectrum by the eye (i.e., spectro-photo-meter). A modern spectrophotometer consists of a source of radiant energy, a dispersing system to provide monochromatic radiation, and a detector system to measure the amount of radiation through the instrument. Generally today the path of radiation is split into two parts within the instrument to provide a sample beam and a reference beam. When a sample is placed in the sample beam, the equality of the two beams is broken and the detector senses the difference and relates that to the transmittance or reflectance of the sample at that wavelength. Because the illumination in the instrument is monochromatic, the spectral-power distribution of the source is not a factor as long as the intensity is sufficient to allow a satisfactory signal-to-noise ratio in the response. Any attenuation of radiation other than that of the sample, such as lenses, mirrors, prisms, gratings, and so on; are canceled out because of the relativity of the measurement.

A. Mode of Operation

All spectrophotometric measurements are made relative to some standard; that is, they are a ratio of the light transmitted or reflected by the sample to that from some reference or standard. With transmission measurements, the references are usually well defined and obtainable. For many optical materials, such as lenses, filters, and so on, air is an appropriate reference. That is, simply the optical path of the instrument minus the sample. In the measurement of transmitting solutions, the reference is usually the solvent or vehicle in which the colorant is dissolved or dispersed. Under these circumstances the reference is usually readily available and unique to the sample conditions at hand.

For a reflection measurement the ideal reference would be a perfectly reflecting, spectrally nonselective material. This would be a rather unique material; the difficulty is in finding such a material in real life. Such materials as calcium carbonate, magnesium oxide, barium sulfate, and most recently, HALON have been used. However, they are plagued by the factors that affect all physical standards, since, in addition to being as close as possible to 100% reflecting at all wavelengths, they should also possess physical durability and permanence, exist in a form that is usable, be readily available, and be economically affordable. Barium sulfate is presently the best compromise of these conditions (see Grum and Luckey, 1968; Budde, 1970, 1976; Erb, 1979; Terstiege, 1974).

Because there is no physical material that perfectly fulfills the desired conditions of a reflection standard,* the concept of absolute reflectance has been developed. Using this concept, reflection measurements are mathematically corrected and reported as if they were measured against a perfect nonselective reflector. This is accomplished by multiplying the measured value of the sample by the absolute value assigned to the reference material used in the calibration of the instrument making the measurement. The difficulty here lies in assigning the absolute reflectance values to the physical standards that are used in the operation of spectrophotometers. Listed values for most of the commonly used materials are available in the literature (Ishizaki, 1956; Erb, 1979; Grum and Wightman, 1977). Although there is not complete international agreement as to what these values should be, work is progressing toward this end (CIE, 1977).

In those cases where fluorescence is present in the sample, the usual procedure of illuminating the sample by monochromatic light must be replaced by heterochromatic illumination from a source of known spectral-power distribution, since under these conditions the spectral-power distribution of the source is a direct factor in the resultant measurements. This is necessary for the fluorescence to be evaluated in the same way as it would be during normal viewing of the sample. The resultant spectrophotometric measurements must, therefore, be defined in terms of the spectral distribution of the source by which they were made, usually the correlated color temperature, for the case of tungsten sources.

B. Geometry

Another set of circumstances that makes reflection measurement more complex than transmission is the influx and efflux geometry of the sample

---

* Reflection (transmission) is the process by which electromagnetic flux incident on a stationary surface or medium leaves that surface or medium from the incident (opposite) side without change in frequency; reflectance (transmittance) is the fraction of the incident flux that is reflected (transmitted).

"Reflectance factor" is defined as the ratio of the radiant flux actually reflected by a sample surface to that which would be reflected into the same reflected-beam geometry by an ideal (glossless) perfectly diffuse (lambertian) standard surface irradiated in exactly the same way as the sample.

"Total radiance factor" is defined as the sum of the reflected and emitted flux from a fluorescent sample, relative to the reflected flux from a nonfluorescent reference, when both are irradiated by the same polychromatic flux.

illumination in the measurement. In many transmission measurements both the illumination of the sample and the collection of the transmitted energy are quite specular. In this case the sample must exhibit very little scattering, since the high specularity of the collection beam would cause the response to drop below a usable level very quickly. Materials possessing a high degree of scatter fall into the category of turbid media, and their color is governed by the characteristics of scattering as well as absorption.

In reflection measurements the instruments illuminating the sample with monochromatic light are usually quite specular. In the case of heterochromatic illumination, however, the light is normally diffuse, because such measurements are normally made by illuminating an integrating sphere with the sample as a part of the sphere wall.

The collection geometry for reflection is either diffuse or 45°/0° (or its optical reverse 0°/45°). These geometries are shown in Figs. 1a and b, respectively. In diffuse collection the sample is positioned in such a way with respect to the incident beam as to exclude the first surface reflection from the sample. This is illustrated in Fig. 2. Figure 2a represents a sphere arrangement where the incident beam falls normally on the sample. In Fig. 2b the sample is irradiated at some angle i and the surface reflectance occurs at an angle r equal to i and is trapped, thus excluding the surface reflection. In Fig. 2c the specularly reflected component is not trapped but is included in the measurement. The major advantage of integrating-sphere collection is that measurements made in this way are nearly independent of the surface

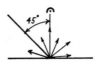

Fig. 1.   Reflection collection geometry: (a) integrating sphere where all reflection from the surface of the sample is collected; (b) 45°/0° where the sample is illuminated at an angle of 45° to the normal and collection is at the normal. The reverse geometry (0°/45°) in most cases is optically equivalent.

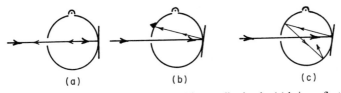

(a)   (b)   (c)

Fig. 2.   The first surface reflection is excluded from collection by (a) being reflected out of the sphere through the entrance port, (b) being reflected into a light trap or included in the measurement, (c) being reflected from the sphere wall.

texture of the sample. This is particularly useful for many textile samples and papers that have a noticeable difference between wire and felt sides.

The use of 45°/0° geometry for reflectance measurements corresponds more closely with the visual viewing of samples and excludes the specular component of reflectance more efficiently than does that of integrating sphere. It is therefore most often used for measurements of colored images and evaluation of color reproductions. Although 45°/0° influx–efflux geometry is closer to visual viewing conditions than an integrating sphere, the true approximation is probably some weighted sum of the two conditions. This is because, although most viewing conditions consist of a directional source, the ambient light level represents diffuse illumination of the subject and in many cases this can be a significant contribution to the total illumination.

For more details of reflectance measurements and integrating-sphere theory, the reader is referred to the work of Wendlandt and Hecht (1966) and to the CIE (1979) Report on Absolute Reflectance Measurements.

For color specification of opaque materials the CIE (1971) has recommended that the total radiance factor β be measured by one of the three following geometries:

(a) 45°/normal (abbreviation, 45/0) The specimen is illuminated by one or more beams whose axes are at an angle of $45° \pm 5°$ from the normal to the specimen surface. The angle between the direction of viewing and the normal to the specimen should not exceed 10°. The angle between the axis and any ray of an illuminating beam should not exceed 5°. The same restriction should be observed in the viewing beam.

(b) Normal/45° (abbreviation, 0/45) The specimen is illuminated by a beam whose axis is at an angle not exceeding 10° from the normal to the specimen. The specimen is viewed at an angle of $45° \pm 5°$ from the normal. The angle between the axis and any ray of the illuminating beam should not exceed 5°. The same restriction should be observed in the viewing beam.

(c) Diffuse/normal (abbreviation, d/0) The specimen is illuminated diffusely by an integrating sphere. The angle between the normal to the specimen and the axis of the viewing beam should not exceed 10°. The integrating sphere may be of any diameter provided the total area of the ports does not exceed 10% of the internal reflecting sphere area. The angle between the axis and any ray of the viewing beam should not exceed 5°.

The radiance factor $\beta_d/0$ is, according to the Helmholtz reciprocity law, identical to the reflectance ρ for directional normal incidence.

C. Spectrophotometric Components

The basic components of a spectrophotometer are shown in Fig. 3.

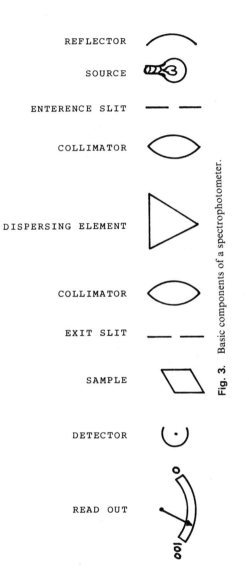

REFLECTOR

SOURCE

ENTERENCE SLIT

COLLIMATOR

DISPERSING ELEMENT

COLLIMATOR

EXIT SLIT

SAMPLE

DETECTOR

READ OUT

Fig. 3.   Basic components of a spectrophotometer.

1. SOURCE

The source must emit continuous radiation over the wavelength range of the instrument and contain sufficient energy at each wavelength to produce a reproducible and accurate response. The spectral-power distribution of the source is not a critical factor in spectrophotometric measurement except for the case of heterochromatic illumination of fluorescent samples. In this case the characteristics of the source must be specified. With the increased use of fluorescent materials the need for instruments with heterochromatic illumination is becoming more and more prevalent. Since correct measurements can be made of nonfluorescent as well as fluorescent materials with heterochromatic illumination, it is reasonable to expect that in the future all instruments will be capable of heterochromatic illumination.

The illumination due to the source may be increased by the use of lenses and reflectors that image the source on the entrance slit of the instrument and illuminate it as uniformly as possible.

2. ENTRANCE SLIT

The entrance slit is imaged by the monochromator collimator system onto the exit slit and serves as the source for the monochromator. It has the geometry of a slit (which is a long, narrow rectangle), so that its width is as narrow as possible, which gives greater wavelength resolution of the dispersed energy, and its length allows sufficient energy into the system to produce a usable signal. Since wavelength is only a one-dimensional quantity, this is a practical solution for increasing the wavelength resolution while at the same time obtaining enough energy of that wavelength to make the system operative.

3. COLLIMATOR

The collimator renders the rays emanating from the entrance slit parallel so that rays from all positions of the slit strike the prism or grating at the same angle. The second collimator forms an image of the entrance slit, after it has been dispersed as a function of wavelength, on the exit slit.

4. DISPERSING ELEMENT

There are two types of dispersing elements, prisms and gratings. This element disperses the energy of the source as a function of wavelength so that wavelengths of light may be spatially separated. Rotation of the dispersing element or other optical components controls the wavelength falling on the exit slit, or in some cases the exit slit is physically moved across the generated spectrum. In the case of the prism all the dispersed energy is placed into a single spectrum, but the dispersion is not linear with wavelength. This results in a varying bandpass instrument if a constant slit width is used.

Gratings produce several images or harmonic spectra known as orders. Modern techniques in the production of gratings, however, permit a large percentage of the total energy to be placed into a single order, making them quite efficient. In this case, the dispersion is constant as a function of wavelength and therefore with a fixed exit-slit width, the bandpass is constant over the wavelength range of the instrument.

There are two other techniques for selectively producing narrow wavelength regions of the spectrum, neither of which involves dispersion. One is the use of narrow-band or interference filters. This restricts the radiant energy transmitted to the relatively narrow spectral transmission band of the filter. Readings are then at discrete wavelength bands determined by the transmission characteristics of the filters. There then needs to be a specific filter for each wavelength reading to be made. Readings at intermediate wavelengths are not possible. Instruments designed in this manner are called abridged spectrophotometers. The second method is the use of an interference wedge. This is an interference filter made in such a manner that its spectral transmission varies as a continuous function of the physical location of the incident energy across the surface of the wedge. Since the change in spectral transmission is continuous, any wavelength can be selected within the operating spectral range and the bandpass is determined by the effective slit width as with the grating and prism instruments.

5.   EXIT SLIT

The exit slit determines the bandpass in terms of wavelength of the energy reaching the detector from the sample. In most instruments with the provision for heterochromatic illumination and monochromatic collection, the roles of the exit slit and entrance slit are reversed.

6.   DETECTOR

The spectral sensitivity of the detector is not generally a factor in the response of a ratio recording instrument as long as it has sufficient sensitivity over the instrument wavelength range to produce a usable signal. In a single-beam instrument, however, the detector must remain stable between the time of the reading of the standard and that of the sample. With a double-beam instrument the time between readings of reference and sample beams is reduced to fractions of a second and the detector must therefore have a frequency response compatible with that used in alternating between the two optical paths.

Another form of detection that has become available with the modern solid-state detectors is the placing of an array of sufficiently small detectors in the projected spectrum so that each detector responds to a specific wavelength (or narrow band of wavelengths, depending on the size of the detector and dispersion of the spectrum). In this case a given spectral range is projected onto a row of detectors and the spectral information is obtained simultaneously as opposed to sequentially, as when the dispersing element is rotated, presenting a sequential display of wavelengths to the detector.

7. READOUT

In a manual instrument the readout consists of adjusting the response to a value of 100% for the reference or standard. With the sample in place, the response of the meter represents the percentage of the sample's reflectance (or transmittance) to that of the reference. This process is then repeated for all wavelengths in question. With newer instruments this is most often done in a digital display mode. When the readout is displayed graphically with the percentage of reflectance (or transmittance) plotted as a function of wavelength, the spectrophotometric curve of the sample is obtained.

It is also possible to plot the spectrophotometric curve directly using a recorder capable of plotting data in both the x and y dimensions, either attached to the instrument or as an integral part of it. The wavelength information is plotted on the abscissa and the reflectance value (or transmittance) on the ordinate. Many modern instruments have microprocessors as an inherent part of their design so that the spectrophotometric data are maintained in memory and can be read or printed out digitally and can be available to drive a plotter for graphic display. In addition, the data are available for any subsequent calculation that may be desired from the information such as colorimetric values, color differences, colorant formulation calculations, and so on.

D. Methods of Measurement

For most instruments designed to make transmission measurements the sample is illuminated at near normal to its surface or the surface of a cell holding the sample. The collection is along the optical axis and if the detector is some distance behind the sample, the measurement is quite specular.

With instruments designed for reflection measurements or using reflection attachments, two types of influx–efflux geometry are used in evaluating the sample, 45°/0° (or its optical reverse, 0°/45°) and integrating sphere. Using 45°/0° the illumination is at 45° to the normal. This can be either a single specular beam for the illumination and collection or an annular ring of illumination at an angle of 45° to the normal. This geometry is designed to prevent the Fresnel specular reflection off the surface of the sample from affecting the readings. It is also a reasonable approximation to the geometry of most visual viewing conditions.

E.   The Integrating Sphere and Its Functions

Integrating-sphere geometry is used to measure the diffuse or total reflection characteristics of a sample. This minimizes the effects that the surface texture of the sample may have on the type of geometry used in the measurements. With the integrating-sphere geometry the front surface reflection from the sample may be included (total collection) or excluded (diffuse collection). The glossier or smoother the sample surface is, the more the first surface reflection will be a factor. A diffuse or scattering surface will reduce the specular component of the measurement. If the sample is a perfect diffuser there is no difference between a diffuse or total measurement, since there is no specular component. This geometry is particularly useful for the measurement of textiles and papers as well as other materials that exhibit marked directional reflection differences because of the characteristics of their surface texture.

Transmitting samples can also be measured with integrating-sphere geometry by placing the sample at the entrance port to the sphere. This permits either the total or diffuse transmission of the sample to be measured as shown in Fig. 4.

If the transmitting sample is placed at the entrance port and the exit port remains open, the specular component of transmission will be lost through the exit port and the sphere will collect only the diffuse transmission (scattering characteristic of the transmitting material, Fig. 4a). By blocking the exit port with a suitable reflection standard both the specular and diffuse transmittance are collected and measured. This is total transmittance (Fig. 4b). The specular transmittance can be determined by subtracting the diffuse from the total transmittance.

If the transmitting material is scattering, its back reflectance (i.e., the reflected rays from the same side as the incident beam) can be measured by placing it in the exit port as for a normal reflection measurement and either backing it with a black material or no backing at all so that any transmitted light is not collected by the sphere and only the surface reflection and back

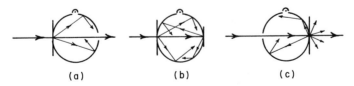

(a)   (b)   (c)

Fig. 4.   Measurement of the components of a transmitting sample by integrating sphere: (a) diffuse transmission, where the specular transmission passes out of the sphere; (b) total transmission, where all transmitted light is collected; (c) backscatter, where only the reflection from the incident side of the sample is collected by the sphere.

scattering are measured. If the collection geometry is set for diffuse measurement the surface reflection will also be eliminated and only the back scattering will be measured (Fig. 4c). For further measurement methods the reader is referred to the CIE (1977) Technical Report and to the work of Webber (1957).

By the proper use of an integrating-sphere measurement and an understanding of the concepts involved, all of the light-attenuating properties of a sample may be evaluated, that is, reflection, transmission, scattering, and absorption. Absorption is readily obtained by subtracting the sum of the total reflection and the total transmission from unity.

The influx–efflux geometries can be achieved in either single-beam or double-beam instruments. In the single-beam instrument there is only one optical path: from the exit slit through the sample compartment to the detector. This means that the sample and reference materials must be alternately placed in the optical path so that the ratio of sample to reference readings can be taken at each wavelength.

As an alternative to this the instrument may store the values associated with the reference and use this information along with that obtained by reading the sample to determine the ratio at each wavelength. In this approach the characteristics of the reference are determined over the total wavelength range to be measured. The reference is then replaced by the sample and as its characteristics are determined, the desired ratios are computed. In the past the information obtained from the reference was retained by some electrical or electronic portion of the instrument; however, with the advent of dedicated computers and microprocessors it is possible to digitize the information and store it in the memory of these devices, where it is then available for calculation with the sample information whenever necessary. One of the assumptions in this technique is that the instrument maintains the same operating characteristics over the time involved in scanning the reference and the sample. Therefore, it must be extremely stable over some period of time that is much longer than the scan times. With modern-day solid-state electronics this is a realistic requirement for most instruments.

One technique for circumventing the problem of time-dependent stability is to make nearly instantaneous comparisons between sample and reference readings. This is accomplished by a double-beam instrument where two optical paths are used, one for reference and one for sample, and the detector responds alternately to these two beams. This condition is shown schematically in Fig. 5. This eliminates physically moving the reference and sample and alternates the sampling of the sample and reference beams instead. Although this eliminates the problem of time stability, it requires that both optical paths be adjusted to have identical output with no sample in the beam.

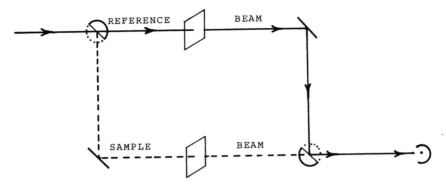

Fig. 5. Double-beam system, where the optical path is caused to alternate between sample and reference by rotating or oscillating mirrors.

Special consideration needs to be given to spectrophotometric analysis of samples possessing optically excited fluorescence properties. There are two processes involved in the mechanism of fluorescence, that of excitation and emission. The excitation spectra is the wavelength region where the material absorbs the energy that excites the electrons to a higher electronic state. As the electrons return to their ground state they emit light of longer wavelength, which is the emission spectra. Since the process is not 100% energy efficient, the emission is of lesser energy than that absorbed and as a result is generally of longer wavelength. This phenomenon is called "Stokes' law." The region where the long-wavelength end of the excitation spectra exceeds the short-wavelength end of the emission spectra is referred to as the "anti-Stokes" region. This relationship is illustrated in Fig. 6.

In the measurement of fluorescent materials with spectrophotometers employing monochromatic illumination, the detector responds to all energy

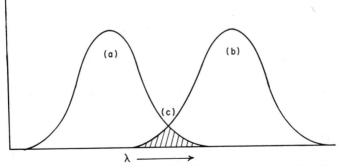

Fig. 6. Mechanism of fluorescence where (a) is the excitation spectra, (b) is the emission spectra, and (c) is the anti-Stokes region where the excitation and emission spectra overlap.

reaching it to which it is sensitive. When the wavelength of illumination is in the region of the fluorescent emission there will be no fluorescence because there is no excitation energy and the detector then records only the true reflectance of the sample at that wavelength. When a fluorescent material is illuminated with the wavelength of radiation in the region of the excitation spectra, however, the sample will fluoresce. This fluorescent energy reaching the detector is read by the instrument as part of the reflectance of the sample at the wavelength of illumination. This result is an apparent reflectance larger than the true reflectance of the material at that wavelength. It is for this reason that the measurement of the total radiance factor was introduced. In this measurement technique the sample is illuminated with heterochromatic light so that all the fluorescence is excited. The instrument then collects the energy from the sample monochromatically and thus measures the true reflectance of the sample at that wavelength *plus* the light that has been emitted by fluorescence at that wavelength. The mathematical expression for the total radiance factor is

$$\beta_T(\lambda) = \beta_S(\lambda) + \beta_L(\lambda) \tag{2}$$

where $\beta_T(\lambda)$ is the total spectral radiance factor, $\beta_S(\lambda)$ the reflected radiance factor, and $\beta_L(\lambda)$ the luminescent radiance factor.

Since the degree of sample fluorescence depends on the energy distribution of the source in the region of excitation, the spectral-power distribution of the source now becomes a factor in the measured value of the sample. If only the true reflectance of the sample is desired, an evaluation using monochromatic incident energy and monochromatic collection is required. See Chapter 6 for further details on the measurement of fluorescent samples.

F.   Types of Spectrophotometers

Spectrophotometers may be categorized according to several different criteria. Single- or double-beam instruments, discussed earlier, constitute one criterion. Also there are single and double monochromators. This classification simply indicates whether the instrument uses one or two dispersing elements with the associated optics. A common classification of spectrophotometers is based on the type of detection system, visual or photoelectric. In the case of a visual spectrophotometer, the photometric part consists of two uniformly illuminated fields, sample and reference, with means for varying the luminance of one or both fields in a controlled manner. The spectrophotometric ratio is determined from the instrument reading at the position where the eye sees an equality of the two fields. The eye is used as a detector; hence, the name "visual spectrophotometer." Numerous methods have been used with visual instruments to form and vary the luminance of

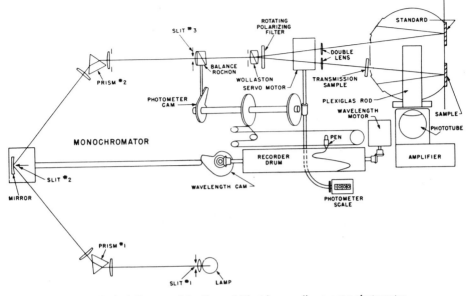

Fig. 7.   Optical diagram of the General Electric recording spectrophotometer.

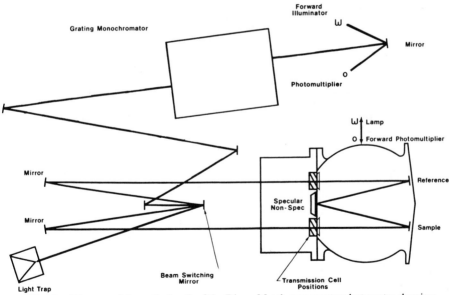

Fig. 8.   Diagram of the optical path of the Diano-Match scan spectrophotometer showing the capability of monochromatic illumination (using forward illuminator and forward photomultiplier) and heterochromatic illumination (using lamp above sphere and photomultiplier after the monochromator).

the photometric fields. These types of instruments were for many years the only spectrophotometers available. Among them are the Gaertner, the Koning–Martens, the Nutting–Hilger, the Huffner, and many others. See Gibson (1925, 1934), Twyman and Allsop (1934), and Walsh (1958) for further description of this type of instrument.

The visual spectrophotometer is no longer in use because the measurements are very tedious and slow to make. They have been replaced by photoelectric spectrophotometers.

The advent of spectrophotometers using photoelectric detectors made a significant contribution to the science of color measurement. There are instruments now available that not only record the data but also provide colormetric specifications computed by built-in microcomputers. Some of these instruments can perform measurements and the computations in a matter of seconds. Examples of these are the IBM 7409 scanning color sensor; Hunter D54P-5 spectrophotometer (1977); Diano Match-Scan; and Macbeth MS-2000.

TABLE I. Several Common Spectrometers and Their Major Features

Instrument	Monochromator	Optics	λ Range	BW
Diano Match-scan	Double grating	DB - reversible (7" sphere)	380-700 250-950	10 10
Diano Hardy	Double prism	DB - reversible (Normally poly- illum.) (7" sphere)	380-700	10 nm
Pye Unicam SP8-200	Master holographic grating (Ebert Monochromator)	DB Sphere accessary	185-950	0.1-10.0 nm
Zeiss DMC-25	Double prism	DB - reversible	200-2500nm	<2.5-10nm -variable
Zeiss DMC-26	Double grating (Ebert) (450nm)	DB	300-900nm	1,5,10nm
Hunter D54P-5	Interference wedge	SB (8" sphere) Interference wedge	400-710	9
B and L Spectronic-505	Double grating	DB	200-650-800	
600	Double grating	DB	200-800	
Beckman 5240	Halographic grating (single)	DD	190-3000	
5270	Halographic grating (double)	DB	190-3000	
Varian Cary 17D	Prism grating	DB Sphere Optional	186-2650	variable
G E Hardy	Double prism	DB	400-700 360-700	10 nm
McPherson EU707	Filter grating Double Monochromatic	DB	200-700 (185-1000)	variable slits
MacBeth MS-2000	Solid state silicon array	DB	380-700	16nm

The optical systems of some of these instruments are given as examples in Figs. 7–11. The pertinent instrumental parameters of some of these instruments are given in Table I.

One of the earliest photoelectric instruments still in use is the General Electric recording spectrophotometer, whose optical system is shown schematically in Fig. 7. The modern version, using basically the same optical system but improved electronics and digital readout, is called the Diano–Hardy. The Diano Match-Scan with integrating-sphere geometry is shown schematically in Fig. 8 and can be used with either monochromatic or heterochromatic illumination. The source is a quartz–halogen tungsten-filament lamp that can be operated without filter as CIE standard source A or optically filtered to simulate D_{65}. The incident beam can be varied from 1 to about 20 mm. A Bausch and Lomb grating monochromator is used, with a 10 nm spectral bandpass. The wavelength range is nominally 400–700 nm. The Hunter D54P-5 shown in Fig. 9 uses a continuous interference wedge as a monochromator. This instrument is a single-beam scanning spectrophotom-

Stated accuracy	Stray light	Scan speed	Read out	Geometry
± 1nm/.06%		16 Pts - 9 sec 161 Pts -30 sec	Graphic/digital	O/d or d/O
0.5%		54 sec fullray "fast"		O/d or d/O
± 0.5nm ± .0015 at 1Å ± .003 at 2Å ± .02 at 3Å ± .2%	<0.02% at 220 <0.01% at 340	0.1-10nm/sec	Graphic/digital	Sphere Specular (optional O/d)
± .1nm		3.5-3500 sec/cm	Graphic/digital	O/d or d/O
± 0.1%	<0.1%	35nm/sec	Graphic/digital	O/d, d/O, or 45/0
± 0.1% >0.5% .005		9.5-14sec	Digital only	d/O
± 0.5nm	<0.1% at 220nm 0.5% or .005 at 4Å	Variable/pen movement	Graphical (digital available)	Specular (optional O/d)
± 0.5nm				Specular (optional sphere)
± 15nm 190-800 ± 2.5nm 800-3000	<.1%	1-400nm/in	Graphical (visual display)	Specular (optional O/d)
1nm 190-800 0.5nm 800-3000	<.0001% 240-500 <.001% 210-690 <.01% 1690nm <0.1%			
<± 0.4nm 0.10 = .0004 1.10 = .001 2.0 = .005 3 = .03	<0.0001%	.5Å/sec	Graphical (digital optional)	O/d, d/O, 0/45, or specular
± 0.1%	2		Graphical	O/d
± .001	<.1% 250-800 <.01% Optional	1-120nm/min	Digital (graphical optional)	Specular
0 - 10% = .01% 10- 30% = .03% 30-150% = .10%		3 sec readout	CRT	d/O or 45/0

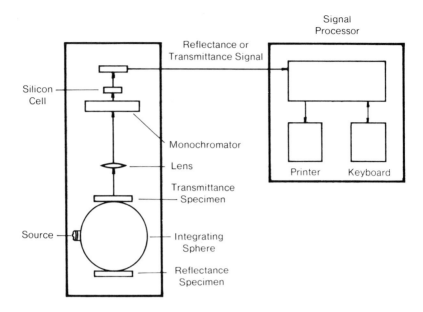

Fig. 9. Block diagram of the Hunter D54P-5 spectrophotometer showing both the optical system and the microprocessor.

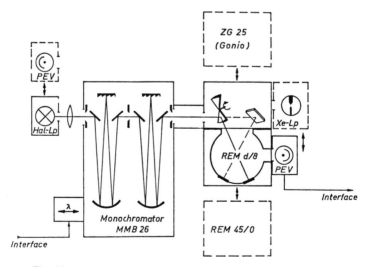

Fig. 10. Optical diagram of the Zeiss DMC-26 spectrophotometer.

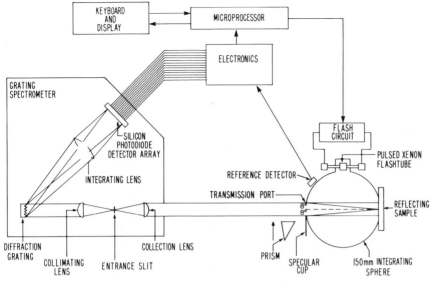

Fig. 11. Block diagram of the Macbeth MS-2000 spectrophotometer.

eter. The source is a quartz–halogen tungsten-filament lamp optically filtered to simulate D_{65}. Integrating-sphere geometry is used with diffuse irradiation and near normal viewing. A silicon photodiode is used as the detector. The spectral bandpass is variable from about 10 to 18 nm at 700 nm. The Zeiss DMC-26 has a double monochromator and covers a wavelength range from 300 to 900 nm with provisions for two types of illumination. It can be used with either integrating sphere or 45°/0° geometry. This instrument is fully automated and is shown diagramatically in Fig. 10. The Macbeth MS-2000 spectrophotometer is an abridged double-beam instrument. A block diagram is given in Fig. 11. The light source is a pulsed-xenon flashtube, optically filtered to simulate CIE standard illuminant D_{65}. The MS-2000 utilizes integrating-sphere geometry with diffuse irradiation and near normal viewing. In each measurement the flashtube is fired four times with the sample viewed during two pulses and the sphere wall, as an internal reference, during the other two. A fixed grating monochromator projects the viewing beam onto an array of 17 silicon-photodiode detectors, spectrally spaced at 20 nm intervals from 380 to 700 nm, with a spectral width of 16 nm each. The 17 signals produced simultaneously at each firing of the flashtube are amplified to produce a logarithmic photometric scale.

III. COLORIMETERS

A. General Concepts

Colorimeters are instruments designed such that tristimulus values may be obtained from their response without the necessity of mathematical integration as is required with spectrophotometric data. In colorimeters the integration is normally accomplished optically by the instrument. Usually the source in the instrument is filtered so that it has the spectral-power distribution of the desired standard source and the response of the detector is also modified with filters so that it matches the color-matching functions of a CIE standard observer.

The spectral-power distribution of the source is usually intended to match that of a standard illuminant recommended by the CIE. In the past this has been illuminant A or C. Since 1964, however, the recommendation has been for the natural daylight series of illuminants, with D_{65} being most commonly used. Illuminant A is still widely used, since it represents incandescent illumination, which is a common source, particularly in the home. There are two standard observers recommended by the CIE for the tristimulus response functions. The 1931 observer for a 2° field of view and the 1964 observer for a 10° field of view. Although they differ significantly in the angle of view, in general the plotted functions of these two observers are similar. These functions are shown in Fig. 12.

Although the colorimeters achieve tristimulus integration optically, they still require the application of simple linear equations to compute the tristimulus values. This is a consequence of the way in which the instrument response is made to match that of the standard observer. The Y tristimulus value is usually obtained directly by achieving a response that is equivalent

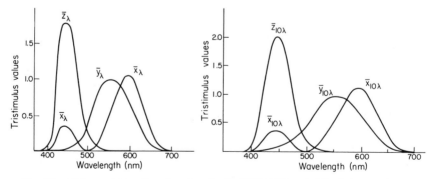

Fig. 12. Color-matching functions for the 2° (CIE-1931) and 10° (CIE-1964) standard colorimetric observer. Although these functions appear quite similar, their differences are colorimetrically significant.

to the $\bar{y}$ color mixture function of the standard observer. This match is usually reasonably accurate because it has already been well developed in the operation of photometers. The $\bar{z}$ color mixture function usually needs only a simple scaling value to achieve a good match. It is the matching of the $\bar{x}$ color mixture function that causes the most difficulty because there are two peaks in the spectral response function, and that makes it difficult to alter the characteristic response of one detector to match this function.

There are two approaches that have been used to achieve the simulation of the $\bar{x}$ color mixture function. One is to assume that the shape of the curve at the secondary short-wavelength peak is proportional to the $\bar{z}$ function. Then it is sufficient to match the long-wavelength portion of the $\bar{x}$ function. If we add to it the proper proportion of the $\bar{z}$ function, the sum is the X tristimulus value, as seen in Eq. (3).

$$X = aR + bB, \qquad Y = G, \qquad Z = cB, \tag{3}$$

where R, G, and B are the responses of the colorimeter and a, b, and c are appropriate coefficients.

The other approach is simply to divide the $\bar{x}$ function into two separate functions, to produce photocell responses that match each one of these, and then to sum them in the right proportions to obtain the tristimulus value, as in Eq. (4).

$$X = aR_1 + bR_2, \qquad Y = G, \qquad Z = cB, \tag{4}$$

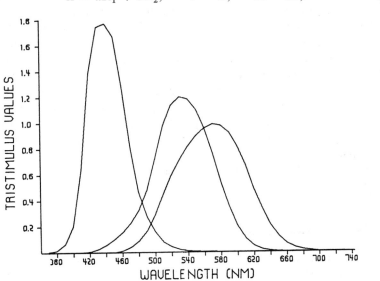

Fig. 13.   A linear transformation of the 1931 standard observer color-matching functions that are spectrally the narrowest and contain only a single maximum for each function.

where R_1, R_2, G, and B are the responses of the instrument and a, b, and c are coefficients. This produces an instrument with four responses instead of three.

It is also possible to use color mixture functions that do not contain any multipeak functions. This is due to the fact that any linear transformation of the normal color mixture functions is appropriate for colorimetry. Such a set of transformations containing not only single-peak functions but also the narrowest bandwidth functions, which minimizes cross talk between responses, is shown in Fig. 13 (Pearson and Yule, 1973).

B.   Computational Methods

Since the advent and incorporation of microprocessors, calculations of tristimulus values as well as many other colorimetric calculations are now made within the instrument. This capability also permits data from previously read standard or reference materials to be stored and compared with the data of current samples in order to calculate differences. It must be remembered, however, that, unlike spectrophotometric data, which can be integrated with any desired source or set of response functions, data obtained from colorimeters are unique to the source and response functions of the receiver designed into the instrument. If information about the sample is desired in terms of other sources or detector responses, other instruments must be used or instruments must be designed to operate in multiple modes.

The same concerns hold for colorimeters measuring fluorescent materials as those discussed for spectrophotometers; that is, in order to measure the fluorescence exactly as it would be during visual examination, the sample must be illuminated with heterochromatic illumination. With colorimeters this means direct illumination from the standard source in the instrument.

The same conditions of influx and efflux geometries that apply in spectrophotometric measurements also apply to measurements made by colorimeters. The geometry is generally either 45°/0° or integrating sphere. Although transmission measurements can be made with most colorimeters, they are usually designed with the reflectance measurement as the primary mode of operation.

C.   Types of Colorimeters

Like spectrophotometers, colorimeters can be divided most basically into two classes, visual and photoelectric. Within these two classes there is a large variety of types and configurations.

3. VISUAL COLORIMETERS

In visual tristimulus colorimeters the human eye is the detector. In this type of instrument the operator looks at two juxtaposed fields and adjusts one of the fields (the comparison side) until it has the same color as the other (the unknown side). The known amount of adjustment required is then used to determine the specification of color of the unknown side. A schematic diagram of a Donaldson-type visual colorimeter is shown in Fig. 14. Visual

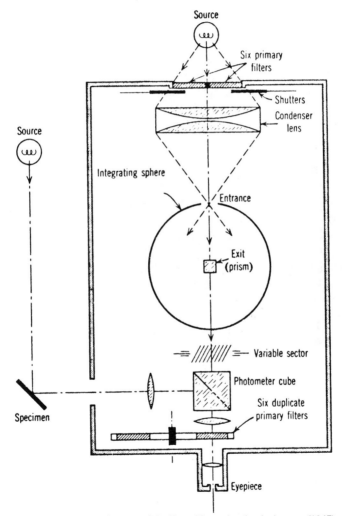

Fig. 14. Optical diagram of the Donaldson visual colorimeter. (1947)

colorimeters may be further classified as additive or subtractive systems according to the manner of producing known colors in the comparison field. However, no matter what technique is used to vary the comparison field, three degrees of freedom are required, since colorimetric specifications are three-dimensional quantities.

2. TRISTIMULUS COLORIMETERS

In the case where the comparison field is filled, in quick succession, by three stimuli of different fixed chromaticities and whose luminances are independently adjusted, we have a tristimulus visual colorimeter. In tristimulus colorimeters the tristimulus values X, Y, Z can be obtained by direct comparison of the unknown stimulus with an optical additive mixture of three primary stimuli in a divided photometric field. The choice of working primaries is important and should be such that they produce as wide a gamut of colors as possible. For further details on this topic the reader is referred to the work of Guild (1931) and that of Wright (1927–1928). The primaries chosen by them were the basis of the CIE 1931 standard colorimetric observer.

The primaries chosen for the CIE 1931 and 1964 standard coordinate systems for colorimetry were empirically derived and are not realizable as working primaries. As a result, tristimulus colorimeters do not yield the X, Y, and Z values directly.

Many tristimulus (visual) colorimeters have been built and are described in the literature (see, e.g., Burnham, 1952; Stiles, 1955; Wyszecki and Stiles, 1967; Donaldson, 1947; MacAdam, 1950; Wright and Wyszecki, 1960; Wyszecki, 1965).

Of all these colorimeters, we should specifically mention two: the Donaldson (1947) and the Stiles (1955) colorimeters. The Donaldson colorimeter makes use of six rather than the customary three primaries to overcome the disadvantages of the large-gamut tristimulus colorimeters. The Stiles colorimeter, sometimes referred to as the NPL trichromator, uses three double monochromators that are mounted vertically one above the other. This instrument was used to determine large-field color-matching functions for more than 50 observers. Portions of these data were used to derive the color-matching functions of the CIE 1964 supplementary standard observer.

Unlike visual spectrophotometers, visual colorimeters are still used extensively in a variety of color vision research projects.

3. PHOTOELECTRIC COLORIMETERS

Photoelectric colorimeters automatically evaluate the tristimulus values of a given sample using a photoelectric type of detection system. As mentioned earlier, the spectral responses of these devices are directly proportional to the color-matching function of the CIE standard colorimetric observer. These instruments are used in various production and quality control operations. They are easy to use, fast, and for most applications, sufficiently accurate. Their accuracy is largely dependent on the degree to which their spectral responses match the standard observer's function. When maximum accuracy is desired, one should turn to spectrophotometry and computation on the basis of the CIE standard observer.

There are several methods available to adjust the spectral response function of a photoelectric detector to that of the standard observer. One of these methods uses a monochromator and three templates (X, Y, Z) to modify selectively an equal energy spectrum in such a way that the spectral response curve of the detector approximates very closely the shape of one of the three color-matching curves of the standard observer. Several such template colorimeters have been built (see, e.g., Harrison, 1952; Mahr, 1958). Template colorimeters, although very accurate, are difficult to use and provide low light signals. They are cumbersome and quite expensive; hence, they offer little advantage over a spectrophotometer.

A more convenient method of modifying the spectral response is by placing optical filters in front of the detector. The filters are chosen so that

TABLE 2. Several Common Colorimeters and Their Major Features

Name	Type	Lamp	Illuminant	Geometry	Output
Color Eye	PT	Incandescent	C	d/0	X,Y,Z
Colormaster Differential Colorimeter	PT	Incandescent	C	45/0	R,G,B
Gardner XL-20	PT	Halogen	C	45/0	X,Y,Z;L,a,b
Harrison Model 4 colorimeter	PT	Fluorescent	D_{65}	45/0	X,Y,Z
Harrison Digital Colorimeter	PT	Fluorescent	6500K	45/0	X,Y,Z
Hunter D25-2	PT	Halogen	C,D_{65}	45/0	$X,Y,Z;L^*,a^*,b^*$
Kollmorgen KCS-40	PT,FS	Halogen	A,D_{65}	d/0	X,Y,Z
Kollmorgen KCS-18	PT,FS	Halogen	A,C,D_{65}	d/0	X,Y,Z
Macbeth MC-1010	①	Pulsed Xenon	D_{65}	45/0	$L,a,b;L^*,a^*,b^*$
Martin Sweets Color Brightness Tester	PT	Halogen		45/0,0/d	X,Y,Z
Ziess Elrepho	PT	Incandescent	A,C,D_{60},D_{65}	d/0	X,Y,Z
Ziess RFC-3	PT,FS	Incandescent Xenon	A,C,D_{65}	d/0,45/0	X,Y,Z

[a] PT = photoelectrec tristimulus, FS = filter spectrophotmeter, and ① = grating/multi-element silion array.

the resultant spectral transmittance function of the combination (correction filters plus tristimulus filters) corrects the spectral response function of the detector to one of the CIE color-matching functions. Close agreement with the CIE color-matching functions using this technique can be achieved, as shown by Robertson *et al.* (1972).

Table II lists some typical photoelectric tristimulus colorimeters. Also given in the table are fundamental parameters of these instruments.

IV. DENSITOMETERS

Densitometers are quite similar to colorimeters in their design and operation. The major difference is that the response of their detection system does not conform to either of the CIE standard observers. They are designed to make color measurements on color photographic materials, (McCamy 1973) and if the color of the sample changes, the measured value of the instrument changes. They were never designed to respond to color as a standard observer would. Regardless of this, the densitometer is still an appropriate and useful instrument for quality control of products, processes, and materials. The major problems often encountered are the attempt to compare data between densitometers and colorimeters or the treating of densitometric data as if they were colorimetric data. These are conditions that the American National Standard Institute (ANSI) is keenly aware of and a great deal of effort is being spent to produce, update, and change densitometric standards. One result is an improved format for specifying a densitometric measurement (ANSI, 1974). This functional notation consists of a symbol representing the type of density measurement being made and a parenthetical expression of four parts describing the spectral and geometrical conditions by which the measurements were made [e.g., $D_R(s;g:g';s')$]. An example of a visual reflection density made with a source operating at 3000 K, with illumination normal to the sample and the reflection collected at 45° would be

$$D_R (3000 \text{ K}; 0° \pm 5°: 45° \pm 5°: V).$$

Densitometers are of two types: reflection and transmission (see Figs. 15 and 16), with some instruments being capable of making both measurements. However, in the reflection mode the recommended influx–efflux is 45°/0° (ANSI, 1976, 1977). The two major applications of color densitometry are in the photography industry and the graphic arts printing industry.

In the photographic industry the primary application of densitometry is in process and product control and the calculation and control of printing exposures both for reflection and transmission printing. For most of these

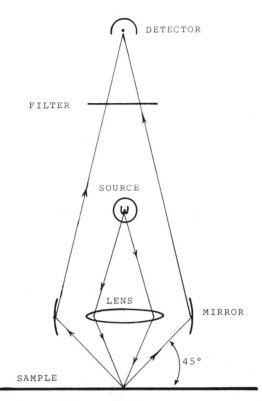

Fig. 15.   Diagram of the optical system of a reflection densitometer using 0°/45° illumination collection geometry.

applications "status" densitometry is used, that is, densitometry where the spectral response of the densitometer is tailored by filters to represent as closely as possible the spectral dye absorption peaks of the photographic materials involved. It has been suggested by Dawson and Vogelsong (1973) that the variability among instruments can be reduced by standardizing the unfiltered response, that is, the spectral response excluding the narrow-band filters and using AA or MM filters to provide the correct response for reading status A densities or status M* densities, respectively.

In the transmission mode there are two types of geometry: (1) diffuse, where the illumination is collimated and normal to the surface and the collection is hemispherical, and (2) specular, where the cone angle of incident and transmitted light is carefully restricted and specified [e.g., $f/4.5$ or $f/1.6$ (ANSI PH2.37)].

* A standard set of densitometer response functions used in photographic applications (see Dawson and Vogelsong, 1973).

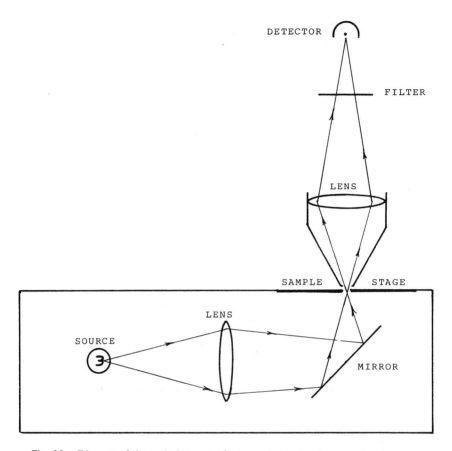

Fig. 16. Diagram of the optical system of a transmission densitometer showing specular illumination and diffuse collection.

In graphic arts printing, densitometers are used for control during printing on the press, evaluation of the final printed image, evaluation of the original copy for determination of separation and masking requirements, and control of the photographic steps encountered in prepress operations. Color transmission densitometry is confined mostly to the evaluation of original transparencies and color duplicates. Control during printing and evaluation of the final product are nearly always done using reflection measurements.

The major differences between graphic arts and photographic densitometry are in the wide diversity of substrates and pigments used in the graphic arts industry and the fact that the scattering characteristics of the pigments are more significant than in photographic dyes.

With the increase in stability and sensitivity of modern solid-state systems, densitometers are now being produced that are capable of dynamic on-line measurements (Kishner, 1977). In the graphic arts industry there are now densitometers that have a response time short enough to make measurements on a moving web of paper during the printing operation. This offers the potential of closed-loop automated control systems for faster, more efficient control.

Some manufacturers of modern densitometers are Macbeth, Welsh, Speedmaster, Cosar, GAM, and Gretag.

REFERENCES

ANSI (1974). PH2.36. Am. Natl. Stand. Inst., New York.
ANSI (1976). PH2.19. Am. Natl. Stand. Inst., New York.
ANSI (1977). PH2.17. Am. Natl. Stand. Inst., New York.
ANSI (1976). PH2.37. Am. Natl. Stand. Inst., New York.
Budde, W. (1970). *Farbe* **19**, 94.
Budde, W. (1976). *J. Res. Natl. Bur. Stand.* **585**, 90A.
Burnham, R. W. (1952). *Am. J. Psychol.* **65**, 603.
CIE (1971). Publ. No. 15 (E-1.3.1). CIE, Paris.
CIE (1977). Publ. No. 38 (TC-2.3). CIE, Paris.
CIE (1979). TC-2.3 Technical Report (in preparation).
CIE Proc. 8th Session, Cambridge 1931, S.19-29 (Empfehlungen 1-5).
Dawson, G. H., and Voglesong, W. F. (1973). *Photogr. Sci. Eng.* **17**, 461.
Donaldson, R. (1947). *Proc. Phys. Soc. London* **59**, 544.
Erb, W. (1975). PTB-Bericht, Opt-3, Braunschweig, Germany.
Gibson, K. S. (1925). *J. Opt. Soc. Am.* **10**, 169.
Gibson, K. S. (1934). *J. Opt. Soc. Am.* **24**, 234.
Grum, F., and Luckey, G. W. (1968). *Appl. Opt.* **7**, 2289.
Grum, F., and Wightman, T. E. (1977). *Appl. Opt.* **16**, 2775.
Guild, J. (1931). *Philos. Trans. R. Soc. London, Ser. A* **230**, 149.
Harrison, W. (1952). *Light Light.* **45**, 132.
Ishizaki, H. (1956). *J. Illum. Eng. Inst. Jpn.* **40**, 23.
Kishner, S. J. (1977). *J. Appl. Photogr. Eng.* **3**, 4.
MacAdam, D. L. (1950). *J. Opt. Soc. Am.* **40**, 589.
Mahr, D. (1958). *Farbe* **7**, 283.
McCamy, C. S. (1973). Color: Theory and Imaging Systems, *in* "Color Densitometry," (R. A. Eynard, ed.). Soc. Photogr. Sci. Eng., Washington, D.C.
National Bureau of Standards (1977). "Geometrical Considerations and Nomenclature for Reflectance," NBS Monogr. No. 160. Washington, D.C.
Pearson, M. L., and Yule, J. A. C. (1973). *J. Color App.* **2**, No. 1, 30–35.
Robertson, A. R., Staniforth, A., Gignac, D. S., and McDougall, J. (1972). NRC Rep. Pro-387. Ottawa.
Stiles, W. S. (1955). *Phys. Soc. Year Book* p. 44.
Stiles, W. S., Burch, J. M, N.P.L. color matching investigation; Final report (1958). Optica Acta 6 (1959), S.I.
Terstiege, H. (1974). *Lichttechnik* **26**, 277.

Twyman, F., and Allsop, C. B. (1934). "The Practice of Absorption Spectrophotometry with Hilger Instruments," 2nd Ed. Hilger, London.

Walsh, J. W. T. (1958). "Photometry," 3rd Ed. Constable, London.

Webber, A. C. (1957). *J. Opt. Soc. Am.* **47**, 785.

Wendlandt, W. W., and Hecht, H. G. (1966). "Reflectance Spectroscopy" Interscience, New York.

Wright, H., and Wyszecki, G. (1960). *J. Opt. Soc. Am.* **50**, 647.

Wright, W. D. (1927–1928). *Trans. Opt. Soc. (London)* **29**, 225.

Wyszecki, G. (1965). *J. Opt. Soc. Am.* **55**, 1319.

Wyszecki, G., and Stiles, W. S. (1967). "Color Science." Wiley, New York.

GENERAL REFERENCES

Billmeyer, F. W., Jr., and Saltzman, M. (1966). "Principles of Color Technology." Wiley (Interscience), New York.

CIE (1971). "Colorimetry—Official Recommendations of the International Commission of Illumination." Publ. No. 15 (E1.3.1). Central Bureau, Paris.

Eynard, R. A. (1973). "Color: Theory and Imaging Systems." Soc. Photogr. Sci. Eng., Washington, D.C.

Johnson, R. M. (1971). *J. Color Appearance* **1**(2), 27.

Judd, D. B., and Wyszecki, G. (1963). "Color in Business, Science and Industry." Wiley, New York.

Weissberger, A., and Rossiter, B. W. (1972). "Physical Methods of Chemistry," Vol. I, Chap. 3. Wiley (Interscience), New York.

Wright, W. D. (1964). "The Measurement of Color." Hilger & Watts, London.

Bausch and Lomb, Analytical Systems Division, 820 Linden Ave., Rochester, NY.

Cosar Corp., 3121 Benton Street, Garland, TX 75042.

Diano-Hardy, Diano Corporation, Optical Systems Division, P.O. Box 346, 75 Forbes Blvd., Mansfield, MA 02048.

GAM, Graphic Arts Manufacturing Co., 2518 South Blvd., Houston, TX 77006.

Gretag, Ltd., Althardstrasse 70, CH-8105 Regensdorf, Zurich, Switzerland.

Hunterlab, 9529 Lee Highway, Fairfax, Virginia 22030

IBM, Instrument Systems, 1000 Westchester Ave., White Plains, NY 10604.

Macbeth, Division of Kollmorgen Corp., Drawer 950, Newburgh, NY 12550.

Speedmaster, Electronic Systems Engineering Co., Cushing, Oklahoma 74023.

Welsh, Sargent-Welsh Scientific Co., 7300 N. Linden Ave., Skokie, IL 60076.

Zeiss, Carl Zeiss, Inc., 444 Fifth Ave., New York, NY 10018.

Index